A CULTURAL HISTORY OF CHEMISTRY

VOLUME 1

A Cultural History of Chemistry
General Editors: Peter J.T. Morris and Alan J. Rocke

Volume 1
A Cultural History of Chemistry in Antiquity
Edited by Marco Beretta

Volume 2
A Cultural History of Chemistry in the Middle Ages
Edited by Charles Burnett and Sébastien Moureau

Volume 3
A Cultural History of Chemistry in the Early Modern Age
Edited by Bruce T. Moran

Volume 4
A Cultural History of Chemistry in the Eighteenth Century
Edited by Matthew Daniel Eddy and Ursula Klein

Volume 5
A Cultural History of Chemistry in the Nineteenth Century
Edited by Peter J. Ramberg

Volume 6
A Cultural History of Chemistry in the Modern Age
Edited by Peter J.T. Morris

A CULTURAL HISTORY OF CHEMISTRY

IN ANTIQUITY

VOLUME 1

Edited by Marco Beretta

BLOOMSBURY ACADEMIC
LONDON • NEW YORK • OXFORD • NEW DELHI • SYDNEY

BLOOMSBURY ACADEMIC
Bloomsbury Publishing Plc
50 Bedford Square, London, WC1B 3DP, UK
1385 Broadway, New York, NY 10018, USA
29 Earlsfort Terrace, Dublin 2, Ireland

BLOOMSBURY, BLOOMSBURY ACADEMIC and the Diana logo are trademarks of
Bloomsbury Publishing Plc

First published in Great Britain 2021
Paperback edition published in 2025

Copyright © Bloomsbury Publishing Plc, 2025

Cover design: Rebecca Heselton
Cover image © De Agostini Picture Library/A. Dagli Orti/Bridgeman Images

For legal purposes the Acknowledgements on p. 10 constitute an extension of
this copyright page.

This work is published open access subject to a Creative Commons Attribution-NonCommercial-NoDerivatives 4.0 International licence (CC BY-NC-ND 4.0, https://creativecommons.org/licenses/by-nc-nd/4.0/). You may re-use, distribute, and reproduce this work in any medium for non-commercial purposes, provided you give attribution to the copyright holder and the publisher and provide a link to the Creative Commons licence.

Bloomsbury Publishing Plc does not have any control over, or responsibility for, any third-party websites referred to or in this book. All internet addresses given in this book were correct at the time of going to press. The author and publisher regret any inconvenience caused if addresses have changed or sites have ceased to exist, but can accept no responsibility for any such changes.

A catalogue record for this book is available from the British Library.

A catalog record for this book is available from the Library of Congress.

ISBN: PB: 978-1-3505-5203-6
Pack: 978-1-3505-5229-6
ePUB: 978-1-3502-5145-8
ePDF: 978-1-3502-5147-2

Series: The Cultural Histories Series

Typeset by Integra Software Services Pvt. Ltd.
Printed and bound in Great Britain

To find out more about our authors and books visit www.bloomsbury.com
and sign up for our newsletters.

CONTENTS

LIST OF ILLUSTRATIONS	vii
SERIES PREFACE	x
ACKNOWLEDGEMENTS	xii

Introduction 1
Marco Beretta

1 Theory and Concepts: The Mythological Foundation of Chemical Theories in Ancient Civilizations 23
Sydney H. Aufrère, Cale Johnson, Matteo Martelli, and Marco Beretta

2 Practice and Experiment: The Conquest of Matter 51
Sydney H. Aufrère, Cale Johnson, Matteo Martelli, and Marco Beretta

3 Laboratories and Technology: From Temples to Workshops: Sites of Chemistry in Ancient Civilizations 85
Sydney H. Aufrère, Cale Johnson, Matteo Martelli, and Marco Beretta

4 Culture and Science: Gods, Myths, and Religions 113
Sydney H. Aufrère, Cale Johnson, Matteo Martelli, and Marco Beretta

5 Society and Environment: The Alteration of the Ancient Landscape 139
Sydney H. Aufrère, Cale Johnson, Matteo Martelli, and Marco Beretta

6 Trade and Industry: The Circulation of Trade in the Mediterranean 161
Sydney H. Aufrère, Cale Johnson, Matteo Martelli, and Marco Beretta

7 Learning and Institutions: The Invention of Chemical Recipes   187
   Sydney H. Aufrère, Cale Johnson, Matteo Martelli, and Marco Beretta

8 Art and Representation: The Iconographic Imprinting of
   Ancient Chemical Arts   211
   Sydney H. Aufrère, Cale Johnson, Matteo Martelli, and Marco Beretta

BIBLIOGRAPHY   236
LIST OF CONTRIBUTORS   286
INDEX   287

LIST OF ILLUSTRATIONS

FIGURES

1.1	Ceiling of the pronaos of the temple of Dendara. Thirty-six decans made of both precious metals and minerals. © Sydney H. Aufrère	26
1.2	Crypt South no. 1. Temple of Dendara. Statue of the god Harsomtus as a serpent. Caption indicating the size and the nature of the metals and minerals used (*Dendara* V: 140, 7). © Sydney H. Aufrère	27
1.3	Names of the temple of Dendara dedicated to the goddess Hathor. Temple of Dendara, northern passage of the hypostyle hall, Ptolemaic period. © Sydney H. Aufrère	28
1.4	One of the names of the temple of Dendara dedicated to the goddess Hathor: "Mansion-of-Electrum" (*Per-en-Djam*). © Sydney H. Aufrère	29
1.5	Cuneiform tablet BM 120960, Middle Babylonian, Tell Umar. © The Trustees of the British Museum	33
1.6	Late Uruk cylinder seal, 4000–3100 BCE. Photograph by Ashmolean Museum/Heritage Images/Getty Images	35
2.1	Flared cylindrical perfume pots set with a palm leaf. Old Kingdom. Cairo Museum. © Sydney H. Aufrère	62
2.2	Three foundation tablets from Khorsabad, Assyrian from the reign of Sargon II (721–705 BCE; gold, silver, and copper). © Bridgeman Images	64
2.3	Vials of kermes and *Rubia tinctorum*. Photograph © Nicole Reifarth	69

2.4	Archaeological remains of dye in Qatna. Photograph © Nicole Reifarth	70
2.5	Fluted glass bottle. © The Trustees of the British Museum	72
3.1	Procession of metals and mineral bearers. Treasure D'. Dendara temple. © Sydney H. Aufrère	91
3.2	*Tinctorium* (dyeing workshop), South-East Dendara temple. © Sydney H. Aufrère	92
3.3	The proto-cuneiform signs representing ovens and kilns: MAH$_a$, AD$_a$, and SIMUG. Drawing by the author, after signs drawn by R.K. Englund	100
3.4	Photograph of a two-chambered oven from Abu Salabikh. Courtesy of J.N. Postgate	101
3.5	Tools from Tell edh-Dhiba'i. Photograph © C.J. Davey	103
3.6	House of the Vettii, Pompeii. Frieze depicting cupids working in a perfumery. Photograph by De Agostini/Getty Images	106
3.7	Distillation equipment in the Byzantine MS *Parisinus* gr. 2327 (fol. 81v) – reproduced in *CAAG* I 161. Wikicommons	109
3.8	Two types of *kērotakis* (from Taylor 1930)	111
4.1	Tayt, goddess of the dyeing process. Eastern staircase of the temple of Dendara. © Sydney H. Aufrère	117
4.2	Text associated with the goddess Tayt. Eastern staircase of the temple of Dendara. © Sydney H. Aufrère	118
4.3	Scene of filling the Udjat-Eye. Eastern Osirian chapels, inner courtyard, southern wall. Detail of the procession. Temple of Dendara. © Sydney H. Aufrère	137
5.1	Pollution visible in a Greenland ice sheet core. Photograph by Jeff Overs/BBC News & Current Affairs via Getty Images	149
5.2	EA 19 letter from Tushratta. Courtesy of www.BibleLandPictures.com/Alamy Stock Photo	151
6.1	One of the Gudea Cylinders, ca. 2140–2124 BCE. Photograph by DEA/G. DAGLI ORTI/De Agostini via Getty Images	172
6.2	Pomegranate vase from the tomb of Tutankhamun, New Kingdom (silver or electrum). © Boltin Picture Library/Bridgeman Images	176
6.3	Stele known as Del Purpurarius. Roman civilization. Parma, Museo Archeologico Nazionale (Archaeological Museum) Palazzo Della Pilotta. Photograph by De Agostini/Getty Images	183

7.1	Text of the manufacture of the statuette of Osiris-Khentymentiu during the Khoiak Festival. Western Osirian chapels, inner courtyard, eastern wall. Temple of Dendara. © Sydney H. Aufrère	194
7.2	Offering of oliban to Hathor of Dendara. Laboratory, northern wall, first register. Temple of Dendara. © Sydney H. Aufrère	194
7.3	The Ur III medical tablet. Courtesy of Science History Images/Alamy Stock Photo	197
7.4	Glass and alabaster jars belonging to the ancient Assyrian King Sargon II. Dated eighth century BCE. Photograph by Universal History Archive/Universal Images Group via Getty Images	201
8.1	Temple frieze from Ubaid. Courtesy of www.BibleLandPictures.com/Alamy Stock Photo	217
8.2	Late Uruk cylinder seal impression of women working. Courtesy of The Metropolitan Museum of Art	218
8.3	Workshops in an Assyrian military camp. Relief from Royal Palaces of Nineveh, ca. 645 BCE. Photography by DEA/G. DAGLI ORTI/De Agostini via Getty Images	219
8.4	Terra-cotta plaque with omega and Kubu. Courtesy of www.BibleLandPictures.com/Alamy Stock Photo	221
8.5	Hephaestus working in his workshop, red-figure *kylix* (foundry painter, ca. 490–480 BCE), Altes Museum in Berlin. Public domain. Courtesy of Wikicommons	226
8.6	Bronze workshop, red-figure *kylix* (foundry painter, ca. 490–480 BCE), Altes Museum in Berlin. Public domain. Courtesy of Wikicommons	228
8.7	A Corinthian black-figure *pinax* (ca. 575–550 BCE) found in Penteskouphia. Public domain. Courtesy of Wikicommons	229
8.8	Fullers at work. Fresco, Pompeii, *fullonica* VI 8,20. Public domain. Courtesy of Wikicommons	232

TABLE

6.1	Local specialties listed in Pseudo-Democritus (*PM* = *Physika kai mystika*; *AP* = *On the Making of Silver*; *Cat.* = *Catalogues*) and in the Leiden and Stockholm Papyri (*P.Leid* and *P.Holm*).	182

SERIES PREFACE

A Cultural History of Chemistry examines the history of chemistry and its wider contexts from antiquity to the present. The series consists of six chronologically defined volumes, each volume comprising nine essays; these fifty-four contributions were written and/or edited by a total of fifty scholars, of ten different nationalities. Of Bloomsbury's many six-volume *Cultural Histories* currently in print, this is the first in the physical or natural sciences; it is also the first multivolume history of chemistry to appear since James Riddick Partington's four-volume *History of Chemistry*, concluded more than fifty years ago. It is distinguished, among other qualities, by its endeavor to take the subject from antiquity right to the present day.

This is not a conventional history of chemistry, but a first attempt at creating a cultural history of the science. All cultures, including the various branches of natural science, consist of mixed constructs of social, intellectual, and material elements; however, the cultural-historical study of chemistry is still in an early stage of development. We hope that the accounts presented in these volumes will prove useful for students and scholars interested in the subject, and a starting point for those who are striving to create a more fully developed cultural history of chemistry.

Each volume has the same structure: starting with an interpretive overview by the volume editor(s), the eight succeeding chapters explore for each respective era in chemistry its theory and concepts; practice and experiment; laboratories and technology; culture and science; society and environment; trade and industry; learning and institutions; and art and representation. Readers therefore have the option to read multiple chapters in a single volume, thus learning about the cultures of chemistry in a single era; or they may prefer instead to read corresponding chapters across multiple volumes, learning about (e.g.) the art and representations of chemistry through the ages. Though the scope is global, major emphasis is placed on the Western tradition of science and its contexts.

Whether read synchronically or diachronically, in any multiauthor undertaking like this one readers will inevitably notice overlaps and repetitions, conflicting historical interpretations, and (despite the magnitude of the project) occasional gaps in coverage. These are inescapable consequences, but they actually offer advantages to the reader, both in making each chapter closer to self-contained and in demonstrating the dynamism of the discipline; like science itself, the study of its history is ever contested and incomplete.

Chemistry has been called the "central science," due to its fundamental importance to all the other physical and natural sciences. It is the archetypical science of materials and material productivity, and as such it has always been deeply embedded in human industry, society, arts, and culture, as these volumes richly attest. The editors and authors hope that *A Cultural History of Chemistry* will be of great interest and enjoyment not just to chemists and specialist historians of science, but also to social, economic, intellectual, and cultural historians, as well as to other interested readers.

Peter J.T. Morris and Alan J. Rocke
London (UK) and Cleveland (USA)

ACKNOWLEDGEMENTS

This publication is part of the research project *Alchemy in the Making: From Ancient Babylonia via Graeco-Roman Egypt into the Byzantine, Syriac, and Arabic Traditions*, acronym *AlchemEast*. The *AlchemEast* project has received funding from the European Research Council (ERC) under the European Union's Horizon 2020 research and innovation programme (G.A. 724914).

Introduction

MARCO BERETTA

THE CONTENDED GEOGRAPHICAL BOUNDARIES OF ANCIENT CHEMISTRY

Historians of chemistry have long discussed where and when chemistry became a science. Both questions have crucial implications for its identity, though they have been addressed in very different ways. Was the cradle of chemistry in Egypt or in the Near East? Did chemistry originate in the development of artisanal chemistry or in alchemists' speculations about the nature of matter? Did the Greek philosophers provide a theoretical framework from which all successive theories stemmed? Did the intense commercial exchange between the Mediterranean civilizations and India influence the spread of chemical ideas and practices? Did Chinese alchemical theories on the transmutation of gold reach the West?

Not surprisingly, the answers to these questions have often produced biased and inaccurate reconstructions. At the same time, these efforts, the first of which date back to early modern times, helped European chemists to find their own epistemological identity. Unlike the exact sciences, until the end of the eighteenth century chemistry embodied the practical arts, occult and religious beliefs, as well as professional guilds. Because of this confused picture of different and, at times, antithetical interests, its academic status and public image suffered from a poor reputation long after the introduction of the first university chairs in the subject. Writing the history of the science, then, became a powerful means of endorsing a specific image of chemistry, at times privileging its experimental background, at others its metaphysical, religious, and philosophical principles, and at yet others its economic value. It should be stressed that the chemical arts have always played a diffuse and vital role in ancient and modern economic systems. Many products of the Mediterranean trade and exchange, now on display in the principal archaeological museums, bear the signs of chemical manipulations. This pervasive influence, however, did not lead to the construction of an image of chemistry that equaled the prestige and reputation of other scientific disciplines. Indeed, the practical and economic value of the chemical arts conflicted with the ancient idea that science was primarily a contemplative and speculative form of knowledge.

Numerous and highly diverse histories of chemistry were published in the early modern period that attempted to elevate the subject beyond a mere corpus of useful knowledge. Already in the second half of the seventeenth century, chemical historiography established the philological foundations of later works; this was due, in particular, to the controversial theories of the chemist Ole Borch (1668, 1674) and the erudite physician Hermann Conring (1669; Abbri 2000). In an impressive display of scholarship, Borch vindicated the antediluvian origins of chemistry and situated its most significant development in ancient Egypt and in the putative work of Hermes Trismegistus, a god who created the art as an emanation of his metaphysical power. In his reconstruction, Borch regarded ancient chemistry as a holy art that combined experimental skills with a religious foundation. Thanks to the rediscovery of this combination, he argued, early modern alchemists were able to enhance the art and restore the prestige it was afforded in ancient times. Although Hermann Conring concurred with Borch's idea that Egypt was the region where ancient chemistry and alchemy made the most significant progress, he believed that the art actually originated in the teachings of Moses (Conring 1669). Borch's and Conring's views were contemporaneously contested by Johann Heinrich Ursin, who asserted the importance of the Zoroastrian tradition and introduced the Near East into the geography of chemistry (Ursin 1661). Regardless of these differences, these scholars sought evidence to show that the chemical arts embodied a superior form of knowledge.

Behind the controversy about the historical and geographical origins of chemistry there was much at stake. Those who supported its mythical and ancient origins defended the relevance of the theoretical and philosophical background provided by alchemical writers. Already during the sixteenth century, however, several authors, such as Vannoccio Biringuccio and Georg Agricola, contested the antiquity of the chemical arts and invited their contemporaries to abandon vain speculation about the superior knowledge of the ancient sages and to support instead the recent progress of the practical arts, such as mining, metallurgy, and glassmaking.

During the eighteenth century, Borch's position began to lose consensus and Jean Baptiste Senac in his famous *Nouveau cours de chymie suivant les principes de Newton et de Stahl* (Paris, 1737) adopted a position that would prevail throughout the century (Beretta 1991). He maintained that "at the beginning chemistry was the art of working metals for the satisfactions of human needs," and that any theoretical assumptions concerning the nature of matters came only after the chemical arts were sufficiently developed. Such a position principally aimed at undermining the influential role still played by alchemy, which was also defended through the noble history of transmutation in Diderot's and d'Alembert's *Encyclopédie* (Beretta 2014).

During the nineteenth century, when chemistry became an established academic discipline and alchemy was left behind, historians of chemistry began to adopt a more balanced position concerning the origins of the science. A few of them, notably Justus Liebig, Ferdinand Hoefer, Hermann Kopp, and Marcellin Berthelot, tried to show the relations between the ancient chemical arts, now regarded as the first steps towards a sound philosophy of matter, and alchemical beliefs (Beretta 2011). The studies published during the nineteenth century, in particular those by Kopp and Berthelot, were often supported by alchemical manuscripts and, for more than a century, these works contributed to designating the perimeters of the history of ancient chemistry. In fact, after Berthelot's and Ruelle's edition of the Greek alchemical corpus of texts (1888), interest in the question of the origin of chemistry was lost, and even scholars like Edmund Lippmann, who heavily criticized the philological reliability of this edition, accepted the idea that the dawn of chemistry occurred in the shadow of Hellenistic and Byzantine alchemy, and that Alexandria in Egypt, the Middle East, and Constantinople were the principal geographical areas where, between the second century BCE and the fourth CE, alchemy and chemistry emerged. Thus, rejecting Borch's claim that chemistry began in antediluvian times, nineteenth-century historians focused their attention on a more recent epoch and on an extremely circumscribed tradition of literary texts. This shift depended both on the importance attributed to literary evidence and on the prevailing idea that Western science originated in Greek culture.

This approach was first questioned in 1935 by James R. Partington, one of the most important historians of chemistry of the twentieth century, in his monumental *Origins and Development of Applied Chemistry*. Partington occupied the chair of chemistry at London's Queen Mary College from 1919 to 1951 and wrote several important historical works, culminating in his four-volume masterpiece *A History of Chemistry*. In *Origins and Development of Applied Chemistry*, he gave a detailed account of the chemical sources and materials "in Egypt, Babylonia, Assyria, the Aegean, Asia Minor, Persia, Syria and Palestine, from the earliest time to the end of the Bronze Age" (Partington 1935: v). He presented in a descriptive but informative way an immense array of material, not least his own analysis of the archaeological findings of the British Museum made available to him. The disproportion between the number of archaeological objects and the scarcity of literary sources conditioned the narrative of this work, which inevitably focused more on the material background of ancient chemistry, leaving aside all the issues of the origins of the chemical and alchemical theories that had attracted the attention of earlier historians. When Partington tried to take into consideration the history of the philosophies of matter in his *History of Chemistry*, before his death he was only able to complete the part related to Greek and Byzantine authors (Partington 1970). The complexity and variety of sources was a major obstacle to the compilation of a homogeneous narrative encompassing the whole of antiquity.

Most histories of chemistry published before Partington's emphasized the central importance of the Mediterranean civilizations and of the Near East, alternatively giving primacy to the Egyptians, the Mesopotamians, and, more often, the Greeks. However, already during the first half of the nineteenth century, the physician and lexicographer Ferdinand Hoefer showed that in China and in India the chemical arts as well as alchemical theories of matter were developed in ancient times and accumulated discoveries that warranted more thorough attention (Hoefer 1866: vol. 1, 9–30). While Hoefer ultimately expanded the map of ancient chemistry to the Far East, he did not advance any serious hypothesis concerning the relations and exchanges between China, India, and the Mediterranean civilizations. It took until the twentieth century before key studies devoted to the history of ancient chemistry in India (Ray 1903–9; Ray and Ray 1956) and China (Needham 1983) shed light on the historical and cultural importance of both traditions and posited the existence of cultural exchanges with the West. However, all efforts to identify relevant mutual influence between Mediterranean civilizations and the Far East have so far failed to produce any compelling evidence. While we cannot exclude that further study will provide us with a different scenario, currently a conservative view of the absence of substantive influences seems closest to what both the material and literary sources are telling us.

The organization of the present reconstruction is limited to the analysis of ancient chemical arts in the Mediterranean and Near East civilizations. In addition to the limitations of space, this choice was driven by two principal reasons. First, we know that chemical artisans, recipes, and ideas were exchanged between Mesopotamia, Egypt, Phoenicia, Greece, Rome, and Byzantium over a very long period. This interconnected history of chemistry has so far escaped scholarly attention, offering the opportunity for the present volume to provide the first synthesis of this epoch of intense exchanges and to explore the strength of these connections in depth. Second, the sources, techniques, materials, and instruments that were transmitted from antiquity to medieval and Renaissance alchemy and chemistry mostly stemmed from the Mediterranean region and the Near East. The growing exchange between the West and the Far East following the opening up of the Silk Road appears not to have significantly contributed to the exchange of practical and philosophical chemical knowledge. Eloquent evidence of this is the rediscovery in mid-eighteenth-century France of the "secret" formula for Chinese porcelain, which had been produced there for over a thousand years.

BETWEEN ART AND NATURE: THE PRIMORDIAL POWER OF FIRE

The knowledge and techniques that have progressively accumulated in the scientific discipline that we now call chemistry are so varied that it is almost impossible to trace back with precision the exact origin of the science. Chemistry became a stable academic discipline by the end of the eighteenth century and it was only during the subsequent century that it acquired a relatively homogeneous and standardized professional curriculum. In sharp contrast to this late academic development, the chemical arts likely preceded all other forms of technical and scientific knowledge. Although the exact origin of the human appropriation of fire technology remains a controversial issue, recent archaeological studies date it to 1.9 million BCE (Gowlett and Wrangham 2013). The first use of fire was for making food, thus altering the chemical properties of foodstuffs. Subsequently, fire was used in agriculture and, with the development of smelting techniques, in metallurgy. While it is not the aim of this introduction to deal with prehistoric epochs, it should be emphasized that already in antiquity the discovery and routine control of fire was regarded as a major achievement in the evolution of human civilization. The appropriation of the transformative power of fire was, in fact, regarded as the first and most important conquest of human civilization, the key to transforming the status of mankind and to opening up a path towards its radical biological and cultural transformations.

In a vivid exposition of the evolution of mankind, Lucretius (first century BCE) used the prehistory and history of man to illustrate this process of biological transformation. Primitive man was naturally equipped with a robust physique that "was built up on larger and more solid bones within, fastened with strong sinews traversing the flesh; not easily to be harmed by heat or cold or strange food or any taint of the body" (Lucr. *DRN* V 925–30). Thanks to these biological characteristics, adapted to resist a hostile environment, primitive men "prolonged their lives after the roving manner of wild beasts" (932), and lived without the assistance of any arts, simply following their instincts, "trusting in their wondrous strength of hand and foot" (966). When they were tired, "like bristly boars, these woodland men would lay their limbs naked on the ground ... wrapping themselves up around with leaves and foliage" (970–2). The condition of these wretched beings – at the mercy of the most disparate adversities and perils – was only made bearable by their ignorant state, unconscious of events and incapable of imagining how they might compete with the overwhelming forces of nature.

Everything changed once man discovered fire, because "... their chilly limbs could not now so well bear cold under the roof of heaven" (Lucr. *DRN* V 1015). This discovery came about by sheer serendipity – a lightning bolt or the friction of tree branches that resulted in a burst of flames, demonstrating to our curious ancestors how heat could be propagated. "Either of these happenings may have given fire to mortals. And then the sun taught them how to cook food and soften it by the heat of the flame" (1101–2). The mastery of fire brought about a gradual change in the morphology of man, who lost his primitive constitution and could no longer survive like the other animals in the wild, where he now risked extinction. It was, therefore, not nature that dictated man's evolution in a particular direction; it was the discovery of fire and the possibility of recreating it artificially that, according to Lucretius, influenced man's biological evolution even before his cultural development. The softening of his constitution by the warmth of fire compelled man to first construct a hut for shelter, then to protect his body from the inclement weather with the skins of animals, and, finally, to recognize himself as a social animal.

Fire was endowed with another quality that was greatly admired by Mediterranean and Near East civilizations. By appropriating the force of this element mankind could reproduce artificially the infinitely creative power of natural fire. Artisans working with fire reproduced the force that gods attributed to nature. When it came to fire and its multiple uses, the characteristic conflict between art and nature that so deeply pervaded the culture of ancient civilizations ceased. We have an echo of the admiration of the ancients for the power of fire in the following passage from Pliny the Elder's *Naturalis historia*:

> We cannot help marvelling that here is almost nothing that is not brought to a finished state by means of fire. Fire takes this or that sand, and melts it, according to the locality, into glass, silver, cinnabar, lead of one kind or another, pigments or drugs. It is fire that smelts ore into copper, fire that produces iron and also tempers it, fire that purifies gold, fire that burns the stone which causes the blocks in buildings to cohere. There are other substances that may be profitably burnt several times; and the same substance can produce something different after a first, a second or a third firing. Even charcoal itself begins to acquire the special property only after it has been fired and quenched: when we presume it to be dead it is growing in vitality. Fire is a vast unruly element, and one which causes us to doubt whether it is more a destructive or a creative force.
>
> (Plin. *NH* XXXVI 68)

This beautiful image of the extraordinarily creative power of fire combined the Heraclitean doctrine of fire, the relatively recent Stoic notion of *pyr technikòn* (artificial fire), and the extensive attention that Hellenistic natural philosophers paid to the prodigious progress of the chemical arts.

By assigning to fire a central role in the chemical arts, Pliny's remarkable assessment is deliberately ambiguous, because it is not clear if he is referring to natural fire or to the flame artificially controlled in the artisanal workshops of the chemical arts he enumerates. By keeping the boundaries between art and nature extremely vague, Pliny regarded the chemical arts as a bridge that could narrow the tension and, at the same time, could enhance the theoretical value of artisanal knowledge. Pliny's ingenious position was, in fact, the outcome of a long-standing tradition dating back to Mesopotamian and Egyptian crafts.

The effort to bridge the tension between art and nature is also shown by the attempt of various Hellenistic philosophers to reconstruct the origins of metallurgy. Pseudo-Aristotle wrote the following regarding a fire that had broken out in an Iberian forest:

> In Iberia they say that when the undergrowth has been burned by shepherds and the earth heated by wood, that the ground can be seen to flow with silver and that after a time earthquakes have occurred and the ground split, that much silver has been collected, which supplied the Massaliots with considerable revenue.
>
> (Aristotle, *On Marvellous Things Heard*, 87)

Seneca says that Posidonius (ca. 135–51 BCE), too, believed in the natural origin of metallurgy on the basis of a similar reconstruction. "Philosophers discovered iron and copper mines, when the earth, burnt by forest fires in molten form cast surface veins of ores" (Kidd 1988: 964).

Lucretius presents a more detailed explanation in the *De rerum natura*:

> ... copper and gold and iron were discovered, so also heavy silver and massive lead, when fire upon the great mountains had burnt up huge forests with its heat: whether by some lightning stroke from heaven, or because men waging war in the forests had brought fires upon their foes to affright them ... which flaming heat with appalling din had devoured the forests deep down to the roots and parched up the earth with fire, through the hot veins into some hollow place of the earth would ooze and collect a stream of silver and gold, of copper also and lead ...
>
> (*DRN* V 1241 ff)

The creation of this myth – the fusion of metals as the result of occasional natural events, subsequently appropriated by man to imitate nature through the introduction of chemical technology – enabled writers to show a parallel between nature and human activity. The rivers of metal that ancient man could see imprisoned within the smith's unhealthy workshop were nothing but the repetition and emulation of that which nature had indicated at the origins of civilization. In this regard, the vivid images of Vulcan's workshop should be remembered, which Empedocles situated inside an actual volcano. Also here the mythological origins of metallurgy were depicted as in nature. This analogy offered a cultural legitimization of these activities and, more importantly, allowed products of metallurgy to be seen as things of philosophic and scientific inspiration. Ancient artisans engaging in the manipulation of matter revealed a skill that enabled them to imitate nature, and their dexterity was appreciated by Mesopotamian and Egyptian priests as well as by Greek philosophers. In the field of chemistry, the coexistence of art and nature was stressed repeatedly. Theophrastus, speaking of a sand with special properties, noticed a means of imitating nature efficaciously in experiments and in technology, thereby outlining a theoretical principle that would later be continued in classical alchemy. In this regard, Theophrastus wrote:

> It is clear from three facts that art imitates nature and creates its own peculiar products, some of them for use, and some only for show, such as paints, and others for both purposes equally, such as quicksilver.
> (*On Stones*, 58)

Among the chemical arts, glassmaking is perhaps the most notable example of the harmonious balance between art and nature. Pliny the Elder reconstructed the history of this material as follows:

> There is a story that once a ship belonging to some [Phoenician] traders in natural soda [*nitri*], put in here [on the beach near the mouth of the river Belus] and that they scattered along the shore to prepare a meal. Since, however, no stones suitable for supporting their cauldrons were forthcoming, they rested them on lumps of soda from their cargo. When these became heated and were completely mingled with the sand on the beach a strange translucent liquid flowed forth in streams; and this it is said, was the origin of glass.
> (Plin. *NH* XXXVI 65)

The story told by Pliny, whose original source is unknown, has often been considered unlikely by archaeologists, even though the passage is cited in many histories of glass. Pliny's reconstruction, however, is of great interest not only because it emphasizes the seminatural origin of glass, but also because it evokes, albeit implicitly, its typical characteristic of being a product of trade. Pliny underlines the marvelous event witnessed by the merchants, who, by submitting the siliceous sand and the soda to the action of fire, saw, after an almost spontaneous chemical reaction, the formation of glass filaments.

Phoenicians, as is well known, have often been credited with many ancient technical inventions whose actual paternity is either Mesopotamian or Egyptian. However, their connection to the introduction of glass, unfounded as it now seems, was reasonable both because of the technical skills achieved by Phoenician glassmakers in the era when Pliny was writing his work (first century CE) and because of the fundamental role played by these merchants in exporting glass products throughout Mediterranean civilizations. In this regard, it is interesting to note that when glass was introduced in Egypt during the XVIII dynasty the Phoenicians increased their trade, thus becoming an essential medium of contact between the Near East, the Minoans, and the Egyptians (Partington 1935: 433). The trade of increasingly large quantities of glass products stimulated the interest in appropriating a technique that promised the possibility to imitate nearly any mineral and precious stone. While the Minoans and other Mediterranean civilizations showed a limited interest for glassmaking and were content to trade the products they brought from the Near East with the Phoenicians, the Egyptians aimed to create an independent industry. But Pliny's history of glass is important because it reveals a view of matter that was shared by the Mesopotamian, the Egyptian, and the Hellenistic technical cultures, namely that there were no differences between the glass produced by nature and the glass produced by craftsmen. This assumption, expressed in very different ways within the three civilizations, enhanced the creative power of the chemical arts and simultaneously increased the value of their artificial products.

The progress made in the chemical arts inspired pre-Socratic philosophers to adopt metaphors taken from techniques that could better explain their philosophy of nature (Mondolfo 1982: 35–50). The tinsmith's bellows were used by Anaximander to explain the fire emitted by the sun and the stars; the felting of textiles and the boiling of water were evoked by Anaximenes to illustrate the properties of matter; glass technology was used by Empedocles and later cosmologists to identify the nature of celestial bodies and spheres. Echoing the Egyptian religious tradition, Empedocles associated the four elements with the gods: "shining Zeus, life-bringing Hera, Aidoneus and Nestis who with her tears waters mortal springs" (Kirk et al. 1983: 296). Zeus represented Fire, Hera Air, Aidoneus Earth, and Nestis Water (Kingsley 1995). Empedocles explained the mechanism supporting this quadripartition of matter by resorting to a particular craft:

> As when painters are decorating offerings, men through cunning well skilled in their craft – when they actually seize pigments of many colours in their hands, mixing in harmony more of some and less of others, they

produce from them forms resembling all things, creating trees, and men and women, beasts and birds and water-bred fish, and love-lived gods, too, highest in honour.

(Kirk et al. 1983: 293–4)

Interestingly, the analogy with how painters mixed the four elementary colors to evoke both material and metaphysical entities explained how the immense variety of things composing the universe could be created from only four elements. Moreover, colors did not merely represent the thing painted; they also evoked its essence, thus participating in its identity even when such an identity was metaphysical.

We should not be surprised that the origin of Greek philosophy was inspired by technological arts that, until recently, historians have regarded as extraneous to the history of science. The chemical arts, in fact, accompanied the everyday life of Mediterranean civilizations; every innovation introduced represented a landmark that not only deeply affected the material conditions in which people lived but also, in a few significant cases, inspired the imaginations of natural philosophers.

However, from the outset, the evolution of the chemical arts was marked by an ambivalent reception of their cultural value. On the one hand, the possibility of imitating and surpassing nature through the technical control of the manipulation of matter elevated the status of craftsmen to the role of superior men. Recent historiography shows that the prestige of the ancient alchemist cannot be properly understood without considering the prominent role played by the ancient chemical arts (Martelli 2019). On the other hand, the chemical arts, especially mineralogy and metallurgy, were often regarded as dangerous and dirty activities, better to be performed by slaves. Moreover, the skill showed by many craftsmen in counterfeiting natural products (gems, precious metals, and minerals) was also associated with fraudulent activities at a very early stage. This is a reason why alchemy, from its earliest appearance, was regarded both as a holy art and as a deception.

ANCIENT CHEMISTRY: AN INVISIBLE SCIENCE

Before examining the social structure and organization of the chemical arts in ancient culture, it is necessary to provide a brief etymology of the word "chemistry." In fact, in antiquity there was no such thing as chemistry, only technical arts manipulating matter; this detail has not hindered historians' extensive exploration of the subject. The origin of the word "chemistry" has been a matter of heated dispute among philologists and historians, who have not yet agreed upon a single etymology (Lippmann 1919: 282–327; Lagercrantz 1938; Hermann 1954). Does "chemistry" derive from the Egyptian term *kemet*, the black pigment used to paint the eyes, or is it the result of the evolution of the Greek verb *kymeia*, indicating fusion? Does it originate from the Akkadian verb *kamû*, which means to bake or to roast? Furthermore, to denote obsidian, a naturally occurring black volcanic glass, the Egyptians used the word *Aner chem* (black stone). Is this material connected to the origins of the chemical arts?

To complicate the picture further, it is interesting to note that one of the earliest associations between chemistry and gold-making is a report, compiled around the tenth century CE, in the Byzantine lexicon *Suda*. Under the heading *kymeia* we read:

> The preparation of silver and gold. Diocletian sought out and burned books about this. [It is said] that due to the Egyptians' revolting behaviour Diocletian treated them harshly and murderously. After seeking out the books written by the ancient [Egyptians] concerning the alchemy of gold and silver, he burned them so that the Egyptians would no longer have wealth from such a technique, nor would their surfeit of money in the future embolden them against the Romans.
>
> (*Suda* 2014, heading: chi, 280)

Despite the interesting insights it produced, these etymological researches have not found any conclusive evidence on the existence in ancient civilizations of a structured and professionalized scientific discipline encompassing the chemical arts. Moreover, the list of etymologies produced in these scholarly studies showed that the occurrence of terms that seemed to be at the origin of our idea of chemistry were all exceedingly rare.

Indeed, the scarcity of ancient chemical and alchemical sources on the one hand, combined with the corruption of those transmitted from the third century CE onwards on the other, makes it difficult to reach conclusive arguments on the origins of ancient chemistry and its disciplinary identity. This difficulty is further underlined by the lack of consensus among historians on the theoretical purposes and the practical contents of chemical research. Partington (1970) attempted to identify ancient chemistry with the theoretical explanations set forth by the Greek philosophers of nature to explain the changes of matter, but in his earlier book he pointed out that the greatest technological breakthroughs made in ancient civilization were achieved within the chemical arts rather than in philosophical circles (Partington 1935). Other historians have insisted on the central role of Hellenistic and late antiquity alchemists, who were regarded as the first to devote works on subjects almost exclusively focused on the manipulation of matter and the use of chemical apparatus (Lindsay 1970). However, the identity of alchemy is also far from being straightforward. The tendency to reduce it to the sole operation of transmuting base metals into gold has led to a focus on the few passages of ancient literature that clearly identify such an intention, leaving aside many interesting texts containing other kinds of transmutation, the process of imitation, and several interesting detailed descriptions of chemical techniques (Halleux 1979). But projecting alchemy's dominant focus in the Middle Ages onto ancient chemical technology has resulted in interpretations that bear only a few literary traces in ancient texts. After all, gold was not seen as the most valuable material until late antiquity; rather, precious gems and lapis lazuli were often regarded as the most valuable minerals, and these stones were easily and often imitated in glass. The fact that glassmaking could already count on a rich ancient tradition of imitating precious stones by the Hellenistic period alone certainly merits the attention of historians of ancient alchemy, as it points to an important field that facilitates a better understanding of the historical background of the *Chrysopoeia* (gold-making).

The introduction of religious rituals before performing chemical operations in the Mesopotamian and Egyptian culture has led some historians to conclude that the chemical arts responded to a metaphysical need to connect matter and spirit (Jung 1944; Eliade 1956).

A recent and more perceptive approach, which is also represented in this volume, has called attention to the pervasive diffusion of recipe books in Mesopotamian, Egyptian, and Hellenistic cultures, thus showing an interesting line of transmission of knowledge that connects these three civilizations. However, the reader should always be aware that the scarcity of ancient sources on the one hand, and the corruption of those transmitted from the third century CE onwards on the other, makes it still difficult to reach conclusive arguments on the origins of ancient alchemy and chemistry and on their epistemological identities.

PRACTICAL ACHIEVEMENTS

The historiographical controversy surrounding the presumed antiquity of the science of chemistry cannot be resolved without taking into consideration two distinct elements that help to illustrate the meaning of this debate. Firstly, it should be noted that ancient craftsmen were masters of technical achievements and discoveries that were significantly surpassed only during the Renaissance. Secondly, the extremely efficient socioeconomic organization of the chemical arts that occurred in ancient cultures deeply influenced their appropriation during the Middle Ages and early modern times. This is not the place to elaborate the chemical discoveries made in ancient times, because their chronology and geography are still fluctuating. There are, however, a few achievements that help us to understand the focus of this volume.

Thanks to the large natural deposits of a dried lake situated in proximity to the Nile delta, the use of *natron* (soda) is attested in ancient Egypt from the fifth millennium BCE. With the addition of two equally common substances, salt and gypsum, Egyptian craftsmen perfected the art of embalming, food preservation, and cleansing (Multhauf 1982: 17–18). The introduction of metallurgy, dating back to the fifth and fourth millennia BCE, originated in the Near East, where gold, electrum, and copper were used at a very early stage. During the third millennium BCE, in the same area, the use of lead, silver, and bronze became equally widespread. Tin was discovered around 1700 BCE. Hittites began to smelt iron around 1500 BCE but, being a relatively rare metal, its diffusion was slow before it was replaced other metals. The geographical area of all these discoveries – the Near East – was crucial to the development of metallurgy in both Mesopotamian and Egyptian civilizations. Alum, a mineral used on a large scale by the Egyptians since the second millennium BCE, was used to fix natural dyestuffs onto cloth.

Glassmaking and glassworking technologies were introduced by Mesopotamian craftsmen in the third millennium BCE. They were perfected by Egyptians and culminated in the first century BCE somewhere in Palestine with the invention of glassblowing, a technical breakthrough that remained unsurpassed until the end of the eighteenth century (Beretta 2009). The early control of glass technology in Mesopotamian technical culture led to the imitation of precious stones such as lapis lazuli. Different recipes to imitate lapis lazuli using glass ingredients were later developed in Egypt with the successful discovery of so-called Egyptian blue in the area around Mount Vesuvius and the large-scale production of the pigment that the Romans called *caeruleum*.

Another area in which the chemical manipulation of matter proved to be extremely productive was that of medicine and pharmacology. Egyptians mastered the art of making medicines to such a degree that their superiority, as shown in the case of Cleopatra, was still acknowledged in Hellenistic times. The heirs of Near East civilizations, too, such as King Mithridates, were regarded as keepers of pharmacological secrets that, once revealed, remained extremely popular until early modern times.

While the Greeks outlined the most systematic effort to provide a theoretical account of the changes occurring in matter, they did not contribute to any major discovery in practical chemistry. This might at least partly be explained by the relative scarcity of mineral resources available in their territories. By contrast, thanks to the rapid geographical expansions of the Republic and the Empire, and the subsequent need to organize large-scale exploitation of the natural resources, the Romans contributed to important achievements and discoveries. Mining techniques made unprecedented progress, metallurgical processes were significantly improved, and the combination of these two factors led to the discovery of mercury, brass, caustic soda, and a wide range of previously unknown salts. As mentioned, the revolution brought about by glassblowing occurred in Palestine in the second half of the first century BCE, at that time a Roman territory. This innovation offers perhaps the best example of the high degree of organization introduced by the Romans in controlling and enhancing a chemical art once it was regarded useful.

The introduction of glassblowing enabled not only the production and manipulation of larger quantities of glass, but also an extraordinary variety of new colors among the objects produced. According to Pliny (Plin. *NH* XXXVI 198), "there is no other material nowadays that is more pliable." Thanks to this double advantage, the new technology enabled craftsmen to attain a hitherto unknown level of achievement and to create imitations of nearly any solid material. Such results were possible thanks to the combination of two factors:

1. The construction of furnaces that reached high temperatures (above 1000°C) and that made raw glass liquid;
2. The use of the blowpipe, which enabled easy handling of the glass melted in the crucibles.

The introduction of glassblowing radically transformed the traditional glassmaking craft over the course of a few decades, and it developed into such a prosperous industry that it is estimated that at the beginning of the second century CE, when the Roman Empire reached its greatest expansion and a population of 54 million people, "glassworkers had to turn out close to 100 million items annually just to keep pace with current demand – production on an industrial level indeed" (Fleming 1999: 60). Glassmaking's qualitative revolution was no less impressive than its quantitative one. Less than a century after the introduction of the new glassblowing technique, glass and glass paste were used in architectural decorations, wall mosaics, and windows to illuminate interior spaces; glass was also used in lamps, mirrors, tableware, aquariums, for unguent jars and the preservation of foods, for panels used in greenhouses, cinerary urns, sarcophagi, and ornaments, for the imitation of gems and the most precious stones, and the glass frit was even employed to produce certain colors for fresco painting, such as blue.

The remarkable discoveries achieved by ancient civilizations in practical chemistry were fostered by the socioeconomic organizations of craftsmen, which appreciated and enhanced their skill. Since Mesopotamian and Egyptian civilizations regarded precious minerals as material emanations of godly power, the professions involved in their manipulation were strictly controlled by the priesthood and their practice involved religious rituals. Although the locations of most ancient chemical laboratories remain unknown, a few archaeological findings, such as those at Dendera, have revealed the presence of chemical workshops in temples (Derchain 1990). We find a remarkable confirmation of the social prestige attributed by the Egyptians to specific crafts, such as the production of artificial lapis lazuli, in a famous passage from Theophrastus' *On Stones* (55), in which he remarked that the inventor of artificial blue (*kyanos*) was a king. This discovery had a crucial religious meaning because

Egyptians believed that the hair of gods was made of lapis lazuli and the formula recreating such a precious material had, therefore, to be controlled by the highest religious authority. Moreover, under royal supervision, the secret of the invention could be secured. The importance of keeping secret the ingredients of the discoveries made in Egyptian chemical workshops further enhanced the prestige of craftsmen and implied an encoded and regulated transmission of knowledge that envisaged the communication strategies later adopted by the Greek alchemists, many of whom were of Egyptian origins, and the organization of Roman *collegia* of arts, where the secrets were transmitted within guilds.

Roman sources provide us with a more complete picture of the rich social organization of the guilds involved in what we may call practical chemistry. Among them, we may recall the guilds of the glassmakers (*vitrearii*), the pearl- and gem-makers (the *gemmarii* and the *margaritarii*), the mosaicists (the *laquerarii* and the *diatretarii*), the pharmacists (*pharmacotribae*), the makers of medical spices (*aromatarii*), the makers of pigments (*pigmentarii*), the perfume-makers (*seplasarii*), and the smiths (*ferrarii*). The increasing scale of the trade towards the end of the Republic and the beginning of the Empire contributed to the specialization of crafts and to a sophisticated division of labor and transmission of knowledge that partially survived during the Middle Ages and the Renaissance.

Ancient chemical arts made important contributions to the equipment and instruments used in chemical workshops. While relatively little is known about the apparatus used in Mesopotamian and Egyptian workshops, the extraordinary results obtained by craftsmen in these contexts show a high degree of experimental knowledge. The most important practical achievements were related to the control and increase of the temperatures of furnaces, the use of durable crucibles, and the specializations of tools. The most important breakthrough, which stabilized the chemical laboratory for over a thousand years, was the introduction in the second half of the first century BCE of blown-glass instruments and receivers. The historical role of glass in these early chemical laboratories can hardly be overstated. Its chemical inertness and resistance to high temperature made it an ideal material for numerous operations and reactions. During the first centuries CE, we find few but significant references to glass apparatus in literary, scientific, and alchemical texts. But as early as the first century CE, the appreciation of glass's chemical inertness was very common. In his *De materia medica* (V 95), Dioscorides acknowledged the properties of glass receivers as resistant to the volatile action of mercury, showing that he was acquainted with the alchemists' technical expertise in the handling of chemical substances.

In the surviving alchemical texts published by Berthelot and Ruelle at the end of the nineteenth century, one finds several references to glass receivers and other vessels (Berthelot and Ruelle 1888), the shape and function of which are often difficult to identify. This difficulty is due to both the lack of archaeological correspondence between a given name and the item to which it refers and to the frequent references to the use of vessels and receivers in alchemical and pharmaceutical texts, of which we know only the ordinary uses established by archaeologists. Examples of the latter are vessels, also made of glass, such as the following typologies: *amphora, ampulla, aryballos, askos, calyx, catinum, crater, guttus, matula, oinochoé, patella, patera, phiala, poculum, pyxis, rython, scaphium*, and *schyphus*. These types of vessels were used both in daily life and in more specialized and scientific contexts. In a few instances, however, the description and nomenclature are more precise and specialized, providing us with useful indications about the progress achieved in constructing glass chemical apparatus. One example of this is the *botarion*, a glass instrument shaped like a breast (*mastarion*), which was used as a receiver for an alembic described by Synesius in the fourth century CE in his description of a process of distillation. In order to overcome the volatility of arsenic, Olympiodorus (fourth century CE) suggested the use of a special glass apparatus coated with earthenware, called *asympoton* by Africanus (third century CE), the aim of which was to cover the receiver in which the sublimation of arsenic was performed. The *angeion* often mentioned by the Alexandrian alchemists was probably a test tube. Glass vessels called *poteria* and *igdis* were used during the coloring of gems, in their turn products of glassmaking. And large glass jars called *bikoi* were intended as components of distillation apparatus. In a work that Berthelot and Ruelle (1888) attribute to Zosimos (third century CE), the difference between the male and female components of a glass alembic is briefly mentioned.

Mary the Jewess (first or second century CE), an author probably from Alexandria or the Syrian–Palestinian coast, wrote a treatise entitled *Perì kaminon kai organon* (*On Furnaces and Instruments*), which dealt with experimental practice and was destined to have an important influence on the history of alchemy during the following centuries. In fact, this is the first known treatise on chemical instruments. In this and other works, Mary mentions more than eighty pieces of apparatus, thus showing the high degree of specialization achieved by Alexandrian alchemy. Among the instruments attributed to her, Zosimos mentions the *tribikos* (still used for distillation), which was connected by three tubes, along with three glass *bikoi*. The invention of the *balneum Mariae* (the bain-marie, water bath, or double boiler) is also attributed to Mary, although it was already known in Theophrastus' time. Mary also introduced many new instruments made of metal, clay, and, above all, glass. Among the latter category, the most important was the *kerotakis*: a cylindrical instrument for softening metallic foils and for the production of compounds with chemical colorants, and thus capable of making artificial gold and silver.

The apparatus in the Hellenistic alchemical laboratory showed a remarkable degree of specialization, which remained unsurpassed until the beginning of the early modern era. Similarly, the experimental skill shown by ancient craftsmen remained unequaled, and the chemical secrets behind the realization of many beautiful artworks preserved in museums were, until very recently, a mystery.

THE LITERARY TRANSMISSION OF CHEMICAL KNOWLEDGE

The literary sources concerning the chemical arts are scant, and they cover only a few moments in a history that, as pointed out in the preceding paragraphs, was extremely rich in discoveries and technical inventions. This contrast between the results obtained in practical chemistry and the rare surviving written testimonies devoted to the methods and interpretations behind the experimental procedure that led to them is probably explained by the central role played by secrecy in the transmission of knowledge. Secrecy was ensured either by relying on the oral transmission of recipes to a limited number of adepts or by compiling recipe books in a coded language. The practice of keeping chemical recipes secret had at least three justifications:

1. As already pointed out in the previous section, the artificial production of precious stones was endowed with metaphysical meaning and was therefore under rigid control by religious authorities.
2. The only way to protect and control a new discovery was to keep it secret within the strict social environment of the guild. The preservation of such secrets allowed sites of production of chemicals and stones to maintain a prosperous monopoly over a long period of time.
3. With the diffusion of chemical texts in classical and late antiquity, secrecy and coded language became the means to enhance the transcendental meaning of chemical practice.

In the Hellenistic epoch, a community of philosophers of nature, which only later became known as alchemists, explored ways of communicating their knowledge that remained successful throughout the Middle Ages and early modern times. While alchemists acquired literary experience and wrote several works, their practice of keeping the art secret led them to create an allusive and often obscure language. Alchemical literature, however, was not the only literary genre circulating in ancient times. Long before the diffusion of alchemical texts another extremely successful genre imposed its authority.

The first written texts concerning chemical practice were produced in Mesopotamian cultures and concern the art of glassmaking. Mesopotamian texts about glass from about the twelfth century BCE resemble medical tablets in their literary form: they prescribe instructions in the form of recipes, and some even recommend religious rituals and prayers, invoking the need to perform certain experiments on propitious days. The reading of the extant recipes reveals a few notable aspects:

1. Glassmaking seems to have been perceived by Mesopotamian authors as a craft suspended between art and nature, which enabled the imitation of precious stones.
2. The literary style of using short and instructive recipes to transmit the secrets of the art appears to be the point of departure for a literary tradition that experienced great success among ancient alchemists.

Even if the technical literature of Mesopotamian glassmaking did not appear to rely on an alchemical philosophy of matter, many of the experimental procedures of coloring or the theoretical distinction between genuine and artificial stones set forth by these texts paved the way for a set of concepts that would be more systematically developed by the Egyptians.

Historians and archaeologists do not agree on the technical value of recipes that were written by scribes and not by craftsmen, and the reader is invited to explore the following chapters presented in this volume in order to obtain an overview of their cultural contexts. However, what is beyond dispute is that this literary genre became extremely successful; it survived the decline of Mesopotamian culture, was adopted in Hellenistic epochs, and became, during the Middle Ages and early modern period, the most popular way to transmit chemical knowledge. The style that was adopted in the first printed book on glassmaking, Antonio Neri's *L'arte vetraria* (1612), would have certainly be understood by Mesopotamian glassmakers. Here, too, as in Mesopotamian glass texts, second-person verbs succinctly prescribe the operations to be followed to produce different kinds of glass. Even today, cookery books echo the great success of this ancient literary genre.

Ancient recipes, both chemical and alchemical, were circulated widely, and a few of them were transmitted almost unaltered to early modern times. It is, however, extremely difficult to track the ways in which these recipes were transmitted from one civilization to another. We know very little about how cuneiform recipe books were translated into other ancient languages, nor the degree to which the migration of craftsmen exporting their technical know-how in different contexts contributed to the oral transmission of recipes. What we do know is that many technical recipes circulated through time and space without encountering major obstacles.

Ancient recipe books bear the signs of a corpus of layered knowledge that reflected a long-lasting tradition of practice. This is the case of two papyri dating from around the third century CE, better known as the *Papyrus Leidensis* and the *Papyrus Holmiensis*. They offer lists of recipes with technical instructions on how to fabricate (by coloration) silver and gold and other metallurgical operations;

how to produce and fix dyestuffs; how to imitate precious stones with the help of glass, rock crystal, and metal oxides; and how to identify fraud in the art of assaying. Many of these recipes surely predated these papyri by centuries, their record being transmitted across generations without major alterations. While the instructions were supposed to be technical, no reference was made to crucial information concerning weights and measures; moreover, sometimes the phrasing was deliberately obscure or encoded, and explicit invocations of secrecy reminded the reader that this kind of knowledge contained information that should be kept within the circles of initiated adepts. Recipe books of this kind established an interesting connection between technical know-how, theoretical aspirations, and useful knowledge of chemical processes. The intersections of these three elements formed the basis of alchemical literature that we have references to during the first century CE, in Pliny the Elder's *Naturalis historia*.

Thanks to the combination of a new passion for gems and the state-of-the-art Roman glassmaking, Pliny mentioned several methods in subsequent chapters of his work that were used during his time to counterfeit all kinds of precious stones. Related to this discussion, he mentioned the publication of technical treatises devoted to the imitation of natural products (Halleux and Cannella 1998: 45–6). In the *Naturalis historia* (XXXVII 75), while testifying to the recent discovery of "a method of transforming genuine stones of one kind into false stones" and lamenting "considerable difficulty in distinguishing genuine stones from false," Pliny mentioned the existence of treatises (*commentarii*), whose authors he preferred to omit, which gave "instructions [on] how to stain crystal in such a way as to imitate smaragdus and other transparent stones, how to make sardonyx of sarda, and other gems in a similar manner" (Plin. *NH* XXXXVII 75).

The circulation of literature of this kind is very important because it attested to the existence of presently unknown authors who wrote about topics that, for centuries, had either been kept secret or treated in general and encyclopedic works such as Pliny's. The amazing dexterity that Roman artisans demonstrated in imitating precious stones justified the diffusion of this literature. While the practice of imitating stones must have been as ancient as chemistry itself, the publication of treatises explicitly devoted to it must have been relatively recent to Pliny's day, as no similar references can be found in earlier sources.

The fact that Pliny deliberately chose to omit the names of the authors of these treatises on glassmaking reveals that the debate about the relationship between natural and artificial stones must have been particularly lively, and that the ambition to create gems using the chemical arts was regarded with contempt by traditional naturalists. Inspired by a conservative philosophical standpoint, Pliny, like Seneca, despised the pretentious attitude of craftsmen who contended with nature over the act of creation. The ancient philosophers' position seemed to be incompatible with the proliferation of opinions and practices that, in Pliny's and Seneca's eyes at least, revealed the cultural and moral decadence of their contemporaries. The high social status of both these authors justified their negative attitude towards the *commentarii* and their authors, but one wonders if their perceptive attention to the recent technical progress in glassmaking was not itself a sign of the power such products exerted on the intellectuals of the epoch. But there were other authors, namely the alchemists, who held a different opinion, and by taking the chemical arts as their point of departure they developed new theories on the properties of matter.

Another ancient literary genre that exerted a considerable influence was represented by lists of minerals, stones, and pharmaceutical remedies. Pliny, in his books dedicated to metals and stones, mentions the main sources for his work, which, besides Theophrastus, amount to forty authors, for most of whom nothing but their names remain. From what one can gather by browsing Pliny's bibliographic references, it is clear that many of the lapidaries he commented upon were written by Egyptian and Persian authors; moreover, the nearly 2,000 observations (*observationes*) and data (*historiae*) he gives in the mineralogical books (XXXVI and XXXVII) of his *Naturalis historia* reflect a richness and variety of approaches – of which Theophrastus' represents the most authoritative, but not necessarily the most popular and influential, work of the field.

According to Halleux and Schamp (1985), the ancient lapidaries derive from four distinct literary traditions. The first was still indebted to the Theophrastian model privileging a descriptive method of classification and, notwithstanding some concessions to superstitious beliefs, placed stones within a taxonomic perspective. The second tradition, inspired by the spread of texts supposedly coming from the Orient, openly adopted a magical and esoteric approach. The third, connected to this, associated mineralogy with astrology – an approach that would be adopted with great success by Paracelsus in the Renaissance. The last tradition, directly inspired by Judeo-Christian beliefs, discussed stones and the mineral world by means of allegories. Although references to glass are not absent in the magical–esoteric tradition, the approach inaugurated by Theophrastus would have greater scientific relevance and was destined to have the most enduring influence. Lapidaries, however, were not the only books that treated the properties of stones and gems, and the tendency to the production of specialized texts led to compilations of lists of minerals as parts of pharmaceutical textbooks. An example is Dioscorides' *De materia medica* (first century CE), the earliest surviving pharmacological treatise and the only complete text belonging to the catalog tradition. It lists remedies taken from the vegetable, animal, and mineral worlds, not presented in alphabetical order, but rather classified according to scrupulous descriptions without concessions to occult or magical beliefs. This approach must have been quite an important tradition, since Pliny almost always stops to describe the therapeutic properties of stones and gems – an aspect completely absent from Theophrastus' work. The therapeutic purpose of the treatise put the different methods of preparing substances at the forefront, although, not unsurprisingly, glass is not mentioned among the remedies. Dioscorides shows that he was familiar with the latest techniques for handling chemical substances. Moreover, the preparation of remedies derived from mineral substances presupposed a familiarity with such chemical operations as calcination and the use of furnaces.

The variety of literary genres and media by which chemical recipes circulated in ancient civilizations and their subsequent enduring influence provide us with a valuable indication of the cultural importance of the chemical arts. The chapters that follow aim to illustrate their historical contexts.

While following the same outline of the other five volumes of the *Cultural History of Chemistry*, the chapters that follow have been written by three different authors (Sydney H. Aufrère, Cale Johnson, and Matteo Martelli), and each ends with a general conclusion (written by myself). The reason for this choice of format is due to the peculiar nature of this volume, which covers over 3,000 years of history and the interactions of three distinct civilizations. The philological skills required to deal with Egyptian, Mesopotamian, and Greco-Roman chemistry necessitate highly specialized competencies that nowadays are impossible to expect from one scholar alone. By subdividing the chapters into three parts, we hope to provide the reader with an easily understood and more homogeneous reconstruction.

Ancient theories of matter were heavily dependent on religious and mythological assumptions, and experimental practice as we know it was rudimentary and still lacking the spaces where it could be practiced in a methodical manner. When it comes to economic structure and growth, the economy was only partially based on trade and industry, but the invention and use of chemical processes did increase the profits of many of the civilizations treated in this volume. This increase was not exclusively positive, as the predominant role of slavery in the exploitation of natural resources also hindered an in-depth reflection on the polluting effects of the chemical arts on their environment.

Thus, when it came to reconstructing the history of ancient chemical arts, the contributors to this volume reflected on problems that hitherto have not been sufficiently explored and that therefore could be addressed only in an indirect way. As ancient chemistry obviously cannot be compared to modern chemistry, the reader will be invited to explore its multidisciplinary ramifications through an attentive reconstruction of the historical and cultural contexts from which they stemmed.

CHAPTER ONE

Theory and Concepts: *The Mythological Foundation of Chemical Theories in Ancient Civilizations*

SYDNEY H. AUFRÈRE, CALE JOHNSON,
MATTEO MARTELLI, AND MARCO BERETTA

EGYPT

Sydney H. Aufrère

Among surviving Egyptian religious texts and culture, no treatise on the spectrum, the origin, the nature, and the uses of precious and raw substances and of other products exists. However, important texts from the New Kingdom (royal panegyrics, funeral reviews, and mythological compositions) as well as from the Ptolemaic and Roman periods give heterogeneous and dispersed information testifying to the permanence and sophistication of these uses as part of the philosophical approach of the hierogrammats (priests specialized in religious texts) of ancient Egypt. In the absence of such a treatise, it is nonetheless possible to draw the outlines and classify a wide range of raw substances and complex products containing ingredients coming from the vegetable, mineral, or animal kingdoms by their origin, their destination, and their use (Aufrère 2016c). In this approach, no substance is considered inert. Even if confined to a minor role, to ancient Egyptians each substance appears endowed with a more or less marked divine potential. Their properties are given by the very essence of the divinities to whom they are closely connected, properties that can be positively or negatively connoted, or presented on a quality scale.

Such a current of thought permeates the priestly circles, even if their expression fluctuates from one religious tradition to another. Far from being watertight, these traditions remain under the control of prestigious Houses of Life – places of intellectual education, of learning and transmission of texts, perhaps even of experimentation (Gardiner 1938) – where "thinking" is the food of life. Priestly libraries, meetings between clergy (synods), and the circulation of scholars and books contributed to the transmission of this line of "thought."

This line of "thought" led the Egyptians to imagine that gods were represented not only by metals, more or less noble, or compounds of metals and minerals, but also by earths, dyes, resins, and plants. These were no longer considered as mere commodities, but as substances with a rich potential. Each substance originated from a distant divine source and had an intrinsic divine signature. Hence the divine statues were made according to a tradition consisting in establishing correspondences between gods and substances, especially since many etiological myths made them divine biological fluids, and even revealed a hierarchy based on their presumed origin.

Indeed, if testimonies initially ascribed this line of thought to the *Texts of the Pyramids*, it in fact asserted itself in the New Kingdom during the economic peak of Egypt, when it imposed its dominance on Asia and on Africa, controlling both remote mining areas and caravan and maritime trade routes. Gold, silver, lapis lazuli, turquoise, carnelian, red and green feldspar, and hematite were then the most valuable materials known in the Egyptian world, since the so-called precious stones (diamond, sapphire, ruby, and emerald) did not become available in the lower Nile Valley until the Ptolemaic period. It is from the latter period onwards that the priestly discourse perpetually emphasized the mythification of the source of noble substances and their close link with the gods. By becoming the elements of a theology that ascribed thought to the divine rather than to a purely poetical or metaphorical approach, precious substances, whatever be their nature, connoted the divine.

Let us take, for example, the enlightening description of the sun god Rēʿ-Atum in the *Book of the Celestial Cow* (Guilhou 1989; Hornung 1997), a myth attested to for the first time in the tomb of Tutankhamun (1345–1327 BCE). The myth explains to us the transformation undergone by the sun, having become extremely old: its bones become silver, its flesh becomes gold, and its hair becomes lapis lazuli. In this process of mineralization of the aging body of the sun, silver prevails over gold – for a long time, in fact, silver – called "white gold" (*hedj*) – was considered more valuable than gold (*nebou*) because of its rarity. This same process applies to deposits of ores and minerals following exposure over a long geological cycle, the "setting" sun corresponding to the start of a paradoxical regeneration, the newborn child representing the start a new life cycle.

The three parts of the divine body (bone, flesh, and hair) then echo an etiological theory, expressed in the Papyrus Jumilhac (XII, 23–XIII, 1; Vandier 1963: 124; Aufrère 1991: 384–6), according to which the flesh and the skin of the newborn child are bequeathed by the mother and his bones by the father. Thus, the golden flesh of this child, delivered by his mother Nut, the celestial vault, is of solar origin, his silver bones are of lunar origin, and his lapis lazuli hair is reminiscent of the darkness in which he was immersed and from which he will spring to life.

Nevertheless, mineral qualities can be observed among the gods. As early as in the *Texts of the Pyramids*, it is said that "metal emanates from Seth," a divinity known for his titanic strength and outstanding fighting abilities. His bones are of "heavenly metal" (*bia-en-pet*; Lalouette 1979; Aufrère 1991: 432–8); that is to say, meteoritic iron, the first known source of iron and from which the dagger of Tutankhamun was made.

This idea prevailed, while it also evolved, because as late as in the second century CE, priestly manuals (Petrie 1889: pl. X, fr. 16; Osing 1998) confirmed a theory by which Manetho of Sebennytos (third century BCE; fr. 79: Waddell 1980: 190–1; see Plutarch, *De Is. et Os.* 62, 376B: Froidefond 1988: 233) associated Seth-Typhon with iron, while Horus, who had defeated him, held his power from magnetite (*beqes-ânkh*). The victory of Horus stemmed from the observation of the properties of attraction of the magnet for iron, transposed in a divine mode. In other words, the essences of Seth and Horus respectively announce their own defeat or victory (Aufrère 1991: 433–4).

Moreover, the Greek transliteration of an Egyptian name Petosorphmus (i.e. "The-One-that-gave-Osiris (*Pet-Osor*)-the-salt (*p-hmus*)") indicates that *natron* (sodium carbonate decahydrate), called "salt" (*hmus* = *hema*) in demotic, was associated with Osiris (Aufrère 1991: 607). The denomination of *natron* was originally represented by the combined sign (*neter*), a divine mast depicted above the prospector's bag, which indicates that it had for a long time been associated with the idea of the god closely connected to it (i.e. Osiris) and took its name from the god. Indeed, the latter, like Hathor, was believed to supply water to the nitraries, namely those of the Wadi el-Natrun, whose efflorescent alkaline salts were used for the desiccation and preparation of bodies for mummification (Aufrère 1991: 607–9). By contrast, sea salt (sodium chloride), considered to be a toxic and harmful substance, was associated with Seth, a negative deity (Aufrère 1991: 636–7).

Taking into account variations in the interpretation of texts, substances defined, more or less, the essence of the gods who cannot escape the material source of their origin. Here lies the beginning of the doctrine of sympathies, established in the *Treatise of Sympathies* of the Egyptian Bolos of Mendes (second century BCE), for whom the planets of the solar system and the stars exerted an influence on both metals and plants (Bidez and Cumont 1938: 189–90; Halleux and Schamp 1985: xxiv–xxviii).

Such an association is supported by the representations and texts on astronomy found on the ceiling of the pronaos (vestibule) of the temple of Dendara (ancient *Tentyris*), which associate the silhouettes of the thirty-six decans (star groups) with pairs of metals and minerals (Figure 1.1). By disappearing under the ecliptic plane (i.e. under the horizon) for seventy days – the duration of the embalming period for Egyptian nobles – and then rising again, following closely the apparent movements of the sun, moon, and other planets, these stars or groups of stars would deposit their cosmic ferments on terrestrial mines and thus affect the type of minerals they contain.

FIGURE 1.1 Ceiling of the pronaos of the temple of Dendara. Thirty-six decans made of both precious metals and minerals. © Sydney H. Aufrère.

These beliefs led to a balance between texts and objects – such as divine statues, coffins, and even embalmed human and animal bodies. Thus, the composition of the divine statues elaborated by the sacred goldsmiths is in agreement with the tenants of the Egyptian liturgical tradition.

The "Mansion-of-Gold" (*Hout-noub*), commonly called "Workshop-of-the-Goldsmiths" (Traunecker 1989; Aufrère 1991: 374–6), in the temple of Hathor of Dendara is considered as a liturgical space in which a series of eight glosses explain the exact designation of substances used at an earlier time. These glosses were written to avoid losing the thread of tradition and are sometimes interpreted as the origin of an alchemical thought (Daumas 1980; Derchain 1990). One of them mentions the following: "If one says of a god that matter is the true stone, one means that it is magnetite" (*Dendara* VIII: 41, 13–142, 12), a substance referring to the very nature of Horus (cf. above). Quite often, captions relating to representations of the divine statues of temples indicate the size and the nature of the metals and minerals used to make the object. Examples are given in the Crypt no. 1 of the temple of Dendara (cf. *Dendara* VI: 68–71, 74–5, 777–8, 81, 84, 87–90, 93–6, 98, 100; Cauville 2004) (Figure 1.2). It is not

FIGURE 1.2 Crypt South no. 1. Temple of Dendara. Statue of the god Harsomtus as a serpent. Caption indicating the size and the nature of the metals and minerals used (*Dendara* V: 140, 7). © Sydney H. Aufrère.

FIGURE 1.3 Names of the temple of Dendara dedicated to the goddess Hathor. Temple of Dendara, northern passage of the hypostyle hall, Ptolemaic period. © Sydney H. Aufrère.

surprising that one of Dendara's names was "House of Electrum" (*Per-en-Djam*) (Figure 1.3), because "Electrum of the Goddesses" was one of the designations of the goddess Hathor (Figure 1.4).

The *Ritual of Opening the Mouth*, attested since the Old Kingdom for making (divine and royal) statues and coffins (substitutes of the deceased), made use of a variety of natron (Aufrère 1991: 606–37), galena and chrysocolla (Aufrère 1991: 581–8), ointments and fumigations. Their use was meant to vivify, by "catalysis," divine and human effigies, and to simulate the restitution of their vital functions (ritual opening of the mouth, nose, ears, and eyes; Goyon 1972: 85–182).

These beliefs had an impact on the preparation of two liturgical ointments composed of a cocktail of mineral and vegetable ingredients, respectively called

FIGURE 1.4 One of the names of the temple of Dendara dedicated to the goddess Hathor: "Mansion-of-Electrum" (*Per-en-Djam*). © Sydney H. Aufrère.

"divine mineral ointment" and "precious ointment." According to its recipe, the first ointment was obtained after a well-codified series of operations, which associated aromatics, bitumen taken from an oil field near the coast of the Red Sea (Gebel el-Zeit), vegetable tar, metals, and crushed precious minerals. Originally, it was specifically used for the statues of the god Min of Coptos, the god of the Eastern Desert boundary, where important mineral resources were exploited. This mixture of aromatics and minerals of divine essence gave it a specifically long-lasting effect. The recipe of the first ointment was intended for the preparation of a blackish coating that when ritually applied with a spatula on divine statues and coffins gave them a particular efficacy. The second ointment, whose recipe is less elaborated, probably was a variant of the first one. Both were used to give divine meaning to the objects they coated. In his writings, Clement of Alexandria (*Protrepticus* 4–6, 48) mentioned such ointments, saying that in the time of the legendary King Sesostris, a Greek artist named Bryaxis took aromatics from the embalming of Osiris and Apis, mixed them with crushed metal and mineral substances, and spread the mixture on a statue of Sarapis (Sauneron 1962a).

Likewise, the Osiris-Khentymentyu figurine was made following a well-detailed ritual procedure during the festival of Khoiak (fourth month of the Flood season in Egypt; i.e. *Akhet*) and buried in the subterranean crypts of the Osirian mounds. It highlights the use of vegetable substances with a mixture of metals and precious minerals to confer a universal character and announcing the resumption of a new vegetative cycle, called germination (i.e. *Peret*; Chassinat 1966–8, 2: 779–88, 822; cf. 379–477, 814–5). On the basis of the same principle, during the Roman period, all the dead of the Egyptian aristocracy were mummified following special rituals whose purpose was to give them a divine appearance, as if they were Osiris (for men) or Hathor (for women).

According to the *Embalming Ritual* (Papyrus Boulaq 3 and Papyrus Louvre, inv. no. 5158), the ceremonialist calls out the names, as if they were fully-fledged beings, of oils, substances of mineral origin (metals, precious minerals, and bitumen), and even chemicals (natron, orpiment, earths, and dyes), which came from Egypt or foreign countries, all attested to in the liturgical preparations, to regenerate and exalt dead bodies with the example of divine personalities through a *mimesis* (Goyon 1972: 17–84).

In the Greco-Roman period, the funerary use of the *First* and *Second Books of Breathings* spread (Goyon 1972: 183–317). The first book announces characteristics of hermeticism (Quaegebeur 1995). The second, more traditional book reminds one of a characteristic of the *Embalming Ritual* (cf. *supra*): the ceremonialist in charge gathers the necessary products to ensure the deification of the body. According to the *Embalming Ritual of the Apis Bull* (Papyrus Vindobonensis, inv. no. 3873; Vos 1993), the same rituals were used, leading scholars to think that they could also have been used for other sacred animals, such as the Bukhis bull of Hermontis (Grenier 2002; Grenier 2009), the Mnevis bull of Heliopolis (Porcier 2012; Porcier 2014), or the crocodile Petesuchos of Crocodilopolis (Widmer 2003).

These beliefs were greatly influenced by a continuous mythopoeic reflection led by high-ranking priests who explained how to use the substances, which, by their nature and origin, induced specific transformation processes. Such reflection allows us to understand the great Royal orders addressed to miners, stipulating that gold, representing the flesh of Re, could not be diverted from its essential purpose – to serve the gods – such diversion being considered as a sacrilege (*Nauri decree* under the reign of Sethi I; see Edgerton 1947). This idea was still echoed in the work of Chenoute (348–466 CE), an archimandrite (superior of a monastery) who dominated Coptic intellectual life in Panopolis (Akhmîm; Daumas 1956; Aufrère 1991: 381). The exploitation of minerals with a divine status was a royal privilege that could be carried out only by those from within the priestly class. The looting of royal tombs at the end of the New Kingdom shows the fragile character of these beliefs and prescriptions.

These rites highlight an approach that can be characterized as mytho-scientific. The Egyptian priests who had access to the House of Life considered the emergence of substances, processes, and reactions, whether natural or activated by an external agent, through the prism of mythology. This mytho-scientific approach used etiological myths for primary or transformed substances. The recipes for complex liturgical preparations are reproduced on the walls of the so-called "laboratories" (*is*), as if they had been composed there, although these chambers only play a liturgical and not a practical role (see Chapter 7).

One of the best examples of this view, a ritual text (Papyrus BM 100090 + Papyrus Salt 825; Derchain 1964; Fermat 2010; Herbin 2010), explains, under cover of secrecy (see Chapter 2), when (at the beginning of every Egyptian winter, during the month of Khoiak; cf. above), how, and in what context the substances could be used to make the figurine of Khentymentyu. The origin of this rite is reported as resulting from the widespread weakness and affliction of the gods when the murder of Osiris by his brother and rival, Seth, was announced. Raw substances – trees, useful plants, herbs, metals, minerals, honey – or products (*shedeh*, an alcoholic beverage; Tallet 1995; Guasch-Jané et al. 2006; Gabolde 2009), were considered as emanating from biological fluids, including tears (Caron 2014), saliva, blood, sweat, or even the sputum and vomit of gods and goddesses. This was due to the psychological trauma caused by Seth, a god of chaos (Aufrère 2021a: 179–80). In another context, Seth was also considered as the one who had disseminated precious substances on the surface of the earth, just as he had dismembered Osiris and scattered all the parts of his body in the desert, a sacred and metaphorical representation of the extent of productions of terrestrial origin (Aufrère 2007: 160–6; Aufrère 2021a: 192).

This mode of production of minerals associated with bodily fluids was a model used for other substances as well. The turquoise, metaphorically called "stone of festival" (*heb*), is said to have come from emanations of the goddess Bastet-Horit when she appeared on the mountain of Heliopolis (Papyrus of the Delta, IX, 2–3: Meeks 2006: 19). We have here a geographical metaphor used for the Sinai Peninsula, where this cerulean stone, considered as representing dawn and sunrise, was exploited from the mines opening in the flanks of the Sarabit el-Khadim plateau, south of the Sinai. Likewise, sea salt, known to sterilize arable land, was called "the sputum of Seth" (*ishesh en Sutekh*). Plutarch, in his treatise *Isis and Osiris* (32, 363E: Froidefond 1988: 205; Aufrère 1991: 636–7), gave the name of "Typhon scum" (*Tuphōnos aphrón*) to sea salt, because it emanated from places dedicated to Typhon such as Lake Sirbonis, east of the Nile Delta (Aufrère 2017b: 42). However, a more detailed approach would demonstrate that by virtue of a sustainable concept, certain gods and goddesses with close links to substances, valuable or not, would in turn become producers of them. Such is the case of Hathor, goodness of metals and precious minerals (also of gums), who was venerated at the entrances of turquoise mines.

It is debatable whether or not there was a hierarchy of materials on the basis of their presumed origin, going by the lists of aromatics engraved on the walls of the so-called laboratories of the temples of Horus at Edfu and of the goddess Triphis at Athribis (Baum 1994a). These show that there were (a) first-choice gum resins, which were cosmic aromatic entities, including myrrh, oliban, and styrax, considered to be falcons' tears spread across Eritrea and Arabia and dropping from the Eye of Horus or of Osiris, and (b) second-choice gum resins of Sethian origin (Aufrère 2017a). This belief prevailed in Eritrean legends about fragrances (Katz 2009).

To conclude, this mytho-scientific approach of the Egyptians allows us to better understand the manufacturing procedure of drug substances to relieve pain (Aufrère 2021a). The idea of the active principle contained in a product is explained, from an Egyptian point of view, by an efficiency based on a mimetic scheme. The principle is based on an analogy: the disease to be treated is considered equivalent to that of the myth. This equivalence generally affects the healing process when the active principles of the *pharmakon* are presented as emanating from an explanatory etiological myth. Such is the case of the chaste tree according to the Louvre Papyrus, inv. no. E 32847 (Bardinet 2013; Bardinet 2017; Aufrère 2021a: 198–9). The qualities attributed to this tree are closely dependent on an etiological legend in which Osiris plays a role. In the field of Egyptian magic, illustrated by many documents (papyrus, magic stelae), if a product is shown to be effective in the treatment of a snakebite or of a sting in a divine patient, then the same product can be given to a human patient (Jelinková-Reymond 1956).

MESOPOTAMIA

Cale Johnson

Salts, lime, gypsum, and bitumen, not to mention the numerous chemical compounds that could be extracted in one way or another from the plant and animal life that naturally and copiously filled the Mesopotamian alluvium: these were not raw materials, naturally occurring in purified form, but rather materials that had to be extracted from complex natural forms through processes of refinement and purification. More important than any of these other complex materials from the Mesopotamian plain, however, is the ubiquitous material that defines Mesopotamian culture: clay. Carefully chosen from natural deposits, then worked and reworked into a smooth and properly resilient surface, the clay tablet was the primary writing surface throughout Mesopotamian history. It is largely through recipes, prescriptions, and instructions, written in cuneiform script on these clay surfaces, that we have access to the procedures that were used to arrive at chemically complex materials, whether arsenical copper, glass, or cosmetics.

FIGURE 1.5 Cuneiform tablet BM 120960, Middle Babylonian, Tell Umar. © The Trustees of the British Museum.

The alluvial sorting of different sediments throughout southern Mesopotamia meant, above all, that a wide variety of clays – suitable for clay tablets as well as the long history of ceramics and other mineral substances – were always available to Mesopotamian scribes, technicians, and inventors. Just as textile was the privileged medium of the Incas, or divinatory shell or bone in early China, it is perfectly appropriate that clay was the privileged material in ancient Mesopotamia.

Much has been made of the absence of certain kinds of "raw material" from this region, whether stone, precious or base metals, or the lengthy and strong beams of timber needed to span monumental buildings, but we should focus on the manifold procedures known from the Mesopotamian textual record that speak to the extraction or production of useful chemical compounds from materials that were regularly found in the land between two rivers. Standing alongside the local materials of the alluvium, imported raw materials also play a central role in the history of Mesopotamian chemistry. We find a wide variety of semiprecious stones and metals imported from neighboring regions, often via age-old supply chains. In terms of Mesopotamian cultural significance, the most important of these imported materials were semiprecious stones such as lapis lazuli, which were fashioned into the cylinder seals that become a centerpiece of administrative life, particularly with the invention of writing around 3300 BCE. Rolled over the entire exterior surface of a clay bulla, the cylinder seal makes the clay bulla into a tamper-proof legal device. It is little wonder, therefore, that the full technological prowess of the last phases of Late Uruk society (ca. 3500–3000 BCE) was focused on importing semiprecious stones and developing technologies that allow for the carving of increasingly hard stones.

These seal impressions often depict economic activities that supported southern Mesopotamian urbanism such as animal husbandry, the storage of grain, and the domestic forms of production in which many chemical processes first develop (weaving, dyeing, pottery production, and tanning). The need for a broad definition of chemistry in early Mesopotamia is made abundantly clear by a material process that was used to produce the designs on hard, imported semiprecious stones: the use of emery in combination with copper drill bits or lead mounts in abrasive solutions. This is a thoroughly mechanical – rather than chemical – process, but the Mesopotamians may well have thought about it quite differently.

The carving of minute designs like these on hard stones proved to be one of the most difficult technical challenges in Mesopotamian history. In the earliest phases, microliths made of flint were used to drill and bore, but the real leap forward came toward the end of the third millennium BCE, when particles of emery were combined with drill bits made of copper. As Gwinnett and Gorelick (1987) have demonstrated, the copper of the drill bit did not itself serve as the cutting element. Instead, the relatively soft copper of the drill bit served as an excellent holder or mount for microscopic bits of emery. Each speck of emery acted as a separate microscopic drill bit alongside hundreds of others, particularly in an abrasive solution made of emery particles in oil (see Simko 2015 and references therein). And while the production of the copper used to make these drill bits certainly represented a chemical process, the interaction between microscopic particles of emery and the copper of the drill bit is entirely mechanical. Can we really take an entirely mechanical

FIGURE 1.6 Late Uruk cylinder seal, 4000–3100 BCE. Photograph by Ashmolean Museum/Heritage Images/Getty Images.

process like this as an example of chemical practice? If for no other reason than counteracting present-day Whig histories of chemistry, in which we pick and choose ancient practices that happen to line up with modern-day scientific practice, we really must allow for the possibility that mechanical procedures that we might deem nonchemical were, in fact, conceptualized by their inventors and users in Mesopotamia as no different from the smelting of ore or the formulation of a medical prescription.

By far the most important means of classifying different material substances in ancient Mesopotamia was the use of cuneiform semantic determinatives (unpronounced cuneiform signs that indicate the substance from which an object was made), which were largely oriented to distinct types of social, economic, or technical activity rather than to a philosophical paradigm or universal categorization of the natural world. Semantic determinatives in cuneiform are much fewer in number and used for a much narrower range of functions than in Egyptian hieroglyphic. The full repertoire of nearly two dozen determinatives only came into regular use in the middle of the third millennium BCE. As a rule, these determinatives group objects made of the same or similar materials (wood, reed, stone, aromatics, plants, vegetables, and several types of metal), a type of manufactured object (pots and textiles), or human professions and offices, geographical features, and stars (for a traditional account, see Edzard 2003: 9; Selz et al. 2017 combine the traditional set of semantic determinatives with a group of quantificational elements such as Sum. didli that have not traditionally been included in the category, so caution is in order).

One might assume that these semantic determinatives provided the key principle of organization for the earliest lexical lists in the Late Uruk period (ca. 3300–3000 BCE) and consequently the earliest systematic descriptions of the different substances from which specific objects were made (lapis beads as opposed to copper axes, for example), but this is only partially and imperfectly true (see Wagensonner 2010 for an overview). The groundbreaking work of Englund and Nissen on the earliest proto-cuneiform lexical lists still provides the essential starting point for any discussion of classification in ancient Mesopotamia (Englund and Nissen 1993, now contextualized by Veldhuis 2014). Although the proto-cuneiform lists of birds, fish, and domesticated animals often receive the most attention, largely because we have a better grasp of the logograms involved, the lists dedicated to wood, (ceramic) vessels, and metals (including textiles), along with designations of offices and social roles, are more frequently attested and demonstrate the centrality of economic management and craft production rather than the classification of birds, fish, and animals as a purely intellectual pursuit. The early lists that are most relevant to the history of chemistry ("metals," "wood," and "plants") do not use a single recurring logogram in every entry in the list (as we see incipiently in "birds" and "fish"). For the "wood" list, to take up a relatively well-attested example, certain manuscripts (such as W20044,45) only include the "wood" determinative (GIŠ) in certain entries and not in others, whereas other manuscripts (such as W20327,2) add the "wood" determinative to every item in the list.

Where we first see specific and robust material or economic domains, clearly defined in the form of a lexical list, is half a millennium later in the Early Dynastic Practical Vocabularies (EDPV) found in Fara and Abu Salabikh in Mesopotamia and Ebla in present-day Syria. That these vocabularies represent a new moment in the lexical list tradition – the seed of a tradition that culminates in the Old Babylonian ur_5-ra = *hubullu* lists – has been repeatedly emphasized in the work of Miguel Civil (1987; see Veldhuis 2014: 71–139 for an up-to-date overview). In his edition of EDPV A, a decade ago, Civil was able to identify clearly demarcated sections, many of which are relevant to us here (Civil 2008):

- Stones (sect. 1 = entries 1–101 = pp. 51–79)
- Metals (sect. 2 = entries 102–68 = pp. 79–92)
- Textiles (sect. 3 = entries 169–95 = pp. 92–8)
- Oils (sect. 4 = entries 196–204 = pp. 98–9)
- Perfumes (sect. 6 = entries 216–25 = pp. 102–6)
- Skins and leather (sect. 9 = entries 262–83 = pp. 118–25)
- Colored clays and pigments (sect. 16 = entries 344–50 = pp. 137–9)
- Tanning materials (sect. 19 = entries 366–70 = pp. 142–5)

These seemingly "craft" or even "industrial" sections are interspersed with sections devoted to musical instruments, equids and harnesses, weapons, and boats, so we have good reason to suspect that the materials in themselves were less important than the finished products being fashioned. And a closer look at each section confirms this: the "stone" section actually begins with a lengthy section of lapis lazuli collars, necklaces, beads and other luxury goods, and many decorative figurines (stones bowls for offerings and stone tools come later and are few in number). The first items in the "metals" section are, in fact, metal statues and figurines, with metal vessels and tools only at the end of the section. The dozen entries dedicated to pigments and materials for tanning toward the end of the list (lines 344–52 and 366–80) are more promising, but only a few of these items can be readily identified. So even here, in the middle of the third millennium BCE, when we can see clear sectors of artisanal or craft production demarcated in a "practical vocabulary," specifically chemical practices are still entirely in the background and do not enter into the lexical tradition explicitly.

If we want to separate out material substances and the properties, real or imagined, from manufactured or finished objects in early Mesopotamia, the nonexistence of Greco-Roman style treatises in the cuneiform textual record has often led moderns to erroneously suppose that Mesopotamians had no such theories. Postgate (1997) provides us with a particularly useful discussion of how Mesopotamian approaches to substances and properties were actually put into practice in nonmythological texts and the archaeological record. The key evidence that Postgate brings together are small collections of raw materials that are occasionally found in the foundation deposits of temples and, in the particular case that gives rise to his paper, in an intramural burial in Early Dynastic Abu Salabikh containing "a neat arrangement of small items: a black pebble and a white pebble, a flint blade, a piece of bitumen, and nine other pebbles of nondescript colour and three or four cockle shells" (Postgate 1997: 206).

Contemporary with the EDPV that we were just looking at, this mortuary assortment is paralleled by archaeologically recovered foundation deposits: Ellis identifies a number of these assortments in excavated temple foundations, such as the "small rectangular pieces of gold, copper, lapis lazuli and slate ... in each corner" of the Oval Temple at Khafajah, with additional materials salted into specific corners (Delougaz 1940: 85–6 and figs 78–9; Ellis 1968: 132). Likewise, in the corners of the main ziggurat at Assur:

> two sets were found, one directly on the bedrock, and the other six courses higher ... The lower deposit consisted of round masses ... of small shells and of glass, frit and stone beads ... The upper deposits were similar; each one consisted of about 1,000 beads, laid on and covered with reeds. Small fragments of iron and lead were included (as were) a pair of small disks of gold and silver, each one bearing an inscription of Shalmaneser III (reigned 859–824 BCE), dedicating the ziggurat to Aššur.
>
> (Ellis 1968: 132)

The second of these foundation deposits, the one found in the ziggurat of Assur, is quite similar to descriptions of assemblages of substances in royal inscriptions running from Šamši-Adad (reigned 1809–1776 BCE) down to the famous statement of Sargon II (reigned 722–705 BCE) that "I aligned its masonry on gold, silver, copper, precious stones, cuttings (of fragrant resins) from Amanus; I laid its foundation and made its brickwork firm" (App. A, No. 15; Ellis 1968: 134).

Postgate's query to this rare instance of parallel bodies of evidence from the textual and archaeological record is clear enough: "What was the thinking behind this age-old practice? ... The texts tell us that these deposits were made, but refrain from explaining why" (Postgate 1997: 211).

These parallel assemblages in the textual and archaeological record point to a self-conscious effort to assemble a group of substances (and thereby their respective properties) *rather than* the finished objects made from these substances. Postgate cites a Hittite text in which it is said that "just as copper is secured, (as) moreover it is firm, even so let this temple be secure!" Postgate goes on to say:

> The point here is that copper is attributed properties, security and firmness or safeness and solidity, which are to stand for the properties the royal builder wishes to impart to the building itself. It is recognised as partaking of a *substance* and is put into the foundation for the *properties* which its substance confers, not for its appearance or its function.
>
> (1997: 211)

In the latter phases of the first millennium BCE, we see this type of systematic concern with substances and properties only in the Late Babylonian effort, in the second half of the first millennium BCE, to align medicinal substances with astrological configurations. This type of astrological medicine first becomes visible in the so-called "stone, plant, tree" lists in which four classes of material (stones, plants, trees, and a fourth category including things like scorpion carapace and dust from various locales) are crossed with three manners of preparation (phylactery, fumigation, and anointment) to arrive at twelve combinations (Heeßel 2005; Heeßel 2008). These twelve combinatorial possibilities could then be aligned with the twelve months of the year or, after its invention in 500 BCE or so, the zodiac (on the microzodiac algorithms deciphered in the last couple of decades, see Brack-Bernsen and Steele 2004; Wee 2016; Schreiber 2018). Unmoored from its astrological roots (where "blood of a ram" is a straightforward cover term for a plant associated with Aries; Steele 2011), this kind of schematic approach to substances and properties led ingredient names like "mongoose tail" (Akk. *zibbat šikkê*), which actually stands for the plant/drug "liquorice" (Akk. *šūšu*), to be systematically misunderstood, especially in later translations, as actually referring to mongoose meat. And this may well have had a significant impact on the "properties" literature that culminates in both Pliny the Elder's collection of exotic recipes (Rumor 2015: 87–95) and in Pseudo-Democritus's four alchemical books (Martelli 2013; Martelli and Rumor 2014).

The most famous and comprehensive of the cuneiform lexical lists, known as (Sum.) ur$_5$.ra = (Akk.) *hubullu*, was compiled in the heyday of the Old Babylonian period (ca. 1800–1600 BCE). In it we see the dynamic variation in the use of semantic determinatives and the gathering of different terms within a specific domain rooted out and replaced by relatively frozen sequences of Sumerian lexical items. Here instead we have extended lists of items, frequenting running for hundreds of entries without interruption, in which each entry begins with the same semantic determinative. These lists have been admirably summarized and the subsequent history described at length in Niek Veldhuis's *History of the Cuneiform Lexical Tradition*, from which I will only highlight a few key points here. The standard list of objects in Old Babylonian ur$_5$.ra = *hubullu* consisted of six large tablets or chapters (Veldhuis 2014: 150):

1. Wood
2. Crafts, including raw materials and manufactured items
3. Animals and cuts of meat
4. Natural entities and objects made from them
5. Geographical names
6. Food

Starting in the Middle Babylonian period (ca. 1400–1000 BCE), the same overall structure is renumbered (though not reconfigured in a substantial way) into a larger number of tablets so that the standard edition in the first millennium BCE ultimately consists of twenty-four chapters, usually designated as "Hh" followed by the tablet number (Veldhuis 2014: 156). The standard editions of these texts are published in the series *Materials for the Sumerian Lexicon* (MSL). Of these, the most relevant for our purposes here are as follows:

- Hh 3–7 ("Wood"), edited by Benno Landsberger in MSL 5 and 6 (1957 and 1958)
- Hh 8–9 ("Reed"), Hh 10 ("Vessels"), Hh 11 ("Leather"), and Hh 12 ("Metals"), edited by Benno Landsberger in MSL 7 (1959)
- Hh 16 ("Stones"), Hh 17 ("Plants"), and Hh 19 ("Textiles"), edited by B. Landsberger, E. Reiner, and M. Civil in MSL 10 (1970)

These editions are now subsumed and expanded online in Veldhuis's *Digital Corpus of Cuneiform Lexical Texts* (DCCLT) website, containing individual editions of each manuscript rather than the mixture of synthetic and "forerunner" editions found in the MSL series.

In each of these tablets or chapters we now find the stereotypical picture of hundreds of entries, each beginning with the same semantic determinative, a pattern that never appears so plainly in the earlier phases of the lexical list tradition. To give some scale to the enterprise, there are nearly 2,000 entries beginning with Sum. giš "wood" in Hh 3–7, ca. 350 entries beginning with Sum. gi "reed" in Hh 8, ca. 335 beginning with Sum. dug "pot," as well as a smaller group of ovens, kilns and related equipment, and ca. 125 entries beginning with Sum. im "clay and minerals" in Hh 10. In Hh 11 there are ca. 280 entries beginning with Sum. kuš "leather" and ca. 140 entries beginning with Sum. an.na "tin" or Sum. uruda "copper," and in Hh 12 there are ca.

350 entries beginning with several types of metal, including statues and tools. Hh 16 includes ca. 450 entries beginning with the general determinative for "stone" (Sum. na₄), Hh 17 consists of ca. 250 entries beginning with Sum. u₂ "plant/drug" and another 150 or so that end with the determinative sar or nisi(g) "vegetable," and Hh 19 comprises ca. 100 entries beginning with Sum. sig₂ "wool," ca. 200 entries beginning with Sum. tug₂ "textile," and a largely destroyed final section on threads and linen. It is really only in the Old Babylonian thematic lists like ur₅-ra = *hubullu* that a normative classification of entire fields of material composition is applied to the lexical list tradition. And yet, even in the Old Babylonian thematic lists there are numerous classificatory outliers such as a variety of manufactured minerals and pastes determined with Sum. im "clay" or the humble "spiderweb" (Sum. aš₅ = Akk *ettūtu*) classified as a type of Sum. u₂ "plant" because of its use in Babylonian pharmacology (Scurlock 1995; Postgate 1997).

The so-called *šiknu* lists are a distinct literary form that expands the basic pattern of the lexical list with a fixed formula consisting of (a) the characteristic name of the overall semantic class and the word *šikinšu* "its feature(s)," as in, for example, *šammu šikinšu* "the plant, its feature(s)," (b) a description or illustrative comparison, (c) the name of the specific entity in question, and (d) the expression *šumšu* "(is) its name" (see Schuster-Brandis 2008: 17–47; Stadhouder 2011; Stadhouder 2012; Mirelman 2015 for recent editions and discussion). This pattern was applied to a limited number of lexical fields: "stone" (Sum. na₄ = Akk. *abnu*), "plants" (Sum. u₂ = Akk. *šammu*), "snakes" (Sum. muš = Akk. *ṣēru*), and what is most likely a cryptographic writing for birds or their calls (BALAG, perhaps equivalent to Sum. mušen). More elaborate forms of this pattern, particularly for *šammu šikinšu* (the descriptions of plants), are known as well, and there may even be one manuscript of this descriptive pattern that dates back to the last centuries of the second millennium BCE (Schuster-Brandis 2008: 18, text E = VAT 9587). The schema outlined above is similar to a few other types of text, notably the so-called *Göttertypentext* (Mirelman 2015: 174), which are descriptions of the different parts of divine statues, a descriptive paradigm that has affinities with so-called Body Description texts and other similar forms (Reynolds 2002; Reynolds 2010; Pongratz-Leisten 2015). These descriptive lists of the properties of plants, minerals, and so on were not the idle speculations of a lone scholar, but rather a regular part of the first-millennium BCE scientific tradition, cited or alluded to in royal inscriptions and wedged into sequences of other compendia through the use of incipits and catch lines (Reiner 1995: 119–32; Robson 2001: 52–3). It is telling that the descriptive lists of plants (*šammu šikinšu*) almost always use other plants as their comparanda, including occasional *Decknamen* (cover names) such as "sailor's feces" (Akk. *zê malāḫi*) or "dog's tongue" (Akk. *lišān kalbi*), but these were clearly recognized as plants (Rumor [2018] argues that the textual structure of Theophrastus' *Historia plantarum* IX is based on the *šammu šikinšu* tradition). As we move into the other descriptive lists, the comparanda become far more diverse, vivid, and nonpharmacological: in the stone list *abnu šikinšu* we find items of comparison ranging from "a swirl of red wool" (Haupttext line 10) or "uncooked bull blood" (line 33) through "dragonfly wing" (line 84) and "sulfur flame" (line 46).

GRECO-ROMAN WORLD

Matteo Martelli

Expressions like "chemical element" or "chemical reaction," at least in their modern meanings and formalizations, are alien to ancient classical philosophy and science. However, Greek and Latin authors certainly made various attempts to describe, conceptualize, and explain the material transformations that they experienced in the natural world. In the framework of different systems, all classical philosophers dealt with natural substances, investigating their properties, compositions, and mutual interactions. Modern scholarship on this subject is vast, and one can potentially find relevant information in every piece of a massive literature dealing with ancient "physics" (i.e. *ta physika*, "natural phenomena") and ancient explanations of what, in Aristotelian terms, may be identified with the sublunary world. J.R. Partington's first volume of his monumental *History of Chemistry* (published posthumously in 1970) is devoted to the "theoretical background" of ancient alchemy: this volume, in fact, is a monograph-length survey of ancient philosophy, from early pre-Socratics to late Neoplatonic schools. Partington also described Plato's *Timaeus* as "the first treatise on theoretical chemistry" (1970: xi), whereas similar definitions were applied by other scholars to certain Aristotelian treatises, such as the fourth book of *Meteorology*, explicitly referred to as Aristotle's chemical treatise (Düring 1944; Viano 2002), or *On Generation and Corruption*, defined as the manifesto of Aristotle's physical chemistry (Giardina 2008; see already Joachim 1903).

Scholars have also addressed the question of the possible influence that classical philosophers, Plato and Aristotle in particular, had on the writings of Greco-Egyptian and late antique alchemists, such as the works by Zosimos of Panopolis (third to fourth centuries CE), Synesius (fourth century CE), and Olympiodorus (sixth century CE; Viano 1996; Viano 2005; Dufault 2015). Byzantine alchemical collections do include the names of Plato and Aristotle in the lists of the fathers of alchemy, which are handed down in Greek manuscripts (*CAAG* II 25–6). In his commentary on Zosimos' (lost) work *On the Action*, Olympiodorus – perhaps to be identified with the writer of that name who was a Neoplatonic commentator of Aristotelian works (Viano 2006: 199–206) – draws a close comparison between early Greek philosophers and the fathers of alchemy, namely Hermes, Chymes, Agathodaimon, and Zosimos (*CAAG* II

80–5; Viano 1995). In this doxographical section, Olympiodorus discusses the different "first principles" (*archai*) proposed by nine pre-Socratics, such as Thales' water, Heraclitus' fire, or the air of Diogenes of Apollonia. Then, he compares these principles with those of the alchemical art. He writes, for instance:

> Anaximander referred to "the intermediaries" [as first principle], namely smoke (*kapnos*) and vapor (*atmos*). Indeed, Agathodaimon's [words are]: "in short, it is the sublimate (*aithalē*)," as Zosimos says. Most of those who philosophically treated this (alchemical) art have mainly followed these (men). Hermes spoke about smoke as well; indeed, speaking about *magnēsia*, he says: "let it be burnt against a furnace with flakes of purple *kōbathia* [arsenic ores?]." For, being white, the smoke of *kōbathia* whitens the (metallic) bodies. Smoke, in fact, is between hot and dry; the same is the sublimate and all the substances made of it; vapor is between hot and wet (substances), and he means wet sublimates, such as those produced with alembics and the like.
>
> (*CAAG* II 84–5)

With Olympiodorus, we are at the end of the chronological period under review in this volume. The alchemist discusses ideas inherited from the earlier philosophical tradition, such as the distinction between dry smoke and wet vapor, which recalls the Aristotelian theory on the composition of minerals (see below). However, rather than being purely philosophical principles, these elements coincide with operative objects, namely wet and dry sublimates that ancient alchemists produced and manipulated in their operations. Indeed, the alchemical art is a combination of theory and practice, as already pointed out by Festugière, who defined Greco-Egyptian alchemy as a *mélange* of Egyptian traditional goldsmithing and Greek philosophy, mainly Platonic and Aristotelian ideas (Festugière 1944: 218–19). This relationship cannot be simply conceived as the application of theories made up beforehand to the explanation of alchemical operations. Since the classical period, in fact, "chemical" arts (e.g. dyeing procedures, metallurgy, perfume-making; see Chapters 2 and 3 on these arts) provided empirical observations that fostered the formulation of various theories on natural phenomena. Relevant examples will be discussed below, with a particular focus on those elements that had an impact on the later development of Greco-Egyptian alchemy.

In Olympiodorus' doxographical section (see above), the name of Democritus is missing. On the other hand, Democritus is mentioned throughout Olympiodorus' commentary (as well as in almost every Greek and Byzantine alchemical work), not as the founder of atomism, but as the author of four pseudo-epigraphical books on alchemical dyeing that date back to the first century CE (Martelli 2013). The later history of the concept of the atom does not need to be emphasized here, let alone the impact that the discovery of Lucretius' *De rerum natura* in the Renaissance had on the works of early modern scientists such as Pierre Gassendi or Robert Boyle (Luthy 2000; Beretta 2015: 219–64). In the same period, other ancient texts providing discussions of ancient atomism were rediscovered, edited, and translated into Latin, such as Diogenes Laertius' *Lives of Eminent Philosophers* and Galen's *On the Elements According to Hippocrates*. Atoms, on the contrary, do not play any role in Greco-Egyptian alchemical writings, where the historical figure of Democritus the atomist is overshadowed by "Democritus" the alchemist, the pseudonym used by a first-century CE alchemical author whose historical identity cannot be reconstructed. Democritus' name seems to have been used by this writer because of the atomist's expertise across several arts or *technai* (Martelli 2013: 34–6), as emerges, for instance, from the portrait given by Petronius (*Sat.* 88): "Democritus extracted the juice of every plant on earth, and spent his whole life in experiments to discover the virtues of stones and twigs" (Democritus, fr. B300,6 DK; transl. by Heseltine 1913: 173).

Indeed, in the framework of his atomistic theory, this pre-Socratic philosopher did focus his attention on arts or *technai*, and he emphasized their pivotal role for the advance of human civilization (Cole 1967). According to Democritus, infinite and eternal atoms, which differ in shape and size and move and combine in an empty space (the void), are the ultimate constituents of the natural world. Atoms of the same kind tend to congregate, thus forming various materials, from the sea to earthy substances, such as salt, soda, alum, and bitumen (fr. A99a DK). Moreover, the philosopher speculates on the atomic composition of metals. Iron is lighter than lead because it contains more void, while its atoms are densely packed in specific areas; lead, on the contrary, is formed of atoms more regularly arranged (Theophr. *Sens.* 62 = fr. A135 DK; see Halleux 1974: 74–6). When discussing the nature of the four primary colors (white, black, red, and yellow), Democritus refers to the shapes of the atoms and their disposition: white, for instance, depends on atoms whose shape is similar to the inner surface of shells (Theophr. *Sens.* 73 = fr. A135 DK). Likewise, the color of metals depends on the mixture of different primary colors: gold and copper, for instance, are bright because they are a combination of white and red – that is, the result of the combination of atoms with particular shapes (Theophr. *Sens.* 76 = fr. A135 DK).

A complete system of four elements – earth, water, air, and fire – recognized as the fundamental constituents of the natural world was fully developed by Empedocles (fifth century BCE), who used the term "roots" (or, sometimes, the names of Olympian gods) to refer to these elements. If early Ionian philosophers identified a single dynamic substance with the cosmic principle (*archē*) undergoing cyclic transformations, Empedocles "posits a plurality of substances of fixed natures that interact in different proportions to produce mixed substances" (Graham 1999: 165). These substances (i.e. the elements), eternal and unalterable, are mixed together and separated by two opposite cosmic forces, Love and Strife, whose work is often assimilated by Empedocles to the work of craftsmen (Wright 1981: 39). For instance, the combination of the four elements is explained in Galen's *Commentary on (Hippocrates') Nature of Man* (I 3 in XV 32 Kühn = Mewaldt 1914: 19) as follows:

> Empedocles maintained that the nature of compound bodies derives from the four unchangeable bodies: these are the first (elements) that mix one with the other, as if someone finely ground verdigris, copper ore (*chalkitēs*), cadmia (zinc oxide) and *misy*, and produced a fine powder by mixing them together, so that none (of its components) can be handled without the others.
> (Empedocles, fr. A34 DK)

The comparison with pharmacology might not belong to Empedocles' original writings (Halleux 1974: 68), although we must observe that the philosopher referred to how painters mix pigments of various colors (*polychroa pharmaka*) when he tried to explain how things come to be from the combination of the four elements (fr. B23 DK; text fully quoted in M. Beretta's introduction, pp. 9–10). Scholars usually agree that, rather than referring to the blending of pigments, Empedocles meant that colors were put side by side to depict a picture (Ierodiakonou 2005: 5–8). Colors (like elements) touch one another rather than fusing one into the other: they may be compared to bricks and stones that are placed side by side to build a wall, as Aristotle explains in *On Generation and Corruption* (II 7, 334a). On the other hand, if the four elements do not change in Empedocles' theory, they are conceived as capable of being broken into very small parts (Wright 1981: 37–8); thus miniaturized, they can intermingle with one another in compositions like the medicinal powder mentioned by Galen. The nature of these compositions seems to depend on the structure of their components. In *On the Generation of Animals* (II 8, 747a–b), Aristotle discusses Empedocles' explanation of the sterility of mules (in contrast to the fact that horses and donkeys are fertile). In Empedocles' opinion, the soft seeds of both parents, when mixed together, become hard, because the hollows in each fit into the densities of the other. The hardness of tin-copper alloys was explained in the same way: while the two metals are soft, their "mixture" is hard, presumably because the structures of copper and tin allow them to fit one into the other in tight juxtaposition (Halleux 1974: 69–70).

While Empedocles referred to the four elements as "roots," one of the earliest occurrences of the term *stoicheion* (στοιχεῖον, lit. "letter") as "element" appears in Plato's *Timaeus*. In this cosmological account – a "likely account" (*eikos mythos/logos*), as the philosopher defines his argument in the dialogue (29d; 30b) – each of the four elements has the shape of a geometric figure (or solid) whose main components are right-angled triangles: fire-tetrahedron; air-octahedron; water-icosahedron; earth-cube. If the *Timaeus* is one of the first attempts to bring mathematics (geometry in particular) into a cosmological discourse, this geometric explanation is often colored by images and terminology taken from the world of ancient arts, which Plato used to explain the work of the demiurge, a craftsman god (Brisson 1998: 35–50). For instance, in order to explain the nature of *chōra* (traditionally translated as "receptacle"), one of the central entities in the dialogue, Plato writes:

> That is why, if it is to be the receptacle of all kinds, it must be altogether characterless. Think, for instance, of perfumery, where artisans do exactly the same, as the first stage of the manufacturing process: they make the liquids which are to receive the scents as odourless as possible. Or think of those whose work involves taking impressions of shapes in soft materials: they allow no shape at all to remain noticeable, and they begin their work only once they've made their base stuff as uniform and smooth as possible.
> (*Tim.* 50e; transl. by Waterfield 2008: 43)

The ambiguity of Plato's *chōra* was already stressed and criticized by Aristotle (*Ph.* IV 2, 209b–210a), and it is also mirrored in the different explanations proposed by modern scholars (Miller 2003), who interpreted it either as "space" or as "prime matter" – a meaning that was fully encapsulated in Aristotle's concept of *hylē*. In the quoted passage, Plato insists on the fact that the *chōra* must be completely formless, a feature that he describes on the basis of two examples taken from the "arts": (a) wax must be formless, in order to receive the impression of a seal; and (b) the liquid basis for a perfume must be as odorless as possible in order to receive the scent or fragrance from the aromatic substances that are added to it. The reference to perfume-making (similar to what we read in the technical section of Theophrastus' *On Odors*, 18) is particularly interesting, since it implies liquid mixtures. The blending of liquids, in fact, is evoked in key passages of the *Timaeus*, when Plato describes how the demiurge blended (a) the world-soul (35a1–b1), (b) the human soul (41d4–7), and (c) the vegetative soul (77a3–5). In cases (a) and (b), moreover, Plato specifies that the "ingredients" of these souls were mixed in a *kratēr* (κρατήρ, lit. "bowl"), which usually refers to a large vessel used to mix wine with water. However, we cannot exclude that Plato had a metallurgical model in mind, according to which metals were melted and mixed in a crucible (namely the *kratēr*; see Brisson 1998: 36–8).

In the *Timaeus*, metals are considered as a kind of water: next to "liquid water" (*hygron hydōr*), in fact, one must also consider "fusible water" (*chyton hydōr*; i.e., solidified water that can be melted). Copper, for instance, is gold-like, but lighter, because its particles of water are mixed with earth: sometimes, in fact, the earthy part appears on the surface and is called "verdigris." As for gold, Plato writes:

> To take a few of all the varieties of what we've called liquefiable water, there's one that is extremely dense (because it consists of the most subtle and uniform parts) and unique of its kind (*monoeidos*), and has been endowed with a shiny, yellow colour: this is our most highly prized possession, gold, the solidity of which is due to its having been filtered through rocks.
> (*Tim.* 59b–c; transl. by Waterfield 2008: 54)

Plato's definition of gold as *monoeidos* ("unique of its kind," "unique species") seems to have influenced Zosimos of Panopolis, who describes a method for the making of gold as follows (*Authentic Memoires*, X 8 in Mertens 1995: 41–2): after treating copper with several ingredients (salt, sulfur, and vitriol) and exposing it to the three sublimates, the alchemist will be able "to subdue the matter (*hylē*) and get the unique species (*to monoeidon*, i.e. of gold) out of many species" (Viano 2005: 99–102). Even though he manipulates, mixes, and treats complex species rather than elemental constituents (as the Platonic demiurge does), Zosimos attributes to the product of these manipulations the same qualities of natural gold. This claim was firmly criticized by the Neoplatonic philosopher Proclus (fifth century CE). In his commentary on Plato's *Republic* (Kroll 1899–1901: vol. 2, 234), in fact, he denies the possibility of *chrysopoeia*: by echoing Plato's words, he defines gold as a "unique species" produced by a demiurgic act, which cannot be imitated by the alchemists who simply mix other species (Viano 1996: 202–3; Viano 2005: 105–6).

Centuries before this debate, the concept of mixture was discussed in Aristotle's *On Generation and Corruption* (I 10). Here, as Cristina Viano recently pointed out (2015), Aristotle introduced the new idea of *mixis* as "chemical mixture"

> … to explain the constitution of those homogeneous substances from which all complex beings in the sublunary world, both natural and artificial, are comprised. In such a mixture, the ingredients "react" together to give rise to a new substance that is qualitatively different and possesses its own form. At the same time, the original ingredients remain present in power (*in potentia*) so that they can be separated again.
> (Viano 2015: 204)

Even though the elements of a *mixis* react to give rise to a new homogeneous substance, they remain present *in potentia*; that is, they can be still detected if the substance is properly analyzed. This analysis is the central object of the fourth book of *Metereology*, in which Aristotle discusses the composition of homogeneous bodies by describing eighteen pairs of opposite transformations (or affections) that they can or cannot undergo. These affections depend on the elemental constitution of the bodies, in particular on the different ways their two passive qualities (i.e. dryness and moistness, which are to an extent identified by Aristotle with earth and water) are affected by the action of heat and cold. For instance, bodies that melt when heated, such as metals or fusible stones, are mainly composed of water. More complex analysis, moreover, involves forms of experimentation, where the *technai*, especially the chemical arts, appear "as an instrument for generating knowledge" (Viano 2015: 211). A telling example is provided by the passage in which Aristotle discusses the composition of must (*Met.* IV 7, 384a3–7); its main components (water and earth), in fact, become evident to the experimenter when must is distilled:

> There is a kind of wine, for instance, which both solidifies by cold and may be dried up by boiling – I mean, must. In drying up all substances of this kind the watery part is separated off. That it is their water may be seen from the fact that the vapour from them condenses into water when collected. So wherever some sediment is left this is of the nature of earth.
>
> (Düring 1944: 45)

If we go back to Aristotle's concept of "chemical mixture," this *mixis* differs from a simple "combination" or "juxtaposition" (*synthesis*), where the elements are put side by side without underdoing any transformation. Aristotle specifies that liquids are particularly prone to "chemically" combine (*Gen. corr.* I 10, 328b3), thus anticipating a fruitful idea in the history of chemistry. As Düring (1944: 10) points out, "one of the chemical principles laid down by Aristotle was undisputed until some twenty years ago, namely the theory formulated in the words *corpora non agunt nisi liquida*. Only in 1925 did Arvid Hedvall succeed in proving that solid bodies also were capable of chemical reactions." The reactivity of liquids was also stressed by various Greco-Egyptian alchemists: natural substances, when dissolved, produce wonderful transformations, as Pseudo-Democritus claims (Martelli 2013: 94–5). In Aristotle's account, examples of liquid mixtures are metallic alloys produced by combining melted metals. In particular, he speculates on a copper-tin alloy: when tin and copper are mixed, "the tin almost vanishes, behaving as if it were an immaterial property of the copper: having been combined, it disappears, leaving no trace except the color it has imparted to the copper" (*Gen. corr.* I 10, 328b). Again, we find here a seminal idea that will be further developed by the alchemists, who tried to transfer the color form of the dyeing ingredients to the metal (often copper) to be dyed, as Synesius (Martelli 2013: 134–7) and Olympiodorus (*CAAG* II 91–2) explain (see already Lippmann 1919: 324; Viano 1996: 199–200; Dufault 2015: 221–5, 237–8).

Aristotle deals with metals and minerals at the end of his third book of *Meteorology* (III 6, 378a–b), where he provides a sketchy account on metallogenesis. Here, in order to explain the blending of the elemental constituents of minerals, Aristotle refers to the mixture of two kinds of exhalations: (a) smoky–dry exhalations and (b) vaporous–moist exhalations (Eichholz 1949; Halleux 1974: 98–105; Wilson 2013: 271–7). Vaporous–moist exhalations are trapped inside dry rocks and solidify (probably because of the cold), thus condensing into substances called *ta metalleuta*, that is, "metals" in this context (Aristotle lists iron, gold, and copper as examples of *metalleuta*; see Halleux 1974: 35–44). These substances, in fact, are fusible and malleable – properties that were clearly exploited by ancient smiths and metalworkers. The different amounts of earth mixed with the solidified moist exhalations determines the variety of metals. On the other hand, smoky–dry exhalations burn the earth in which they are trapped and thus produce a second type of minerals, which Aristotle refers to as *ta orykta*, lit. "fossils." He lists among these substances realgar, ochre, minium, sulfur, and cinnabar. Most of these "fossils," Aristotle continues, are colored dust, thus making reference to the production of pigments in antiquity (see Chapter 2).

The exhalations theory was later inherited and reshaped by the Stoic philosopher Posidonius (ca. 135–51 BCE; see Sen. *QNat.* II 54,1), who explained the composition of gemstones as well. In fact, as Halleux pointed out (1981: 50–1), we can assume that Diodorus Siculus (first century BCE), when discussing the origins of Arabian gemstones (*Bibliotheca historica*, II 52), had Posidonius' work in mind, even if the name of the philosopher is not mentioned in the passage (text fully quoted in Chapter 4, p. 134). Diodorus explains that rock crystals are made of water that solidified underground: here, specific kinds of rock crystals, namely emerald and beryl, take their colors from a bath of sulfur (*baphē theiōn*), as well as from the smoky exhalations produced by the sun heating the earth. Various elements of the theoretical explanation derive from empirical observations based on artisanal practices: in fact, the similarity with ancient dyeing techniques, which either used dyeing baths to dye crystals (as explained in the third-century CE Leiden and Stockholm papyri) or exposed metals to dyeing fumes (especially in the Greco-Egyptian alchemical tradition), is evident (see, for instance, the description of the *kērotakis* in Chapter 3, pp. 110–112).

CONCLUSIONS

Marco Beretta

Chemical arts played a central role in the development of ancient civilizations. Many important discoveries and technological breakthroughs remained unsurpassed until early modern times. However, such achievements were rarely the objects of autonomous reflection on the composition of matter and on the nature of its changes. Chemistry as a science simply did not exist. This does not mean that the effects of the manipulation of matter did not generate an effort to embody them in a theoretical frame. In the Egyptian civilization, precious stones and minerals were not just valuable commodities, but were believed to be material manifestations of the presence of gods. For this reason, their treatment was included in religious rites that were carried out under strict control of the clergy. The Egyptian classification of stones and metals depended on the nature of the gods from which they stemmed. The mythological origins of these materials exalted their value well beyond the world of the living. The funerary adornment of the pharaohs, the rite of embalming, and other complex religious rites endowed these precious materials with the properties of regenerating the dead.

The classification of chemicals, minerals, and metals in the Mesopotamian civilization was approached through quite different criteria. The transmission of knowledge by the cuneiform texts favored the systematic use of semantic determinatives, which attested to the centrality of economic management and craft production rather than a theoretical classification of chemical products and operations. In fact, chemical practices were not recognizable in one specific lexical classification, and they were often associated with medical and astrological texts.

The natural philosophies conceived by the Greek and Roman authors created a theoretical framework that was partially independent of the mythological and metaphysical assumptions of previous civilizations. Some of the most successful concepts, however, embodied earlier ideas. Empedocles' theory of the four elements and Pseudo-Democritus' view on the sympathies existing among substances echoed ideas and concepts that had been circulating in Egypt for a long time. On the other hand, by exploring the ideas that matter could be constituted of atoms (Democritus), of solid geometric forms (Plato), or of ever-changing combinations (Aristotle), the Greek philosophers presented new chemical theories that were at the basis of a classification of metals and stones destined to be absorbed, discussed, and developed by the Byzantine alchemists.

CHAPTER TWO

Practice and Experiment: *The Conquest of Matter*

SYDNEY H. AUFRÈRE, CALE JOHNSON,
MATTEO MARTELLI, AND MARCO BERETTA

EGYPT

Sydney H. Aufrère

In his book *Isis and Osiris*, the Greco-Roman philosopher Plutarch, writing about 100 CE, proposed a synthesis between Platonist philosophy and Oriental wisdom (Pleše 2005), explaining the presence of natural elements in Egyptian thought in passages dealing with physical myths. In the first excerpt (Plutarch, *De Is. et Os.* 36, 365C: Froidefond 1988: 209), the first element is the humid substance, or "emanation of Osiris" (*Osiridos aporroēn*; Plut. *De Is. et Os.* 36, 365B: Froidefond 1988: 208–9), equivalent to the Egyptian "humours of Osiris" (*redjou Ousir*; Pantalacci 1982; Kettel 1994; Nardelli 2017: 408). This element and the other three elements (earth, air, and fire) would have had the power to autoregenerate. A second excerpt relates a war that had broken out between Zeus-Amon (Breath or Air) and his brother Apophis (Dry or Igneous, different from the Sun). Osiris (Moist) chose to side with Zeus. After defeating Apophis, Zeus not only adopted Osiris as his son but also gave him – a perfect example of *interpretatio Graeca* – the name of Dionysus. Thus, according to this excerpt, it would follow that the association of Air and Moist prevailed over Dry and Igneous, with Earth in the background. In a third excerpt from the same book (Plut. *De Is. et Os.* 7, 353E), the sea of the Egyptians, considered isolated from the rest of the world, was perceived as a foreign, corrupt, unhealthy, salty body like that represented by Apophis-Typhon.

Plutarch's conception was not isolated in Egyptian thought. Manetho of Sebennytos, an emblematic character created during the Ptolemaic period that probably inspired Plutarch's discussions, evokes a system in which five elements were symbolized by divine names. Thus, Fire was Hephaestus (Ptah), Water Ocean (Noun) or Nile (Hapy), Air Athena (Neith), Earth Demeter (Isis), while a fifth element, Spirit, was Zeus (Amon; Waddell 1980: 196–7).

However, if Plutarch's and Manetho's views were meant to show that the traces of elements mentioned in Egyptian thought were equivalent to those found in Greek philosophy, this did not necessarily imply that this notion corresponded exactly to that of the Egyptians. Whereas the ancient Greek philosophers argued for a terrestrial realm consisting of four elements (Fire, Water, Air, Earth), ancient Egyptians preferred a nomenclature consisting of elements of the universe over which the gods had an influence, for instance: "Sky, Earth, Water, Mountains" (*Wb* V, 213, 4; Barucq and Daumas 1980); "Sky, Earth, Hellish world, Water, Ether" (Cauville 2009; *Edfou* IV 309); or "Sky, Earth, Hellish world, Mountains and Oceans" (Agut-Labordère and Chauveau 2011: 23, 26).

The concept of kingdoms, in the Western sense of mineral, vegetable, and animal, is hard to discern here, as the Egyptian priests thought in terms of up to five classifications. Moreover, the ancient Egyptian classification of kingdoms was a more extensive and variable one, like the one depicted on a bas-relief engraved on the plinth of the door of the East sanctuary of Amon-Rē's temple in Karnak (Aufrère 1991: 307–9). According to the caption, this bas-relief represents animate beings in prayer before the king of the gods, Amon-Rē, represented in its solar form circulating in the celestial vault. The gods (*netjer*), the humans (i.e. the Egyptians; *remetj*), the plebaians (*rekhyt*; Griffith 2006), the trees (*nehet*), and last, combined in a pair, the herbs (*sem*) and stones (*iner*) appear successively. The last three are shown with both arms.

The priests of Amon-Rē – the king of the gods – in Karnak conceived a nomenclature consisting of five classes of kingdoms over which he reigned: divine, human, servile (perhaps also animal), ligneous, and herbomineral. On the basis of an identical growth process, the last two represent a double paradigm, recalling for us that the Middle Ages inherited a tradition of giving the same status to herbs and to stones. While there are even more extensive models for classifying beings, which differ from one document to another (Meeks 2012), this vision is significant for classes of substances considered, religiously speaking, as noninert and endowed with a life of their own, being the subjects of a particular ontological conception.

The nomination and determination of the terms in the hieroglyphic system made it possible to identify concepts relating to matter and its transformations. The hieroglyphic system, combining images of sounds and pictures, enhanced a more comprehensive interpretation of their meaning. If the Egyptian words consist of a skeleton of consonants using phonograms (the Egyptians had a wide range of strong to weak consonants represented by uniliteral, biliteral, or triliteral values), a wide range of classifiers, technically called "determinatives," specify the semantic class to which they belong and their concrete or abstract character. In addition, word signs or ideograms combined the two former functions by attaching a sound to a classifier. A vast vocabulary contained references relating materials and substances to elements and kingdoms.

Moreover, by its very nature the hieroglyphic system enabled the reader to visually perceive the texture of the materials. In fact, to make their intrinsic nature more explicit, a range of determinatives provided access to information about the characteristics of the material used, such as its appearance, its state (liquid, dry, solid, or gaseous), its temperature, the mass or texture of the substances used, the mode of packaging, and even their mode of manufacture. In the absence of a rigorous spelling system, each scribe, depending on his degree of language proficiency, could add a panel of semantic nuances. It is therefore important to examine more closely the interpretation of the most common hieroglyphs in order to obtain valuable clues. For example, placed at the end of the phonetic skeleton of a term, one grain or three grains ° ° °, or even the group such as ιιι, indicated their powdery, fragmentary, or granular character. This is the case for a detrital rock such as sand (*sh'ā*), for an oleoresin such as frankincense (*'ānty*), for a mineral substance such as natron (or niter; *neter*), and also for dry fruits. The viscous or oily nature of a substance resulting from an amalgam of products, such as "oil" (*merehet*), was indicated. The same applied to minerals, dyes, chemicals, and drugs. The same classifier could also be used for the word "metal" (*bia*), which could be broken down by the addition of epithets for various types of metals such as "native copper" (*bia-her-khaset.ef*, literally "the metal on its desert"), "Asian copper" (*bia-Setjet*), "black bronze" (*bia-kem*), or iron, which was called "celestial metal" or "metal of the sky" (*bia-en-pet*).

However, if it was important to insist on the weighty nature of a material, the hieroglyph signified this by the squared stone block ready to be transported: ▭. The word "stone" (*iner*) was used more for building stones than for minerals. The exceptional character of a stone block was connoted by a wooden sledge on which was fastened a squared block sliding on a skidding ramp: , a hieroglyph that suggested the idea of a "finding," of a "marvel": (*biaou*), as if it were an exceptional stone block spotted and drawn from a mineral vein. It will be observed that the notion of "weight," "to weigh" – for example, "to be heavy" (*denes*) or "a block of heavy stone" (*oudenet*) – combined two hieroglyphs: the pestle hitting a material in a block of stone (*supra*) or not in stone 𓏤. The hieroglyph of the stone block could sometimes be represented by a grain, because the word designating "a small stone, a pebble" was indifferently written 𓂝𓂋𓏤 or 𓂝𓂋𓏤 (*'ār*).

In those distant times, when mineralogy was not yet a discipline, a wide range of rocks and minerals were all designated by the same hieroglyphic word 𓂝𓏤 or 𓂝𓏤 (*'āat*), which even embraced resins used to make beads (Aufrère 1991: 101–4). Such rocks and minerals included what we now know as igneous (granite, porphyry, diorite), sedimentary (limestone, greywacke, sandstone, shale), metamorphic (gneiss, quartzite, shale), or metamorphosed (marble, schist, gneiss), as well as minerals (alabaster, calcite, travertine), all of which substances were used to make statues, precious items (coffins, naos), and vases (an exhaustive list of Egyptian mineral products is given by Aston et al. 2000: 21–63).

Concerning metals, there were several signs evoking the shape of the crucible used to melt and pour the molten ore into a mold: 𓎯, 𓎰, 𓎱, etc. Thus, the word "metal" was written 𓃀𓇋𓄿, 𓌃 (*bia* or *hemet*), but the writings could vary, showing an ax-head (𓌃) or a mold and an ax-head (𓌃; Aufrère 1991: 449–50).

The designations of liquid substances were frequently accompanied by either the determinative for liquid (𓈗), used to represent the word "water" (*mou*) itself, or by various containers made of ceramics or of rocks or minerals, the shapes of which connote the products they contain. For example, the abbreviated word for "oil" (𓏺𓏺𓏺 *merehet*) was represented in the form of a cylindrical travertine or alabaster/calcite pot and only used to contain oils or fatty ointments (see Chapter 8). The word "natron" or "niter" (𓊹 *neter*) was represented by an ideogram combining two hieroglyphs: the sign of the god – a mast (𓊹) – and a prospector's leather bundle (𓍊; Aufrère 1991: 606–7).

The name of a vegetable substance was characterized by the determinative for herbaceous plants (𓆰), a species of tree by the silhouette of a sycamore (𓆭), and a woody material by a branch (𓆱). The combination of two or three determinatives increased the degree of semantic precision of the written word.

Whereas the name of a basic product was culturally unambiguous in its local context, the contents of exotic substances had to be specified by their origin and intrinsic quality. Indeed, their potential liturgical role had to be identified according to appearance, color, and smell. These specifications were frequent in lists of aromatics exported from Punt (i.e. the littoral of southern Arabia and the Horn of Africa) and from Asia. They were in conformity with a mytho-scientific model associating information of a religious and physical nature and their affiliation to a kingdom, a genus, or a species. The mythical origin of a variety of styrax, its appearance, and its smell were thus specified:

> The *gaiu-maa* tree, here is what is said about it. Its "variety" is blackish woody, with a pleasant smell. It comes from the pupil of Rē's eye. Its upper extremity is black, its middle part is grey, and its lower extremity is as clear as the resin

of the terebinth tree precipitated on its trunk. When its flank is incised, it takes on the golden color of the wing of the oriole. (Gloss :) Concerning the oriole, it is the *tefnyt* bird (the bird of the goddess Tefnut) with striped gold-colored wings. When scratched, the smell is that of the *tishepes*.

(Aufrère 2005a: 256–7; Incordino 2017)

Colors are often named according to the external appearance (liveries) of living creatures or of products with stable colors (Bardinet 2018). Such precision about substances clearly indicates that the exotic unknown was associated with the diversity of the known materials from the Nile valley.

From the Second Dynasty (2925–2700 BCE) onwards, the Egyptians indicated on the funerary steles – and later, in the Middle Kingdom, on coffins – the list of products for daily use of the deceased, some made with different manufacturing processes (Barta 1963). Thus, on the stele of Princess Nefertiabet (Louvre Museum, inv. no. AE 15591), Fourth Dynasty (2625–2510 BCE), pieces of meat and various types of fruits are depicted alongside perfumes, cosmetics such as the resin of the terebinth tree (*senetjer*; Loret 1949; Espinel 2017), high-quality oil (*hattet*) perfumes, different kinds of pigments such as chrysocolla (*ouadj*) and galena (*semdet*), alcoholic fermented beverages (i.e. beer [*sekhepet*] and wine [*irep*]), and bread of different shapes showing the rising of the dough and the nuanced colors of the baking. Naturally, these lists became ever longer as the standard of living of the aristocracy improved.

Furthermore, a literary genre – the *onomastica* – appeared, attested from the Middle Kingdom until the second century CE, in several papyri (*Papyrus Ramesseum*, *Papyrus Golenischeff*, *Papyrus Hood*, *London Leather Roll*, and *Tebtunis Papyrus I*), giving a didactic nomenclature of the universe. The incipit of the book written by Amenope (*London Leather Roll*) refers to this nomenclature in these words:

Beginning of the teaching to clarify the memory, to educate the ignorant and to learn about all that exists: what Ptah created and Thoth copied, the sky and all that concerns it, the earth with what is inside, what mountains spit, what is flooded by the flow, anything that illuminated Rē', all that grow on the back of the earth, everything written by the scribe of the *Sacred Book* in the House of Life, Amenope son of Amenope.

(Gardiner 1947, 1: 1*–3*)

Some chapters of this document allow us to better understand the nature of cereals and other solid and liquid products used to make beverages, but the contents of other chapters – "the earth with what is inside, what mountains spit" – are missing. However, the *Tebtunis Papyrus I*, in which whole sections of Egyptian artisanal culture are fortunately preserved, gives more information. Indeed, in addition to chapters dealing with the diversity of the animal kingdom, other chapters provide information on the characteristics of minerals, on quarries, on mountains and the products of the mines (fr. J, 21,1–22,6; Osing 1998: 107–10), on leatherwork and basketry (fr. N 6,22–9,2; Osing 1998: 120), on wooden utensils, and on various plants (fr N 1–6,21; Osing 1998: 116–20). In the *onomastica*, Egyptians could view the full spectrum of artisanal activities and production.

Despite the small-scale output of mining operations (Aston et al. 2000: 5–20) and of the harvesting of aromatic plants, as testified to in the texts dealing with mines and temples, extracting any substance of high symbolic and religious value was seen as breaking into a universe with a strong original and divine footprint. Indeed, breaking into the vast, desert-like, and uninhabitable expanses stretching from Egypt to the boundaries of the Red Sea suggested a relationship between the distant geological upheavals in that area and the time when, according to legend, the gods lived on earth. It is reasonable to assume that in order not to offend the gods the extraction of materials from these divine spaces encouraged piety on the part of the prospectors (Aufrère 2008). This was according to a widespread belief that the gods had left, from a poetical point of view, their emanations in mineral veins as well as in gum resins (myrrh and frankincense) resulting from the tapping of the bark of the tree. According to the Edfu texts, these gum resins were considered to be the tears of gods and goddesses taking the form of divine hawks (Aufrère 2017a). Even if it only meant returning the product to the rightful owners of the land, the opening of mineral veins was ritualized to counter unexpected dangers and to obtain the best quality and the highest possible quantity of material (Loret 1928; Kurth 1996; Pantalacci 1996: 87–91; Valbelle and Bonnet 1996).

In extracting and refining techniques there were two distinct phases: in the first, physical means (sieving) were used, and in the second, dealing with ores, a refining procedure exposed the sorted material to a high temperature. Depending on their qualities, the collection of minerals proceeded by sorting and then crushing and washing – as was done for the argentiferous galena of the Gebel el-Zeit (Castel and Soukiassian 1989). Indeed, for obvious reasons of weight, the ores and minerals extracted from the subsoil of the desert adjacent to the Nile valley could not be shipped. The technique of gold extraction is now better known thanks to the study of the Bir Samut mines in the Eastern Desert. The gold ore came from two types of deposits: gold-bearing quartz veins and alluvial gold veins from adjacent wadis. Alluvial gold was simply sieved.

According to Agatharchides of Cnidus (second to first centuries BCE), the gold-bearing quartz was crushed and milled by millstones (Diodorus Siculus, *Bibl. Hist.* Book 3, 12–48; Peremanns 1967; Redon 2016; Faucher 2018). The heavier gold flakes remained on the inclined planes of the washing units.

Archaeology has uncovered traces of important mills dating from the second half of the third century BCE. On account of the lack of local fuel, the purification of the precious metal was carried out in the valley after melting at 1064°C (Rabot and Goncalves 2015). This traditional seasonal way of exploiting the mines, using less sophisticated techniques, peaked twice in the Eighteenth Dynasty (1552 and 1314–1295 BCE), respectively in Samut el-Beda and then in the Greco-Roman period in Samut. Mineral resource maps of the Eastern Desert were kept in the royal and priestly libraries. Preserved today in the Egyptian Museum of Turin, the map of the gold mines (i.e. the map of Wadi Hammamat dating from the reign of Seti I [1294–1279 BCE]), shows the location of the gold and greywacke, also known as "stone of *Bekhenu*" (𓃀𓈎𓈖𓏤 *bekhenu*; Goyon 1949; Sauneron 1967: 143; Baud 1990).

As for copper ore, it may have been available in significant quantities in the Sinai desert, in the Eastern Desert, and in Sudan, but there are few documented traces of its exploitation throughout Pharaonic Egypt. Analyses show that the Egyptians were probably supplied with copper and tin from the ports of the Levant, including copper containing a high percentage of arsenic (8 percent; Garenne-Marot 1984; Garenne-Marot 1985). Following the classical reduction process, the objects manufactured with that copper (Garenne-Marot 1984; Garenne-Marot 1985) were rubbed with an ointment made of a mixture of realgar and orpiment, giving them a silvery shine.

It is only on the site of Bir Nasb, west of Wadi Nasb, in South Sinai, that this activity is testified to by the presence of slag (traces of reduction of the copper ore) and of multiple batteries of furnaces (3,000 in all) in which the temperature could be brought up to a maximum of 1334°C, a temperature necessary even today to obtain copper from malachite. This quasi-industrial-scale production by Egyptian metallurgists in the Old Kingdom (Fifth Dynasty, 2510–2460 BCE) was due to the availability of local fuel in sufficient quantity (Tallet 2000: 19–22; Castel et al. 2008; Tallet et al. 2011; Tallet 2013–18). However, there is no trace of tin mining in Egypt, despite the presence of an association between gold and tin ore, testified to by a number of bronze objects found in the area of the Eastern Desert. From the Middle Kingdom on, tin was probably imported from the Eastern Mediterranean shores (Garenne-Marot 1984: 107–8), while it was always exported during the Achemenid Period (Briant and Descat 1998: 67).

Apart from a few exceptions, the metal refining and casting processes were carried out in the Nile valley, where fuel, namely charcoal, was readily available. While craftsmen in most towns of the Nile valley met the requirements of daily life, capital cities like Memphis, Thebes, or Pi-Ramses had specialized industrial centers. Archaeology and iconography show this concentration of techniques and work to be of Levantine heritage. Excavations in the area of the temple of Amon at Pi-Ramses revealed the presence of copper casting facilities contemporary with the Nineteenth Dynasty (1295–1188 BCE).

To optimize mass production, craftsmen of all trades such as charcoal producers, potters, tanners, wax sculptors, and molders, all very hierarchical, were comprehensively reconfigured (Pusch 1994; see Garenne-Marot 1985; Hampson 2012: 187–230). The manufacture of weapons (swords, knives, arrows, spears, and chariots of the New Kingdom), requiring the mastery of the art of metallurgy (copper, bronze, and iron), was done in Memphis, renowned for the operations of its arsenal (*Pa-Khepesh*; Sauneron 1954) in the Eighteenth and Nineteenth Dynasties. As imported metals – copper, tin, and bronze – were available only through authorized intermediaries, it was important to recycle and recast worn copper tools and other metal fragments, thus avoiding their theft (Valbelle 1982; Allam 1997: 6: Pap. Geneva, inv. no. 15274 v° 1). In view of their importance, goldsmithing and the welding of other materials (Vernier 1907; Hampson 2012: 141–70) will be discussed in Chapter 8. The solder for gold was made using chrysocolla, but the use of silver-on-bronze welds is suspected (Evrar-Derrick and Quaegebeur 1979: 30).

Large granite and granodiorite blocks (Aston et al. 2000: 65–6) for obelisks or statues (Aston 2000: 35–8) were quarried and worked by thermal expansion of surfaces. First weakened by hot embers and then altered, after thermal shock, these surfaces were hammered with dolerite (microgabbro) balls, often equipped with handles, on the still hot surface, before polishing it. This technique was used in addition to that of the mortises – that is to say, notches specially made to receive bronze wedges, which, when hammered, made it easier to cut granite blocks (Goyon et al. 2004: 285–9; Hampson 2012: 271–90; Gremilliet and Delangle 2017).

The manufacture of ceramics is one of the fire arts that best attests to the mastery of various techniques that spread over time. Two types of materials were used: Nile clay (for ritual and domestic use) and limestone clay (to store food). Blocks of these materials were first crushed (to prevent the formation of air bubbles) and then rehydrated to make a malleable paste. Slips were used to nuance the color of the paste, and vegetal and mineral degreasers were used to strengthen its resistance to heat. The sophistication of turning pottery (by using the potter's wheel instead of more rudimentary methods like turntables) and the control of the entry of air into the ovens led to the production of black and red ceramics. These techniques were invented and applied during Nagada I period (3900–3500 BCE; Arnold and Bourriau 1993; Bourriau et al. 2000).

Plaster, obtained by burning gypsum (Goyon et al. 2004: 70–1), and mortar (plaster mixed with sand) are attested in the construction of the pyramid of Giza. These had previously been used to fill irregularities in graves dug in the rocky walls (limestone or sandstone; Goyon et al. 2004: 71). Plaster was also used for coatings (Goyon et al. 2004: 73–4) and to make funerary masks. Significant deposits of gypsum existed in Northern Egypt and in the Suez Isthmus. Others were exploited in the desert at Umm es-Sawwan, near the Fayoum. The gypsum was extracted using hewn flints (Kemp 2005: 318, fig. 111). Lime is obtained by firing the purest calcite available (e.g. limestone) to a temperature of 1000°C. The increase in temperature leads to the release of carbon dioxide and the formation of the product quicklime (calcium oxide). When water is poured onto the fresh quicklime, it becomes very hot, and hydrated (slaked) lime is obtained. Mortars made of slaked lime with added plaster were not used before the Twenty-Sixth Dynasty (ca. 600 BCE; Goyon et al. 2004: 74). Quicklime was used in Roman times to burn corpses in order to stop an epidemic of plague (third century CE). Traces have been observed in the tomb of Harrua (excavations of Francesco Tiradritti). It is likely that this quality of quicklime was associated by the Egyptians with a divine mechanism.

Several glass-like glazing techniques coexisted. The frit or sintered-quartz ceramic, attested from the fifth millennium BCE to the Late Period, was also called "Egyptian faience." The making of "faience" results from the use of different glazing techniques that varied over time and were used to make several types of objects. Its Egyptian name was *tjehnet*, "brilliant, lustrous" (=⟨hieroglyphs⟩). The process consisted of molding objects (figurines, beads, amulets, seals, instruments) made of a siliceous core of fine ground quartz, slaked lime, copper oxide, and sodium carbonate (natron) and submitting them to high temperatures. During the heating, oxides migrate to the surface, which start to melt and form a glaze while the core remains hard and porous. The copper oxide yields a color ranging from green to blue, depending on the temperature reached (Vandiver 1998). Other oxides were used to decorate the objects. Pigments – some synthesized – were strongly associated with precious metals and noble minerals (Aufrère 1998b; Aufrère and Menu 1998; Mathieu 2009), which they imitated. They required elaborate manufacturing processes (Rouchon et al. 1990; Colinart and Menu 1998).

Since blue could not be made with lapis lazuli or turquoise for reasons of expense, as these were high-value imported products, from the Fourth Dynasty (2665–2510 BCE) on Egyptians produced an artificial blue pigment "synthesized by cooking in a closed cup, a mixture of sand, lime, a copper compound and perhaps an alkaline flux," yielding a product with the modern formula $CaCuSi_4O_{10}$ (Blet et al. 1997: 121; Matoïan and Bouquillon 2000). The manufacturing process of this artificial lapis lazuli blue, now called Egyptian blue, is not mentioned in Egyptian texts. It has successfully been reproduced in the laboratory of the Louvre (Pagès-Camagna 1998). This artificial product was called *irtyu* (⟨hieroglyphs⟩; *Wb* I, 116: 11), which, etymologically, signified "the artificial product." However, Egyptian inscriptions also differentiated the "true lapis lazuli" (⟨hieroglyphs⟩ *khesebed ma'āt*) from "artificial lapis lazuli" (⟨hieroglyphs⟩ *khesebedj iryt*; Delamare 2007: 18–28). The name "maker of lapis lazuli" (⟨hieroglyphs⟩ *iru khesebedj*) appeared in the New Kingdom (*Wb* III: 334). While in the Old Kingdom the blue–green range was obtained with the help of copper oxide, it is in the New Kingdom that cobalt blue was invented (Tite et al. 1998). Its use is attested as far away as in Scandinavia (Varberg et al. 2015). It was called "Alexandrian blue" in the Roman period (Delamare 1998a; Delamare 1998b) and *cæruleum* by Vitruvius (VII 12).

Antimony and manganese oxides were used to outline details in yellow and black, respectively. First used in the Middle Kingdom, green glaze (𓎛𓏤𓅓 *hemet*), a name connoting the idea of "crafts," imitated turquoise. Under the reign of Amenothep III (1411–1352 BCE; Tite and Bimson 1989), it was employed to coat minerals such as the soapstone used to manufacture statuettes and beetles to make them resistant to high temperatures.

There is little evidence that glass craftwork could have reached Egypt before the New Kingdom. Small containers made of lapis lazuli-colored glass paste (a material probably called *khesebedj oudjeh*) enriched with a feather-like decoration (plumeti) are attested in the aristocratic milieu as early as the reign of Thutmosis III (1458–1425 BCE), a time that corresponds to the political and economic expansion of Egypt toward the Levant, from where this technique originally came (Lilyquist and Brill 1993). Three glassmaking workshops are known: one workshop in Tell el-Amarna (ca. 1350 BCE); another in Lisht (Smirniou and Rehren 2016: 57–8); and one in Qantir-Pi-Ramses, in the Eastern Delta around 1250 BCE (Rehren 1997; Smirniou and Rehren 2016). In all three workshops, the primary material was obtained using a technique consisting of melting ground quartz pebbles heated to 900°C in the presence of vegetable ash in pottery and then, after removal of the slag, of heating to 1000°C in cylindrical ceramic containers to obtain round ingots of red glass colored with copper oxide, a technique requiring a high degree of control (Rehren 1997; Smirniou and Rehren 2016). Blue glass was obtained, as we have seen, using cobalt oxide, which was exported throughout the Mediterranean Sea basin. This production, which declined in the Twenty-First Dynasty (1096–945 BCE), experienced a final renaissance during the Twenty-Sixth Dynasty (664–525 BCE).

Tanning, currying, and dyeing processes used almost the same techniques (Vartavan 1998). According to archaeological evidence, tanning, which consists in stopping the natural process of degradation of the skin, required the use of urine, animal excrement, or alum for depilation and tannins obtained from acacia seeds (*Acacia nilotica* [L.] Delile 1813). The final touch was the cleaning of the skins with half-moon-shaped flint knives. Red, yellow, and green colors were usually used to dye leather (Lucas 1962: 171–86).

Egyptians wore linen clothing, the whiteness of which was obtained by subjecting it to the action of water, air, and light, followed by an alkaline wash. It was possible to dye the linen, but only with difficulty (Goyon 1996). The study of dyeing in ancient Egypt revealed the knowledge of processes such as using a mordant such as alum from the oases (Lucas 1962: 257–9) and iron acetate (obtained from iron nails dissolved in vinegar). The Egyptians had a long-standing tradition in this field. Egyptian dyeing processes reached such a high reputation that Pliny (*NH* XXXV 42) described as a singular process the mordanting process used by Egyptian dyers. The texts of Dendara referring to liturgical fabrics revealed the use of four substances – red, blue, green, and yellow – associated with the status of the goddess, highlighting the use of the scum resulting from the fermentation of crushed woad leaves (*Isatis tinctorum*). Indeed, the color changes from red to blue and green (Goyon 1980) depending on the level of oxygenation of the bath. Two other reddish tints were used by tanners: one for fatty substances, alkanet (*Alkanna tinctoria*); and the other for blood-red leather, madder (*Rubia tinctorum*), an imported plant (Loret 1930). The dyer's safflower (*Carthamus tinctorius* L., 1753), occasionally used in Egypt before the Roman period, had to be added to obtain a yellow–orange color (Mathieu 2009: 48 and n. 106; Newton et al. 2013: 13).

Cereals were stored in silos to keep them dry. Even in prehistoric times, techniques of watertight construction were sophisticated (Dachy 2014). Silos from the Thirteenth Dynasty (1785–1633 BCE), 5.5–6.5 m in diameter, were discovered in Edfu. For meat products (red meat, poultry, fish), open-air drying was used (e.g. fish were gutted and the heads cut off, sliced, and hung), but they could also be pickled or salted (salt and natron) and stored in jars. This will be discussed in Chapter 8.

Although cereal grains were first eaten as gruel, empirical observations led to the control of several processes discovered serendipitously: dough rose due to fermentation caused by adventitious airborne yeasts, which were also responsible for the alcoholic fermentation of beer (Doyen and Warmenbol 2004). On the names of different kinds breads and cakes, see Salavert and Tengberg (2005) and Schwechler (2017). Beer (*henket*) was made using ground germinated barley (𓏥𓂺𓄑𓏥 *besha*; *Wb* I: 458, 10) mixed with wheat flour. The Egyptians had observed that using germinated barley turned it into malt and changed the taste. Indeed, we now know that germination triggers biochemical changes in the grain through the release of enzymes, which gives it new flavors. In addition, the observation of the fermentation of natural agents on the surface of the liquid in the beer jar led to the empirical discovery of fermentation. The name given of "brewer's yeast," literally "the one on (it)" (𓎛𓂋𓏤𓏥 *herut*; *Wb* III: 148, 18), supports this observation. The Egyptians came to the conclusion that cereals were not inert substances, but possessed intrinsic powers that triggered processes such as fermentation (for bread and beer).

As for perfumes, the Egyptians had recourse to several techniques (Castel et al. 2012). The preparation of vegetable perfumes was particularly important. Batches of selected petals, replaced several times during the process, were placed in refined beef dripping to macerate. The oily macerated mixture was then filtered and the perfume obtained – technically a pomade – was packaged in flared cylindrical pots set with a palm leaf (𓎯).

FIGURE 2.1 Flared cylindrical perfume pots set with a palm leaf. Old Kingdom. Cairo Museum. © Sydney H. Aufrère.

It is possible that the famous lotus pomade may have been prepared in this way. When rubbed on the skin, hair, and clothes, pomades gave a feeling of freshness (Derchain 1975; Cherpion 1994). Another technique consisted in cold-pressing petals with a delicate fragrance in a cloth sack. This is how the lily perfume was obtained (Bénédite 1921). Many other sophisticated preparations using multiple ingredients and different cooking phases existed. For example, the Egyptian perfume-makers described that the gum resins frankincense and myrrh emitted their odors only when warmed by a charcoal flame.

MESOPOTAMIA

Cale Johnson

As recently as the 1960s, specialists in cuneiform could not agree on the linguistic identification of the five most important metals in Mesopotamian history: it was clear from studies of artifacts and several late examples of words written on the material they denote that the five metals in question were gold, silver, bronze, tin, and lead; however, only three could be securely equated to corresponding Akkadian or Sumerian words: gold (Akk. *huraṣu* = Sum. $ku_3.sig_{17}$), silver (Akk. *kaspu* = Sum. ku_3.babbar), and copper (Akk. *(w)erû* = Sum. uruda). Benno Landsberger, by far the most important lexicographer in the history of Assyriology, tells the story of Victor Place's discovery in 1854, during his excavations at Khorsabad (ancient Dur-Sharrukin), of "a stone box containing seven inscriptions on tablets of different materials, (i) gold, (ii) silver, (iii) antimony, (iv) copper, (v) lead, (vi) alabaster and (vii) marble. The first four are currently in the Louvre, the other three heavier tablets fell into the Tigris" (Landsberger 1965, summarizing Lenormant 1878 and Delitzsch 1887). As several more recent investigations have shown, only five tablets were actually excavated (made of gold, silver, tin-bronze rather than copper, lead, and magnesite), and only the lead tablet was lost in the Euphrates (Bjorkman 1987; Brinkman 1988).

On nearly all of the tablets, the inscription includes the telltale phrasing "I wrote the spelling of my name on tablets made of gold, silver, copper, tin, lead, lapis lazuli and alabaster and placed it in their [the palaces'] foundation," giving us the final developmental stage of the placement of substances in foundation deposits in order to imbue the building with their properties (see Chapter 1). There is now a widespread consensus that Sum. an.na, read by some as Sum. nagga, corresponds to the Sumerian loanword in Akk. *annaku* "tin," while Sum. $a.gar_5$ "lead" corresponds to Akk. *abāru*. Due to the fact that tin was difficult to identify in the archaeological record, many objections arose against the identification of Sum. an.na with tin, but Landsberger brings together the relevant evidence for this conclusion, including several inscriptions, ranging from the Gudea Cylinders (ca. 2150 BCE), down to the Neo-Assyrian period (ca. 800 BCE), in which "copper" (Sum. uruda) is combined with "tin" (Sum. an.na) to produce "bronze" (Sum. zabar = Akk. *siparru*). Alloys between copper and

FIGURE 2.2 Three foundation tablets from Khorsabad, Assyrian from the reign of Sargon II (721–705 BCE; gold, silver, and copper). © Bridgeman Images.

lead do exist, but they were never widespread in ancient Mesopotamia or the neighboring region.

Recent surveys of archaeologically attested bronzes suggest that arsenic bronzes arose earlier, perhaps accidentally or through the selection of copper ores that already contained arsenic such as the arsenic-rich copper ores from the Amarak-Talmessi mines on the central Iranian plateau, while tin bronze, which must be a consciously formulated alloy, only begins to dominate in the middle of the third millennium BCE (Moorey 1994: 252–4; De Ryck et al. 2005; Potts 2007: 127; see, however, the discussion in Chapter 5). In the Late Uruk list of metal objects, three identifiable finished products predominate: "knives" (GIR_2), "drill bits" (NAGAR), and perhaps "scrapers" (NU). And as Englund and Nissen argued in their pathbreaking edition of the lexical lists, the primary metal used for these objects was almost certainly copper (Englund and Nissen 1993: 34; Englund 2006: 12). Within the list, however, ordinary objects in copper like GIR_2, NAGAR, and NU are contrasted with the same signs followed by the cuneiform sign AN, which may be related to the later orthography for tin (Sum. an.na): these entries probably refer to arsenical bronzes produced, if intentionally, through the selection of arsenic-bearing ores rather than the alloying of tin and copper (Englund 2006: 12, citing Waetzoldt 1981 against Vaiman 1982).

Until the discovery of the Ebla archive in 1974–5, relatively little progress had been made in making sense of tin bronzes in Mesopotamia. The earliest known cuneiform document to distinguish between copper and bronze (UET 2, 373) dates to 2900–2800 BCE or so, but the Ebla archive (dating to 2380–2333 BCE) presents us with the crucial evidence for tin bronze alloys (Archi 2017). The Ebla texts distinguish between two qualities of copper: "unrefined copper" (Sum. urudu) and "refined copper" (Sum. a.gar$_5$.gar$_5$, a neologism derived, through reduplication, from the standard Sumerian term for "lead," Sum. a.gar$_5$), whereas in contemporary southern Mesopotamia refined copper was termed "washed copper" (Sum. urudu luh.ha; see Waetzoldt and Bachmann 1984). Refined copper and tin appear in Ebla administrative documents in the 6:1 ratio that we might expect in a proper tin bronze (Archi 1993: 618, referencing the tablet TM.75.G.1310). The clearest evidence, however, for the alloying process used in the creation of tin bronze is found in an Ur III (ca. 2112–2004 BCE) problem text known as UIOM 892: it begins with the desired finished product, one mina (= 60 shekels or 500 g) of bronze (Sum. zabar), and then, line by line, sets out the ingredients and material losses that took place during the smelting and alloying of the materials (Jones 1961: 114; Waetzoldt 1981; Waetzoldt and Bachman 1984). The inputs are 56 shekels of refined copper (= 466 g) and 8 shekels of tin (= 67 g), giving a ratio of 7:1, but it also states the amount of material lost during the alloying process: 4 shekels (= 33 g, hence 56 + 8 − 4 = 60 shekels of bronze). The fact that the text begins with a single unit (Sum. 1 ma.na "mina" = 60 gin$_2$ "shekels") and goes on to separate out the individual elements shows its affinity with contemporary mathematical procedure texts, but we already have this type of schematic ratio in an Early Dynastic III administrative document from Girsu (RTC 23), where 80 shekels of refined copper are combined with 13 1/3 shekels of tin, yielding a slightly lower 6:1 ratio, designated as a "seven part mixture" (Sum. 7(dištenû) la$_2$).

If "copper" (Sum. urudu) served as the prototypical base metal, the Sumerian word ku$_3$ "shining (metal)" was used to refer to precious metals, with the distinction between gold and silver made in terms of color: "shining yellow (metal)" (Sum. ku$_3$ sig$_{17}$ = Akk. *ḫurāṣu*) designated gold, while "shining white (metal)" (Sum. ku$_3$.babbar = Akk. *kaspu*) was used for silver. Ostensible gold objects from the Royal Graves at Ur, dating to the Early Dynastic IIIa period (ca. 2600–2500 BCE) have recently been shown to consist of varying percentages of gold and silver, almost certainly due to the co-occurrence of these two elements in nature; similar objects that required greater strength or durability also show substantial amounts of copper (Hauptmann et al. 2018). But soon after this, again in the Syrian city of Ebla (ca. 2350 BCE), we find careful grading and labeling of different qualities of gold, based on its relative value in comparison with silver (Waetzoldt 1985; Paoletti 2016a). Likewise in the Ur III period (ca. 2112–2004 BCE), gold was classified into three grades on the basis of its purity: "yellow brilliant gold" (Sum. ku$_3$.sig$_{17}$ huš.a, valued at 15–21 shekels of silver for each shekel of gold, and often translated as "red gold"), "mixed gold" (Sum. ku$_3$.sig$_{17}$ HI.da, valued at 10–15 shekels of silver), and "normal gold" (Sum. ku$_3$.sig$_{17}$ si.sa$_2$, valued at 6.5–10 shekels of silver; Waetzoldt 1985; Paoletti 2016b: Hauptmann et al. 2018; Kleber 2019). The first decisive evidence for cementation (using vapors from salt or alum to convert silver into silver chloride and thereby remove it from molten mixtures of gold and silver) comes from an Old Babylonian period letter from Mari (ca. 1750 BCE), which reads, in part, as follows:

> Speak to my lord, thus [says] Mukannišum: "The four minas of gold [ca. 2 kg] for the two sun-disks that my lord has sent to me have been powdered. I have taken four shekels [ca. 33.3 g] for each of the four ingots[?], and I purified it in order to determine the fineness. Half a shekel and 10 grains of gold [= ca. 4.63 g] were lost from the four shekels of gold, that [means] 25 grains per shekel [= 13.9 percent] were lost. The goldsmith said "It [viz. the gold] is not red [*ul sām*]!"
>
> (Kleber 2019: 22; cf. Paoletti 2016a)

Here, a fire assay shows that the gold in question – once it is refined – will not be sufficient to make the sun-disks that the goldsmith has been asked to produce. The goldsmith describes the gold as "not red" (Akk. *ul sām*), demonstrating the continued use of "red" as a description of the highest-quality gold. Only much later, in the Neo-Babylonian period, do we see new terminologies arise for the testing of the quality of gold using a touchstone (Akk. *pidānu*), a loanword from Arabic or another language from the Arabian peninsula that enters the Akkadian alongside a distinctive term for Arabian gold (Akk. *nalṭar*; see Kleber 2016; Kleber 2019).

Few textiles other than linen and wool were regularly used throughout Mesopotamian history, and linen could only be dyed in a limited and rather specific way (Gittinger 1982; Dalley 1991: 120); therefore, much of the Mesopotamian textile tradition is based on an ever-growing interdependence between the production of textiles in wool and the dyeing of wool textiles with a wide variety of materials. Contemporary with the development of metal industries at the beginning of the third millennium BCE, we also see in Mesopotamia a so-called "fiber revolution," in which woolen textiles were produced on an industrial scale (McCorriston 1997). It was in the context of this revolution that methods of finishing and dyeing woolen textiles became a central area of technological development in Mesopotamia. Thanks to the unprecedented level of administrative detail preserved in the Ur III textual record, the production of woolen textiles in Mesopotamia is one of the best-documented economic sectors in all of antiquity. Texts like ITT 5, 9996 describe the processing of the materials for large 7 × 7 cubit textiles through several different stages of cleaning (Sum. za.ri$_2$.in) the wool, plucking it apart, airing it out and sorting it (Sum. peš$_5$), combing (Sum. ^{giš}ga.rig$_2$ ak), spinning (Sum. U.NU), warping (Sum. dun.dun), and weaving (Sum. tag or tuku$_5$; Waetzoldt 1972). Once the weaving was complete, a distinct group of workers finished the textiles using a number of substances that are occasionally recorded in administrative documents such as ITT 2, 902+:

> (Five garments): the work-days required for them (equals) 195 work-days. The oil needed for it: 1 liter. The alkali: 5 1/2 liters. The string: 10 shekels.
> (Waetzoldt 1972: 158)

Based on Waetzoldt's work, Potts has described the work of the fullers as follows:

> It is estimated that 7.7 work days were required by the fullers to treat each kilogram of finished cloth. Depending on the type of cloth being treated, greater or lesser amounts of oil and alkali were needed, varying somewhat in ratio between 1:5.5 to 1:4 [i.e. oil:alkali] per kilogram of cloth.
> (Waetzoldt 1972: 159; Potts 1997: 95)

Different oils and fats, both vegetable and animal, two types of alkali (Sum. naga and Sum. naga.si.e$_3$), and substances such as "white earth" (Sum. im.babbar$_2$), "probably fuller's earth" according to Potts, also played important roles in the work of the fullers (Potts 1997: 106; Wasserman 2013: 266). What is usually termed "alkali" is actually ash from an alkali-bearing plant, such as *Salsola kali* or *Salicornia*, which was mixed with oils or fats to make soap, while "white earth" was presumably burned and crushed gypsum that was rubbed on cloth as the last stage in the fuller's work (Forbes 1965: 85; Firth 2011). Other materials, such as barley for the brewing of a kind of beer for soaking garments (Akk. *miḫḫu*) and a type of alum called Akk. *allaḫaru*, appear in the texts as well, but there is no trace, in Mesopotamia, of using sulfur fumes as a bleaching agent.

Unlike the work of the fuller, the production of linen from flax does not seem to have involved any major chemical processes in Mesopotamia, but leather was a different story: many of the same materials that were used for finishing cloth or dyeing linen are also found in the tanning of hides and the finishing of leather. Stol's entry for "Leder (Industrie)" (1983) in the *Reallexikon der Assyriologie* outlines two methods of depilation and tanning: "a 'primitive' method involves using salt, sour milk or flour to remove the hair, and pomegranate skins or plant roots to perform the actual tanning. The 'industrial' method uses calcium [i.e. lime] to take off the hair, and gall nuts, oak bark, sumac, or alum for tanning" (Pott 1997: 96). Both of these techniques are attested, to a limited degree, in cuneiform sources, but in the two recipes we have for tanning leather (one for making a leather poultice, for medical purposes, out of goatskin and the other for making a drumhead out of steer hide), only the later method, involving "madder" (Sum. giš/u_2.hab = Akk. *hūratu*) and, as its mordant, "alum" (Sum. im.sahar.na_4.kur.ra = Akk. *gabû*) is used. The hair was removed using a milky liquid made with fermented flour known as Sum. a.gar, and a depilated hide was spoken of as "leather eaten by a.gar" (Sum. kuš a.gar gu_7.a). It is difficult to separate out tanning, staining, and dyeing, but, as Van De Mieroop notes, there were distinct processes and ingredients for producing white, black, red, and green leathers (1987: 30). The alum known as Akk. *allaharu* seems to have been used to produce white leather, while the ingredients for black leather included, alongside strained oil and pomegranates, a substance named Sum. im.ku_3.GI = Akk. *šarserru* (i.e. "golden" or perhaps "shining earth"), which we might expect to be green vitriol (ferrous sulfate) based on Talmudic and Greco-Roman parallels, but is, in fact, described as a red earth or paste in the lexical tradition. Campbell Thompson (1936: 19) took it to be ferric oxide, but does not recognize that it is used for dyeing leather black; green ferric sulfate becomes red ferric oxide when heated, so the lexical tradition may have been misled by traditional associations between gold (Sum. ku_3.GI = ku_3.sig_{17}) and the color red in the languages of Mesopotamia. Madder (Sum. u_2.hab_2 = Akk. *hūratu*) is often enough found in combination with *allaharu*-alum (Sum. al.la.ha.ru = Akk. *allaharu*) in the Ur III and early Old Babylonian periods, sometimes alongside im.ku_3.GI, and the resulting red leather was known as "madder leather" (Sum. kuš u_2.hab_2; lexical lists make a strict distinction between Sum. giš.hab_2 = Akk. *hūratu* and Sum. u_2.hab_2 = Akk. *būšānu*, but only u_2.hab_2 is found in combination with mordants in the Ur III period). Green leather was known as "*dušu* leather" (Sum. kuš du_8.ši.a), after the stone of the same name, and it was pigmented using copper acetate or verdigris: in BIN 9, 455, for example, we learn that slightly more than 8 shekels of copper (ca. 70 g) was used in the coloring of one skin (Van De Mieroop 1987: 31). As Van De Mieroop puts it: "Three materials were used for coloring: im-KÙ.GI for black, copper for green, and madder for red" (1987: 32).

The earliest reference to dyed fabric or textiles, predominantly wool in Mesopotamia, is from an Early Dynastic list of terms related to lapis lazuli (Biggs 1966), and for much of the early history of dyeing textiles, which must be kept separate from both finishing textiles and dyeing leather, it was the resulting colors rather than the material of the dye that served as the basis for categorization. Besides Forbes' survey (1965), the most important point of orientation for colors in Mesopotamia is Landsberger (1967), while the

FIGURE 2.3 Vials of kermes and *Rubia tinctorum*. Photograph © Nicole Reifarth.

best recent survey of textile dyeing is Thavapalan's new monograph on color terminology and meaning (2020), which offers a comprehensive picture of the different methods for dying fabrics in Mesopotamia but appeared too late to be incorporated here in a substantial way. In line with van Soldt's 1990 overview, it is best to look at dyes in three or four groups: insect-based dyes, plant-based dyes, and copper/verdigris dyes (murex-based dyes were probably not sourced in Mesopotamia, although there may have been some possibilities in the vicinity of Bahrain; see van Soldt 1990: 345–6; Eden 1999). Insects found in the kermes oak were known by the same name: "kermes and cochineal, dyes extracted from the bodies of female insects belonging to the family Coccidae" (Forbes 1965: 100), and they were used to produce a reddish dye known in Akkadian as *huruhurātu* (not to be confused with Akk. *hūratu*, "madder"). The most important of the pharmaceutical lists, Uruanna III 237, references a "red worm" (Akk. *tūltu sāmtu*) in connection with "the red dye extracted from the kermes worm" (CAD T 467a; see already Forbes 1965: 104, n. 19; for Nuzi and Middle Assyrian evidence, see Donbaz 1988 and Soldt 1990: 347). The most important plant-based dye was, as noted above, madder or dyer's madder (*Rubia tinctorum* = Akk. *hūratu* = Ugr. *puwatu* = Arab. *fuwwatu*; see Hoffner 1967: 301; Soldt 1990), but there were other plants involved, namely Akk. *tiyatu* and Akk. *nuhurtu*, and like madder, both *tiyatu* and *nuhurtu* are also found fairly often as ingredients in medical recipes. Other terms, including Akk. *uqnâtu* for woad or Akk. *azupirānu* for saffron, are used to refer to both a material in dyeing processes and, separately, a plant, so we can be fairly confident that these were also plant-based dyes (Soldt 1990: 347–9). The most important mineral dyes were, as noted above, made from copper for leather goods, in particular verdigris and copper acetate (Akk. *šuhtu*).

Much of the Akkadian terminology for dyed fabrics, however, refers to several shades and varieties of colored garments, without specifically describing their color or the dye that was used to produce them: color terms are often used to translate Akkadian words like *hašmanu* "red-purple (garment)," *tabarru* "red (garment)," *takiltu* "blue (garment)," among others, but this can be rather misleading: these are not color terms, but rather words that designate types of garment (or textile) that have a distinctive color. Campbell Thompson raised the possibility that the rare Akkadian term *hazallūnu*, on the basis of a supposed Syriac parallel in *halāzūnā*, might be the term for murex in Akkadian, but this has not been substantiated (Campbell Thompson 1934: 781; Levey 1959: 108). Purple garments dyed with murex must occasionally have been imported

FIGURE 2.4 Archaeological remains of dye in Qatna. Photograph © Nicole Reifarth.

into southern Mesopotamia, and we have solid evidence for their production and distribution in and around Ugarit on the Syrian coast, including surviving traces of purple-dyed garments from a royal tomb in Qatna (James et al. 2011).

It is telling that the one collection of recipes for dyeing fabric that we have from Mesopotamia, a Neo-Babylonian compilation that likely reiterates recipes known from the mid-second millennium BCE, includes various combinations of different dyeing elements and mordants for a variety of colors, but only describes the red-purple known in Akkadian as *argamannu* as "imitation" (Leichty 1979; Finkel et al. 1998). This was the term used in the Eastern Mediterranean for murex-based purple dyes, a term originally borrowed from Hittite, where it referred to "tribute" delivered to one's suzerain, but it is little wonder that this prototypical form of tribute eventually morphed into a designation of the royal dyestuff itself (Soldt 1990: 344, n. 164). In administrative documents from both Nuzi and Assur in the second half of the second millennium BCE down to Neo-Babylonian temple records, we find clear descriptions of the same materials mentioned in the Neo-Babylonian recipe collection, including specifications of Akk. *huruharātu* dye used for Akk. *tabarru* "red" garments in Middle Assyrian sources (Donbaz 1988) and, in contrast, in the Neo-Babylonian period, Akk. *tabarru ša inzahurētu* "red (wool) dyed with *inzahurētu*-dye" (Payne 2007: 137). As Oppenheim already suggested in his brief description of the mid-second-millennium BCE chemical trade in 1967, the terminology for non-murex dyes seems to include Hurrian elements, such as the Hurrian term loaned into Akkadian as *inzahurētu*, and it is likely that both the terminology and the recipes for non-murex red-purple dyes arose in inland Syria, in the state of Mitanni in the middle of the second millennium BCE, likely in competition with the famous murex dye trade in the Eastern Mediterranean (Oppenheim 1967: 242–3).

Glass is the most important and dramatic of the artificial materials invented by humankind in antiquity: it was probably invented in Mesopotamia, and it plays a special role in the early history of chemistry in the ancient Near East. As Henderson puts it, "glass was the first man-made translucent solid," but unlike its immediate precursors, such as glazed steatite and faience, the production process involves careful management of the pyrotechnic environment in which it is produced and of the forming of objects in a molten state (2012: 1). Two archaeologically well-stratified examples, from Tell Brak and Eridu, date to ca. 2300 BCE, suggesting that glass technologies were already in development in the late third millennium BCE in Mesopotamia. This is symbolized by the highpoint of a Sumerian epic known as *Enmerkara and the Lord of Aratta*, in which Enmerkara, the ruler of Uruk, is engaged in a technological competition with his nemesis, the ruler of faraway Aratta. Challenged to produce a scepter of no known material, not made of wood, metal, or stone, Enmerkara comes up with an unfortunately broken and poorly understood recipe for producing a "reed of splendor" (Sum. gi su.lim.ma, lines 426 and 430), which is "poured like oil"

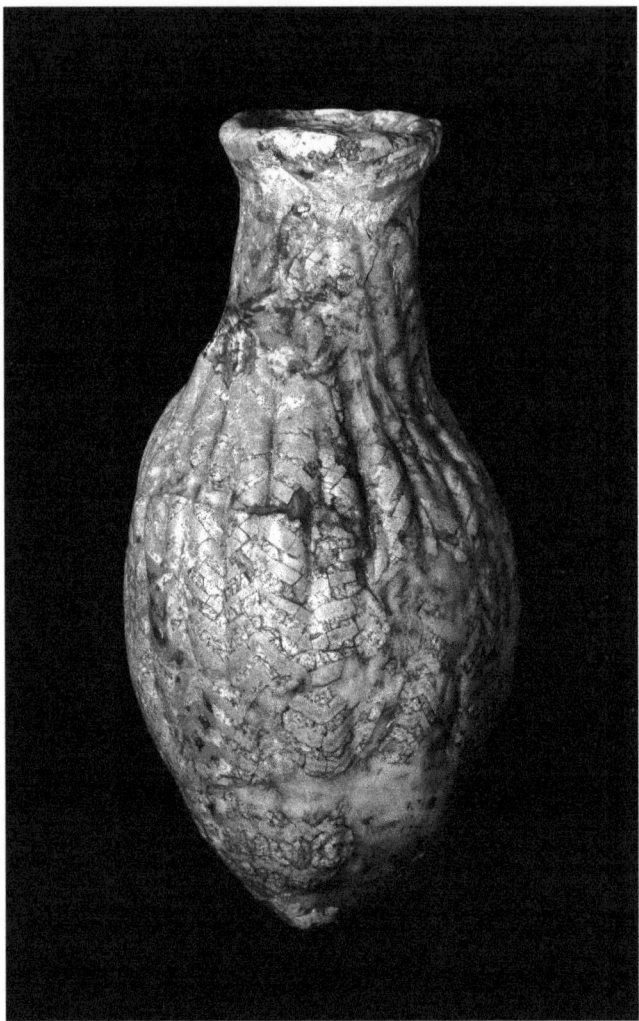

FIGURE 2.5 Fluted glass bottle. © The Trustees of the British Museum.

(Sum. i$_3$.gin$_7$ mu.ni.in.de$_2$, in line 426). The resulting material is characterized by a two-line proverb: "From the sunlight it emerged into the shade, and from the shade it emerged into the sunlight" (transl. after Vanstiphout 2003: 80–1), likely a description of the translucent nature of glass. This crowning achievement in the competition between the technological centers in Uruk and Aratta, namely the creation of a translucent solid, is portrayed as taking place in the Early Dynastic period – Enmerkara is one of the rulers of Uruk preceding Gilgamesh – but was certainly first composed in the Ur III period (ca. 2100–2000 BCE), well within the experimental period leading up to the origins of the glass vessel industry in 1600–1500 BCE. Contemporary Ur III administrative documents use the Sumerian term an.zah (= Akk. *anzahhu*), in all likelihood to refer to faience or the partially processed material known as frit, while the term used in reference to glass in *Enmerkara and the Lord of Aratta* (Sum. su.lim) may be related to the term for "amber" (Sum. $su_3.ra_2.ag_2$ = Akk. *elmešu*) in the lexical tradition (see Cassin 1968 for a survey of the terminology of "splendor" and its links to royal display).

Core-formed glass *vessels* first appear, as part of a regular and well-contextualized archaeological industry, in the middle of the second millennium BCE, largely in cities under the control of the Hurrian state of Mitanni such as Tell Brak, Chagar Bazar, Alalakh, Assur, and Nuzi (Henderson 2012: 129–38). Besides the Kassite glass ax-heads, which derive from a late Seleucid-period hoard, glass mosaics are known from Kassite-period Dur-kurigalzu and, as Henderson emphasizes, "the largest amount of Bronze Age glass of any Middle Eastern site" was excavated in Ugarit, so glassmaking was not a Hurrian monopoly (Saldern 1970: 207, 213–15; Henderson 2012: 136). Flinders W.M. Petrie (1924–5) was the first to suggest that the Egyptian glassmaking industry may have been based on Hurrian artisans and technologies that were extracted from Syria in the course of Thutmose III's campaigns there, an idea that was reiterated by Oppenheim (1973). As Nicholson (2012: 17–19) emphasizes, this would nicely explain why copper and cobalt blue glasses were known as "Menkheperre [i.e. Thutmose III's throne name] lapis-lazuli and turquoise" in contemporary inscriptions. In the Amarna Letters, as Oppenheim first pointed out, two terms are used for glass, supposedly Hurrian *ehlipakku* and another term, perhaps West Semitic, *mekku*; *ehlipakku* is actually glossed with the term *mekku* in RS 17.144 (PRU 6, no. 8), a letter to the ruler of Ugarit about high-value goods, including horses, textiles, and glass, so we can be certain of the equation (Nougayrol 1970: 7–9; Richter 2012: 76 rejects Oppenheim's contention, following the suggestion in Landsberger and Reiner 1970: 26 [= MSL 10] that *ehlipakku* is actually an Akkadianized Sumerian expression, viz. hi.li.ba, subsequently loaned into Hurrian). In spite of the terminological discrepancies, it is clear that a wide variety of glass objects were included in the gifts of Mitanni kings to the Egyptian Pharaoh, reinforcing the general perception that inland Syria was the center of innovation in glassmaking in the second half of the second millennium BCE.

Within a few centuries of the development of the glassmaking industries in Syria and Egypt, at the midpoint of the second millennium BCE, the first recipes for making both imitation semiprecious stones and colored, translucent glass were recorded in cuneiform writing. Unlike the archaeological finds from the Eastern Mediterranean, Syria, and northern Mesopotamia, the glassmaking recipes point to Mesopotamia proper (present-day Iraq and eastern Syria) as the central region for innovations in glassmaking in the ancient world. The cuneiform glassmaking recipes are all written in Akkadian and stem from both northern Mesopotamian Assyrian sources and southern Mesopotamian Babylonian traditions in the fourteenth to twelfth centuries BCE. The three most important blocks of text are (a) the Alpha Group from the Neo-Assyrian period compilation from Assurbanipal's Library in Nineveh, (b) Paragraph U in the Beta Group from the same compilation, and (c) the Middle Babylonian Glass-Making Tablet (BM 120960), published by Gadd and Campbell Thompson in 1936 (Oppenheim 1970). Each of these groups of recipes differs in orthography, vocabulary, and textual format; moreover, the Neo-Assyrian compilation from Nineveh actually includes a widely divergent array of different text types and formats. The procedural texts, such as paragraphs 1–3, 4–6, and 13–15 in the Alpha Group, have garnered the most attention from present-day scholars, largely because they describe a multistage process – with each stage demarcated by a horizontal ruling in the text and named intermediate products such as Akk. *zukû* or Akk. *tersītu* – and distinctive final products: paragraphs 1–3 and 4–6 produce *zagindurû*, a "deeply saturated lapis lazuli blue," while paragraphs 13–15 produce "fast bronze" (Akk. *siparru arhu*; see Thavapalan et al. 2016, especially 201–3, for these translations and the research history). The third paragraph in each of these sequences focuses on opacifying translucent glass, since the goal of all these recipes is to produce imitation stones, which are necessarily opaque. In contrast to the use of the *kūru* furnace and tests for the viscosity of molten glass in the procedural texts, the texts in paragraphs 7–12 of the Alpha Group adopt a simpler formulation – nearly identical in places to the well-known style of medical prescriptions – and exhibit an older technological profile: an Akk. *atūnu* "kiln" is used (instead of the Akk. *kūru* "furnace"), the repeatedly ground materials are placed in "molds" for a week at a time, and if not sufficiently vitrified, they are simply returned to the oven for another week. The Beta Group from Nineveh and the Middle Babylonian text focus largely on the production of different types of red glass, with individual recipes attributed to different regions (Assyrian, Babylonian, and Elamite, in present-day Iran, but no mention of Syria) and even, in two cases, named ancient authorities, although both names are unfortunately broken.

Perfume-making was also an essential part of the technical practice carried out in the palace in order to promote the well-being of the crown and those around him. Already in the Early Dynastic and Ur III periods, in the second half of the third millennium BCE, we find administrative records that quantify more than a dozen different aromatics that were used in the production of perfume (Brunke and Sallaberger 2010). Most of these ingredients reappear in receipts and perfume-making recipes in the second millennium BCE and also form the basis for numerous medical recipes that were presumably first recorded in the Old Babylonian period but only survive in the first-millennium BCE Library of Assurbanipal. The main ingredients throughout this lengthy period of time were the oil that served as the carrier, almost certainly sesame oil (Sum. i$_3$.giš =

Akk. *šamnum*), from at least the Old Babylonian period on, and oils or resins derived from evergreens such as "cedar" (Sum. erin = Akk. *erēnu*), "juniper" (Sum. za.ba.lum = Akk. *supālu*), "cypress" (Sum. šu.ur$_2$.me = Akk. *šurmēnu*), and "myrtle" (Sum. ad$_2$ = Akk. *asu*). In addition, approximately a dozen other ingredients, nearly all classified as "aromatics" in cuneiform through the use of the cuneiform determinative šim (= Akk. *riqqu*), occur in different periods in much smaller quantities. It is fairly clear from the Middle Assyrian step-by-step perfume-making recipes, published by Ebeling in 1948–50, that warm maceration was the primary way of producing perfumes for the crown, including numerous repetitions and quite complex movements of the macerating material from one vessel to another throughout the process. The complexity of these procedures, in combination with his own preconceived notions, led Levey (1959: 36–8) to see in these practices a primitive form of distillation, but there is a solid consensus, at present, that distillation was not known in second- or first-millennium BCE Mesopotamia (Jursa 2005: 335). The fact that these "perfumes" played a central role in religious practices and that the same aromatic ingredients reappear in many medical recipes shows that these substances were not only or even primarily aesthetic in function. Except for Jursa's recent entry under "Parfüm (rezepte)" in the *Reallexikon der Assyriologie* (2005), there has not been much effort to identify broader continuities between the different historical periods, although there have been a number of important studies of individual historical phases in recent years (Jursa 2009; Brunke and Sallaberger 2010; Cousin 2013; Middeke-Conlin 2014; Escobar 2017).

GRECO-ROMAN WORLD

Matteo Martelli

Despite its Greek etymology, the term "chemistry," or at least its earliest cognate, entered the Greek vocabulary only at the edges of the chronological period under consideration in this volume. Indeed, the Greek term *chymeia* (with its various Byzantine spellings, such as *chēmeia, chymia*, or *chēmia*) made its first known appearance in the writings of the Greco-Egyptian alchemist Zosimos of Panopolis at the end of the third century CE. Zosimos interestingly introduced *chymeia* – usually translated as "alchemy" – in the framework of a wider discussion on the boundaries of this art (*technē*): an alchemical book, he claims, must not only explain how silver can be dyed gold, but it must also include a wider spectrum of practices that could produce a variety of chromatic transformations in all kinds of metals (Martelli 2014a: 9–15). The same definition of alchemy and the practices it should encompass seems to be under debate here.

This debate did not take place in a vacuum, but was firmly rooted in a rich technical and artisanal tradition, which alchemy inherited and reshaped when it took its first steps in Greco-Roman Egypt. Already in classical antiquity, a wide range of artisanal practices exploited properties of the natural world that we would call chemical nowadays. One can retrospectively detect a variety of chemically relevant passages, so to speak, which are scattered in many Greco-Roman texts dealing with different technologies, from metallurgy to pharmacy, from dyeing techniques to cosmetics. The variety of the described procedures is so wide and scattered in texts belonging to such different genres that it would be risky and anachronistic to reconstruct some supposed classical view of chemistry as a unified and well-defined field of inquiry in the Greco-Roman world (Healy 1999: 115–16). However, against this fragmented picture, one must observe that some practices started to be grouped together and recognized as a consistent set of technologies by the authors of the earliest alchemical writings. As clearly emerges from the extant works of this literature, in the first centuries CE, alchemical authors constructed their own writings around specific areas of expertise. The four books of Pseudo-Democritus dealt with four main "chemical" subjects: (a) the making of gold, namely a set of techniques to dye metals yellow; (b) the making of silver, that is, how to whiten metals; (c) the making of artificial gemstones; and (d) purple dyeing (Martelli 2013). Likewise, a similar range of techniques is described in the Leiden and Stockholm papyri, two chemical recipe books dating to the third to fourth centuries CE (Halleux 1981; English transl. in Caley 1926 and 1927).

These sources describe procedures that in many cases can be located in other ancient technical writings as well. The Leiden and Stockholm papyri include recipes on ink-making along with various techniques to test artificial products, thus continuing a rich tradition well attested in earlier authors such as Pliny the Elder and Dioscorides (Halleux 1981: 42–3; Greenaway 1986). Pliny (*NH* XXXIV 112) explains that verdigris was adulterated with *atramentum*, a green vitriol (i.e. ferrous sulfate). A papyrus soaked in an infusion of gallnuts could reveal the composition of the drug to be tested: the papyrus, in fact, turns black in the presence of ferrous sulfate (Healy 1999: 136), and ferrous sulfate was indeed an essential component of iron gall black inks. In his *Compendium on Mechanics* (IV 77, in Diels-Schramm 1920: 79), the third-century BCE writer Philo of Byzantium mentions a special ink made of gallnuts dissolved in water, which allowed letters to be written that became invisible as the mixture dried; however, the letters became legible again after being washed with a sponge soaked in a solution of vitriol (Forbes 1965: 236–9; Christiansen 2017: 188–9).

Between the fifth and fourth centuries BCE, technical writings were composed on a variety of arts (Cambiano 1991: 29). Plato refers to cookbooks (*Grg.* 518b), Aristotle mentions books on agriculture (*Pol.* I 11, 1259a), and many medical treatises were grouped around the figure of Hippocrates. The author of *On Ancient Medicine* (§§ 3–9 = I 574–91 Littré) links the origins of medicine to the development of cooking techniques, such as the making of wheat bread and barley cakes. Leavened wheat bread became widespread in the Roman Empire, when different sources of leavening were exploited: next to dough left to sour naturally, Pliny mentions millet or wheat bran fermented in grape must (*NH* XVIII 102) and the yeast produced by the Gauls and Iberians (*NH* XVIII 68), who, when making beer, used the foam that formed on the surface during the process as leavening (Forbes 1965: 97–8; Monteix 2015: 214). Beer, indeed, is often mentioned in Greco-Roman sources: in Hellenistic Egypt, brewing was a royal privilege, and an anonymous alchemical text (*CAAG* II 372) describes the making of barley beer (Forbes 1965: 130–4; Rasmussen 2014: 29–48). Mead and palm and date wine are also mentioned by Roman authors, but grape wine was certainly the most popular alcoholic beverage in the Greco-Roman world: it was often flavored with pitch and resins, which could also prevent the conversion of alcohol into vinegar, thus preserving the wine (Rasmussen 2014: 18–25, 66–7). Vinegar, on the other hand, was the main acid substance known in the Greco-Roman world. As well as being used to treat metals (e.g. iron or gold; see Halleux 1987), it was employed in medicine. Various recipes for the preparation of vinegar are given by ancient authors, such as Columella and Julius Africanus. In a fragment of his *Cesti* (F12 § 19 in Wallraff et al. 2012: 94–7), Africanus lists various kinds of vinegar, produced with the addition of soda (*nitron*), pepper, and burned barley (see also Columella, *On Agriculture*, XII 17).

A wide range of chemicals were used by ancient artisans, such as metalworkers, shoemakers, tanners, goldsmiths, and pharmacists. Even though mainly based on plants, the Hippocratic medical recipes include mineral compounds as well: along with metals (especially copper), one can find litharge (*lithargyros*), verdigris (*ios*), white lead (*psimythion*), and different kinds of vitriol, to name but a few (Goltz 1972: 107; indexes in Totelin 2009; Laskaris 2016: 156–7). Many of these substances were used as pigments as well. Section 8 of Theophrastus' treatise *On Stones* is specifically devoted to mineral colors: both natural and artificial materials are described, according to a useful division that was still used to classify *colores* in the eighth book of Vitruvius' *On Architecture* (Romano 1998). Art imitates nature, Theophrastus claims (*Lap.* VIII 60): technological treatments seem to follow those natural processes that produce minerals under the ground, such as filtering, roasting, or cementing; therefore, if the treatment is correct, what is artificially produced departs minimally from its natural "model."

Some artificial products were discovered by accident. The Athenian painter Cydias (fourth century BCE), for instance, realized that roasted yellow ochre could be used as a purple dyestuff after a fire burned a shop (*Lap.* VIII 53). In some cases, similar treatments were applied to different metals. Pieces of lead or copper were exposed to the vapors of vinegar in order to produce a white or a green pigment, respectively. Theophrastus describes the making of the white pigment (*psimythion* in Greek and *cerussa* in Latin), namely a basic lead carbonate later called "white lead," as follows (*Lap.* VIII 56):

A piece of lead as big as a brick is placed above some vinegar in a cask. When after about ten days, the lead has acquired thickness, the cask is opened and a kind of mildew scraped from the lead, which is repeatedly placed in this way until it is used up. The scrapings are pounded in a mortar and continually strained away; and the white lead is the matter finally left deposited.
(Eichholz 1965: 79; see also Vitruvius, *On Architecture*, VII 12.1)

Modern replications have shown that the treatment produces a thick layer of white crust around the lead, which can be easily scraped off. The jar certainly contained air, an essential source of atmospheric carbon dioxide, which allows the conversion of lead acetate into lead carbonate (Principe 2018: 163–5). According to Dioscorides (V 88.1–3), a net of reeds was put in the jar over the vinegar: it collected the fragments of the white crust that dropped down during the process, thus preventing them from dissolving in the vinegar. We must observe that in his work *On the Capacities of Simple Drugs* (IX 3.39 = XII 253 Kühn), Galen suggests administering *psimythion* after diluting it in strong vinegar. Moreover, white lead often received a second treatment: roasted on hot coals, it was turned into a red substance suitable both as a pigment and as a medicine (called *sandyx*). Pliny prescribed adding red ochre to white lead before roasting (*NH* XXXV 40). Dioscorides (V 88.5) provides a different description of the process, where a new clay container containing ground white lead was placed on hot coals (see also Plin. *NH* XXXIV 176).

Lead oxide or litharge (*lithargyros* in Greek and *spuma argenti*, "foam of silver" in Latin) was produced by treating argentiferous lead ores. The ores were first smelted to obtain metallic lead with different percentages of silver. In order to separate silver, this lead was remelted in open furnaces (a cupellation process): it quickly oxidized and turned into liquid lead monoxide, which overflowed onto a second crucible (Plin. *NH* XXXIII 106–10). According to the cooling time, two allotropic forms of litharge were produced, namely a yellow and a red litharge, which only differ in their crystalline structure. Their possible identification with the three types of litharge mentioned by ancient sources – *chrysitis* ("gold-like"), *argyritis* ("silver-like"), and *scalauthritis* or *molybditis* ("lead-like") – is still debated (Bailey 1929–32: vol. 1, 215; Halleux 1975; Healy 1999: 320–6). These kinds of litharge were further manipulated in order to make medicines of different colors. Broken into pieces and boiled in water with wheat and barley wrapped in linen cloths, *argyritis* turned into a shiny white drug (Plin. *NH* XXXIII 108; Diosc. V 87). Ground in mortars and treated with salt and soda, it could possibly produce white lead chlorides, as suggested by Healy (1999: 325; see Diosc. V 87.11; Plin. *NH* XXXIII 109). Bailey (1929–32: vol. 1, 216) compared this product with "Pattinson's white lead," produced in 1841 by the English chemist Hugh Lee Pattinson, who patented a new process to produce lead oxychloride, commercialized as white lead pigment. The polychromy of mineral products captured the attention of Greco-Egyptian alchemists as well. In a recipe from his book *On the Making of Silver* (§ 5), Pseudo-Democritus explains how to make litharge whiter than *psimythion* by means of different ingredients (sulfur, orpiment, cadmia), but a strong heat – he explains – would turn the product yellow, since "the nature of lead quickly undergoes many transformations" (Martelli 2013: 110–11).

As in the making of white lead, vapors of vinegar were used to treat copper leaves as well, in order to produce verdigris (*ios* in Greek and *aerugo* in Latin). Theophrastus (*Lap.* VIII 57) prescribed the use of wine lees rather than vinegar, which appears, in turn, in the recipe given by Vitruvius (VII 12.1). According to Dioscorides (V 79.1), a jar containing vinegar was simply covered with an inverted copper vessel, whose internal surface reacted with the vapors of vinegar. Sometimes a lump of copper was stashed among old pressed grapes, or copper blades were simply sprinkled with vinegar (Diosc. V 79.2; Plin. *NH* XXXIV 110–11). The production of verdigris is accurately described in the chemical Stockholm papyrus (§ 74): a sheet of Cyprian copper is first cleaned by means of pumice stone and water, then smeared with oil; a cord is tied around it in order to suspend it in a vessel containing strong vinegar.

Along with verdigris, the ancients produced other kinds of rust. Iron rust, for instance, was used in medicine (Diosc. V 80), while Greco-Egyptian papyri mention "gold rust" (*PGM* XII 193–201; Halleux 1981: 163–6; Betz 1986: 160–1). Gold, in fact, was treated with specific cements, namely mixtures of minerals (salts, sulfur, alum, vitriols), which reacted with its impurities (especially copper), thus producing a "rust" to be scraped off (Halleux 1982: 197–8; Halleux 1985: 45–8). In some cases, vinegar was used to "wash" these impurities away, as one can infer from the mention of "vinegar from the purification of gold" in the chemical Leiden papyrus (§ 14; Halleux 1981: 87; Halleux 1985: 55–7).

Cementation processes were frequently used in ancient metallurgy, when metals were heated in contact with different ores or mineral products. In the *Sophistical Refutations*, Aristotle mentions various metals that imitate gold and silver (§ 1, 164b23–4): "silver-like metals produced with lead oxide (*lithargyrina*) or tin (*kattiterina argyrâ*), and gold-like metals dyed with bile (*cholobaphina chrysâ*)." The use of bile (*cholē*) to tinge metals – replicated by Bailey with some success (1929–32: vol. 2, 164) – is recorded in various inscriptions, as well as by Pliny the Elder (Halleux 1981: 41). Lead oxide (*lithargyros*) is added to tin or lead in order to whiten the metals in the papyrus of Leiden (§ 11). As for tin, a chapter of Pseudo-Aristotle's *On Marvellous Things Heard* reads (§ 62, 835a): "They say that Mossynecian copper is very shiny and white, not because there is tin mixed with it, but because some earth is combined and molten with it" (Hett 1936: 263).

The identification of the earth mentioned in the pseudo-aristotelian text is a matter of debate among scholars: it could refer either to arsenic ores, which would have produced a white arsenic-copper alloy, or to zinc ores (Halleux 1982: 194). Indeed, zinc-rich minerals too were added to copper in a crucible to produce brass. An early reference to this process seems to be detectable in *On Stones*, where Theophrastus mentions an earth that, when mixed with melted copper, improves the beauty of its color (*Lap.* VIII 49). The fourth-century BCE historian Theopompus describes a stone from the city of Andeira in Troad, which produces *oreichalkos* when mixed with copper (fr. 112 Jacoby; see also Strabo, *Geography*, XIII 1.56). Even though the meaning of *oreichalkos* is uncertain in ancient sources (Halleux 1982: 195), Theopompus seems to be referring to a zinc-copper alloy produced by treating copper with zinc ores. More certain is the identification of the Latin *aurichalcum* with brass, since Roman writers often stressed the yellow color of this alloy. Pliny specifies that copper *Marianus* (mined in Cordova) "most readily absorbs *cadmea* and reproduces the excellence of the *aurichalcum*" (*NH* XXXIV 4). The term *cadmea* (*kadmia* in Greek) referred either to natural zinc ores (such as calamine) or to zinc oxide artificially produced by heating zinc-rich ores and collecting the vapors that condense in contact with the walls of the furnaces (Plin. *NH* XXXIV 100–3; Diosc. V 74).

The ability of ancient sculptors and metalworkers to produce metallic alloys clearly emerges from Pliny the Elder's description of ancient statuary in book 34 of his *Natural History*. He mentions the Egyptian black *niello*, a black alloy manufactured by using sulfur to darken a silver-copper alloy (*NH* XXXIII 131; see also the Leiden papyrus, § 35 in Halleux 1981: 93). Corinthian bronze was also very precious – a dark blue bronze made by treating the surface of a copper-silver-gold alloy (Plin. *NH* XXXIV 8). Similar techniques for producing black bronzes are described in the alchemical books by Zosimos of Panopolis (*CMA* II 222–32), who explains the preparation of many dyeing *pharmaka* (Giumlia-Mair 2002; Hunter 2002). These could be watery substances as well, such as "sulfur water" or "divine water," a red liquid made of sulfur and lime diluted in vinegar or urine. Ancient alchemists used to play with the term *theion*, which could mean both "divine" and "sulfur" (Martelli 2009). The earliest recipe for the making of this "water" is recorded in the chemical Leiden papyrus (§ 87; Halleux 1981: 104). Lime and sulfur are ground and mixed in vinegar or urine. After being boiled, the liquid becomes a red "water," which is used after being filtered. Modern replications have shown that if a piece of silver is dipped into this "water," the metal quickly changes its color, becoming very similar to gold. As Principe explains, the process leads to the "the formation of extremely thin layers of sulfides on the metal surface" (Principe 2013: 10–11).

Mercury is often mentioned in processes for dyeing metals superficially, such as mercury gilding, in which a gold–mercury amalgam was smeared on copper (Halleux 1981: 97–8; Healy 1999: 290–3). A method for extracting the liquid metal is already attested in Theophrastus' *On Stones* (VIII 60): cinnabar (a mercury sulfide) was ground with vinegar in a copper mortar with a copper pestle. A mechanochemical reaction is here described, in which copper reacts with sulfur, thus liberating metallic mercury (Takacs 2000). Pliny (*NH* XXXIII 123) and Dioscorides (V 95) also explain a second technique, namely a hot extraction, in which cinnabar was placed in an iron spoon and heated in a double vessel (see Chapter 3): drops of sublimated mercury precipitated on the upper vessel, from which they were collected. These two methods were inherited by the earliest alchemists. Maria the Jewess and Zosimos ground cinnabar by using mortars of different metals, such as lead and tin (*CAAG* II 172,12–19; *CMA* II 47). Pseudo-Democritus used to grind cinnabar in oil of soda and then distil the mixture (*CAAG* II 123,3–7). Moreover, ancient alchemists tried to make mercury solid again (thus producing artificial cinnabar). Pseudo-Democritus mentions that mercury is made solid by means of various minerals, such as stibnite, roasted lime, and sulfur, while Zosimos specifies that vapors of sulfur can solidify mercury and make it yellow/red again (see Mertens 1995: 11; Colinet 2010: lxxxiii–xci; Martelli 2013: 86–7).

These procedures may have fostered the idea that mercury could be extracted from various minerals. Pseudo-Democritus, in particular, mentions mercury from cinnabar as well as from other ores, especially orpiment and realgar. We cannot exclude the possibility that ancient alchemists tried to process different minerals with the same methods that were used to extract mercury from cinnabar. This seems to be most evident for arsenic ores. While in Pliny the Elder (*NH* XXXIV 177–8) and Dioscorides (V 104–5) orpiment or realgar is simply boiled in water – a process that might have produced white arsenic oxide (Bailey 1929–32: vol. 2, 207) – an alchemical treatise ascribed to the goddess Isis (first to second centuries CE) describes a different process: orpiment and realgar were ground in a mortar with an ear of corn and a little oil, then sublimated in a double vessel (*CAAG* II 32–3; Mertens 1984: 138). The collected vapor that condensed on the upper vessel (perhaps metallic arsenic) was probably equated with a kind of mercury.

Glass too was included by ancient authors among metallic (or, more generally, mineral) substances (Beretta 2004: 4–16; Beretta 2009: 1–55). In his *On Stones*, Theophrastus both mentions vitreous sand as an important component of glass (*Lap.* VIII 49) and lists the Egyptian *kyanos* among artificial blue pigments (*Lap.* VIII 55). The term *kyanos*, already attested in Mycenaean tablets (Halleux 1969; Delamare 2007: 29–30), refers to different blue substances that, when finely ground, could be used as pigments: natural blue ores, in particular azurite, a deep blue copper mineral (Diosc. V 91; Plin. *NH* XXXVII 119); perhaps lapis lazuli, a semiprecious stone usually called *sappheiros* in Greek (Theophr. *Lap.* IV 23; Diosc. V 139); and an artificially produced pigment, the so-called Egyptian blue, actually a vitreous product containing silica, copper, and calcium. This pigment had been manufactured in Egypt since the third millennium BCE by heating together quartz sand (as a silica source), lime, a source of copper (natural ores such as malachite or copper filings), and soda (*nitron* as the alkali flux; Nicholson and Shaw 2000: 108–10). If archaeological evidence points to the production and use of this pigment in pharaonic Egypt, no Egyptian texts record its recipe. Many centuries later, Vitruvius describes how to make a synthetic blue pigment, called *caeruleum*, which was produced in Pozzuoli according to an Alexandrian method (Vitr. *De Arch.* VII 11.1): sand and soda were finely ground and mixed with copper filings to produce small balls that, put in an earthen jar, were heated in a furnace. However, quantities are not specified, and, more importantly, no source of calcium seems to be mentioned among the listed ingredients (Delamare 2003; Delamare 2007: 40–3).

Pliny deals with glassmaking at the end of book 36 (§§ 190–6). Here, he also skeptically reports the discovery of a "flexible glass" (*vitrum flexile*) during the reign of Tiberius, who ordered the workshop responsible for the invention to be destroyed, since it would have resulted in the devaluation of metals; according to Petronius, its inventor was executed by the emperor (*Sat.* 51). Flexible glass is usually regarded as a technological myth. However, it is worth mentioning a recipe for unbreakable glass (*vitrum quod non frangitur*) that is included in the *Mappae Clavicula*, a Latin early medieval recipe book probably based on a lost Greek model (third to fourth centuries CE): finely ground glass was mixed with lead, white lead (*cerussa*), quartz, and resins, then melted in an iron crucible. The process is described in recipe 162 according to Baroni's edition of the *Mappae Clavicula* (Baroni et al. 2013: 174; see Halleux and Meyvaert 1987: 56 and Tolaini 2004: 203). This text, by contrast, is not included in the English translation by Smith-Hawthorne (1974), where we do find another recipe (§ 69 on p. 37) that describes how to produce a glass stronger than metals (see Baroni et al. 2013: 188).

On the other hand, in the same section, Pliny records the use of glass to imitate obsidian and other colored stones. The topic is fully developed in book 37, specifically devoted to gemstones: in fact, Pliny claims, many precious stones were imitated by coloring glass, such as opals, "carbunculi," and jaspers (*NH* XXXVII 83, 98, 117). Pliny also mentions treatises on the counterfeiting of gemstones (Plin. *NH* XXXVII 197), and Seneca reports that Democritus was credited with the discovery of how "a boiled pebble could be turned into an emerald" (*Ep.* XC 32). As already seen, the third alchemical book of Pseudo-Democritus was devoted to the making of precious stones. The book is unfortunately lost, and the practices described there are difficult to reconstruct on the basis of later sources (quotations in Byzantine recipe books and Syriac translations). While the *Mappae Clavicula* includes recipes on the making of colored glasses for imitating gemstones (Tolaini 2004), the Leiden and Stockholm papyri provide evidence for a different technology. Rock crystal (*crystallos*) or tabashir (*tabasios*) were dipped in dyeing baths (made of mineral, vegetal, or animal dyes) after undergoing a preliminary treatment called *stypsis*. In fact, rock crystals were processed with mordants (astringent substances such as burned copper, alum, or iron oxides), which made them more receptive to the dye. This technique seems to imitate the methods used to dye textiles. Pliny, in particular, credits Egyptian dyers with the invention of mordants: they used to smear fabrics not with dyes (*colores*), but with chemicals (*medicamenta*, lit. "medicines, drugs") that caused them to absorb the colorants (*NH* XXXV 150). In his *On Odors* (§§ 7–8), Theophrastus interestingly compared these techniques for fixing dyes with the making of perfumes. In fact, in order to produce perfumes, the ancients used a variety of aromatics (mainly vegetal substances) that were boiled in oils; fixatives (especially myrrh) were added to the mixture in order to make the perfumes more stable (Forbes 1965: 30–40).

Purple dyeing was particularly valued in antiquity. Aristotle already provided a description of the sea snails (belonging to the Muricidae family) that secrete the dye known as Tyrian purple (*Hist. An.* V 15, 547a; Longo 1998). However, many substitutes for this expensive dyestuff were used; dyeing algae, cochineals, woad, alkanet, and madder are often mentioned. Many of these dyeing substances are described by Pliny and Dioscorides, such as madder (Diosc. III 143; Plin. *NH* XIX 47), kermes (Diosc. IV 48; Plin. *NH* XVI 32), or woad (Diosc. II 184; Plin. *NH* XX 59). Sections on purple dyeing are included in Pseudo-Democritus' alchemical books (Martelli 2013: 78–81) and in the Leiden (§§ 89–99) and Stockholm (§§ 89–159) papyri (Pfister 1935; Halleux 1981: 43–6). Alkanet – also used to dye perfumes (Theophr. *On Odors*, 31) – produced a dyestuff that the ancients had trouble fixing on fabrics, since it requires a specific treatment with iron salts; the complex method for processing madder, on the other hand, is fully described in the chemical Stockholm papyrus (§§ 109–11).

CONCLUSIONS

Marco Beretta

The degree of specialization achieved in the chemical arts in ancient Egypt is evident when you look at both the complex classification of minerals and metals expressed in the hieroglyphics and the vast repertoire of archaeological remains. The Egyptian classification embodied the principles of an approach to experimental practice that was shared by all three ancient civilizations addressed in the present volume. The analogy established between native metals and minerals and those products of the chemical arts that imitated them enhanced the role of artisanal knowledge. Within this framework, it was therefore consequent that metals were sometimes evoked in the hieroglyphic sign of a crucible, alluding to the operation of melting. Chemical operations, therefore, helped to define natural products.

The growing importance of the chemical arts in Egypt and Mesopotamia encouraged the creation of large-scale sites of experimentation. Metals, minerals, and precious stones played important roles in Mesopotamia's economy. Glass, the "first man-made translucent solid" (Henderson 2012: 1), became one of the most valuable commodities, and it was within this art that the most important innovations prior to glassblowing were introduced and exported. The remarkable property of glass of imitating precious stones and minerals demonstrated the idea that chemical crafts could replicate natural products. The fluctuating hierarchy between *physis* and *techne* in the Greco-Roman world was the effect of both the progress made in the chemical arts and the interest they raised among natural philosophers. It is in this context that we see the first attempt to group together the different techniques entailed in metallurgy, pharmacy, dyeing, and cosmetics into a consistent view on the properties of matter. The works of the alchemist Pseudo-Democritus and the surviving Leiden and Stockholm papyri listing chemical recipes offer crucial evidence of the efforts to treat the manipulation of matter within a comprehensive philosophy of matter.

CHAPTER THREE

Laboratories and Technology: *From Temples to Workshops: Sites of Chemistry in Ancient Civilizations*

SYDNEY H. AUFRÈRE, CALE JOHNSON,
MATTEO MARTELLI, AND MARCO BERETTA

EGYPT

Sydney H. Aufrère

From the First Dynasty (3150–2925 BCE) onwards, the reunification of the various territories of the Egypt was under the yoke of a royal power and the authority of one or two nominated viziers (Moreno Garcia 2013), whose task was to establish a central administration and to see to the increase of industrial production. At the regional level and from a political and religious point of view, this administration was carried out by nomarchs (provincial administrators) representing the king. The emergence and availability of written texts facilitated the work of these two viziral administrations and ensured overall management of the agricultural resources of the Nile valley. As the Nile Delta was considered a "Gift of the Nile," according to Herodotus (*Hist.* 2: 5; Sall 2005–6), a policy was established imposing the digging of irrigation canals, exemplified by the Mace Head of King Scorpion (Oxford, Ashmolean Museum, inv. no. AN1896.1908.E.3632; Goyon 1982; Menu 1994; Gauthier and Midant-Reynes 1995; Manning 2002). This irrigation limited the recurrent hunger episodes caused by irregular Nile water levels (Vandier 1936).

Administrative centralization from the south to the north of the country made it possible to balance resources and develop an industrial economy in various fields of agriculture. It not only improved the exploitation of mineral resources in the immediate vicinity of the Nile and in the deep desert but also increased the importation of high-value substances and products. (For a general history of Egyptian economy, see Muhs 2016.) In Egypt there was no currency and all transactions were made with monetary equivalents (Daumas 1977; Kemp 2005: 319–26) standardized by a unit of weight and measures, the *deben* (☰); its value varied more or less with the product weighed, generally 90–91 g, and it was subdivided into 10 *qites* (𓏺 *qedet*) or 12 *shats*, or in volume as the *heqat* (4.8 l). This mode of exploitation of the country's resources led to a redistributive economy under the aegis of the Pharaonic state providing payments in raw commodities (cereals, meat, fish, wood) or processed materials (bread, beer, wine, salt products, clothing; Koenig 1979–80).

The Egyptian economy was based on the collective production of foodstuffs and on the control of storage and modes of transportation. The improved exploitation of quarries and mines in the surrounding valleys increased the number of expeditions for the supply of materials to meet orders coming from royal or local personalities. (See the information on the quarries of Wadi Hammamat; cf. Couyat and Montet 1912; Goyon 1957; Gasse 1987: 207–18; Gasse 1988.) Overexploitation of mines led to the imposition of drastic bans or even convictions (Meeks 1991: 234). The importation of exotic products through traditional channels and from distant lands to meet the requirements of the temples for various materials that were indispensable for the conduct of their rituals (precious metals and minerals, oleoresins, etc.; Grandet 1994) had to be organized and strictly controlled. Treasury-dependent state warehouses and major and minor temples then became places where raw materials were processed. Cult products requiring specific know-how for the treatment of metals, minerals, and aromatics were produced in specific temples. Despite the fact that several designations relating to the mineral world have been debated (Putter and Karlshausen 1992; Klemm and Klemm 1993), it is possible to define the geography of deposits of raw materials (rocks, precious metals and minerals, soils and dyes, resins, aromatics, and chemicals) in Egypt (see the maps of hard rock resources in Aston et al. 2000: 8–11; Gremilliet and Delangle 2017: 9) and in its bordering deserts and distant lands. The Pharaonic state had the capacity to locate these deposits. Large contingents of quarrymen and miners or even personnel of private enterprises could be sent by a central or provincial administration into these deserts, their security ensured by the military (e.g. Chartier-Raymond n.d.). These expeditions included sailors employed to handle cables and move the stone blocks cut from the mountain and craftsmen. The search for mineral veins in the deserts of the East was done by prospectors named *sementyu* (𓋴𓏠𓈖𓏏𓊖; *Wb.* III: 135, 18), identifiable by the bundle (𓋳) they carried on their shoulders at the end of a stick (Yoyotte 1975). According to a scene showing a transport of typical desert products by nomads (Kemp 2007: 317), the idea of a monopoly on desert resources, which would result from the reading of inscriptions relating to major royal expeditions, is disputed. Nonetheless, under the reign of Seti I, based on the idea that gold represented the flesh of the sun, the smuggling of gold nuggets from the mines of Samut was strictly prohibited (Aufrère 2016d). This clearly meant that the Pharaonic state wanted to assert its sovereign rights on the veins.

The two plateaus flanking the Nile valley were the main sources of sedimentary rocks for the building of temples (Aufrère 2001e); limestone of different qualities came from quarries in the region of Gîza (and Masara) and Middle Egypt (Goyon et al. 2004: 142–5), sandstone came from the Gebel Silsila and from many other places (Aston et al. 2000: 54–6; Goyon et al. 2004: 145–6), and sedimentary quartzite came from Gebel el-Ahmar and Aswan (Aston et al. 2000: 16–17, 53–4). Igneous rocks such as granite and granodiorite, destined for the carving of obelisks and columns, came from the Aswan quarries (Aston et al. 2000: 35–7; Goyon et al. 2004: 161–71). Granite and porphyry were respectively exploited in Roman times in *Mons Claudianus* and *Mons Porphyrites* in the Eastern Desert. Anorthosite gneiss or diorite-gabbro (*mentet*), formerly considered as diorite, came from Toshke (Gebel el-Asr) in Lower Nubia (Engelbach 1933; Klemm and Klemm 1993: 423–6; Aston et al. 2000: 30–1). Metamorphic rocks (greywacke, siltstone, shale), for the manufacture of coffins, pyramidions, and small statues, came from the quarries of Wadi Hammamat (Aston et al. 2000: 57–8; Goyon et al. 2004: 173–4). In the first dynasties, minerals such as travertine (or alabaster-calcite), used for the manufacture of large statues and the industrial-scale production of containers such as dishes and jars, came from the Hatnub quarries of Middle Egypt (Putter and Karlshausen 1994: 43–6; Aston et al. 2000: 59–60; Goyon et al. 2004: 172–3).

Although Egypt had abundant supplies of iron ores such as magnetite and hematite (Ogden 2000: 166–8) and sometimes had recourse to meteoritic iron, the Egyptians were not able to master a controlled reduction process (Besançon 1954: 313–14; Gremillet and Delangle 2017: 39). Copper and tin ores and their different alloys (copper-arsenic, copper-tin, copper-zinc; Ogden 2000: 149–61) are well documented in the Eastern Desert, but they were underexploited since copper and tin mainly came from abroad. In contrast, the Eastern Desert and Sinai mines were rich in other minerals such as agate, amethyst carnelian, chrysoprase, chrysocolla, garnet, green feldspar, green and red jasper, malachite, onyx, sard, sardonyx, silicified wood, silver galena, turquoise, various oxides (Aston et al. 2000: 25–30), and particularly bitumen (essential for the preparation of recipes of liturgical ointments; see Chapter. 1, pp. 28–30; Chapter 7, pp. 187–188), found in Gebel el-Zeit (or *Mons Petrolius* of the Romans), the Dead Sea Basin, and the deposits of Syria-Palestine (Serpico and White 2000: 454–6). Antimony came from abroad (Ogden 2000: 149). Aquamarine or beryl (emerald) from the Sikait-Zubara mines (*Mons Smaragdus*; Harrel 2004) or olivine (peridot) from Zabargad Island off the Red Sea coast were used only from Ptolemaic times onwards (Aston et al. 2000: 24–5, 47–8).

Moreover, earths, mineral dyestuffs, and chemicals were processed in local deposits in the Nile valley and its surroundings. This was not case for the red and yellow ochre deposits mentioned in religious texts, for the hematite of Aswan, or for the natron deposits of Wadi el-Natrun, west of the Delta (*Nitria* of the Greeks) and of Elkab (Upper Egypt; Aufrère 1991: 609–37). Natron, used for the dehydration of mummies and the preservation of various foodstuffs, seems to have remained under royal monopoly right through to the Greco-Roman period. There is no archaeological proof attesting to the organization of official expeditions carried out to exploit these resources (see Chapter 6, p. 166). The sea salt coming from the regions of Peluse and Thonis-Herakleion were important for the preservation of meat products.

Aromatics and gum resins – various species of frankincense and myrrh – came from two areas: the plateaus of Southern Arabia and the mountains of the Horn of Africa. Oman frankincense (*Boswellia sacra* Flueckiger, 1867) – the best-quality incense – came from Yemen and Oman. Other species of *Boswellia* came from Yemen and Somalia. The myrrh tree (*Commiphora myrrha*) or basalm tree grew naturally on these plateaus. The method of harvesting of their products varied. Like today, frankincense trees were tapped and the resin collected several times a year. As for myrrh trees, the naturally exuding resin was simply detached from the tree trunk.

Punt was the name given to the region situated on both sides of the Bab el-Mandeb. Recent discoveries have shown that after crossing the Eastern Desert, the Egyptians had access to several harbors on the shores of the Red Sea, from where they could embark for Punt (see Chapter 6, p. 162). Taking advantage of the trade winds, they transported these resinous trees in baskets to acclimatize them gradually and planted them in the Nile valley, as illustrated by scenes depicted in Queen Hatshepsut's (1478–1458 BCE) temple at Deir el-Bahari. Other products, such as styrax resin, balm, and terebinth resin, came from various parts of the Near East (Baum 1994b).

As all manufacturing sectors were under state control, a rationalization of the manufacturing processes was imposed to obtain mass production. This was true for the mastery of ceramics techniques, reflected by the imposing ceramic decoration of aristocratic tombs in the Old Kingdom. The tombs of the common people contained only a basic piece of funerary furniture. The shape of the coffin was often indicative of their content: the number and richness of tombs in archaeological layers is a relative indication of the living conditions of the population.

The demand for standardized ceramic pots necessitated the rationalization of production, the distribution of food resources, and the organization of a society with common norms for all. For example, the evolution of norms in bread-making can be determined by the shapes of bread molds, indicating that there were several types of molds and which mold was most in demand. This mold was then used to meet the ever-increasing demand. The same observation applied to coiled-clay handmade beer jars, water jugs, and plates until the Middle Kingdom. A rudimentary hand-operated potter's wheel was then invented, making it possible to manufacture objects with thinner and lighter walls. Spinning the potter's wheel with the feet appeared much later, allowing the manufacture of larger containers (Bourriau et al. 2000).

The study of multiple kilns found in the fortified city of Qila' el-Dabba, where the palace of governors of the Oasis of Dakhla was located, show that the Egyptians had mastered the firing techniques required for the production of ceramics. These kilns were used not only to meet the daily needs of the palace but also to make the funerary furniture for the mastabas (monumental bench tombs) of dignitaries. The reproduction of an experimental kiln made of raw bricks by researchers of the French Oriental Institute in Egypt has shown that a sufficiently high temperature could be reached using only desert scrub as fuel, since wood resources were not readily available (Soukiassian et al. 1990; Bourriau et al. 2000: 123).

The manufacture of papyrus (*Cyperus papyrus*), a kind of sedge (Leach-Tait 2000: 227–31), was developed very early, perhaps even before the Third Dynasty (2790–2625 BCE). This became generally known when rolls of virgin papyrus ready for use were discovered in the tomb of Hemaka. The Nile Delta contained many large fields where papyrus grew, especially at the mouths of branches of the Nile, in areas where, according to Pliny, the depth of water did not exceed two cubits. It is said that each region produced a different quality of papyrus. Scenes of the harvesting and transportation of bundled stems of papyrus are attested in tombs of the Old Kingdom and Middle Kingdom (Leach-Tait 2000: 231–6). According to the iconography, these bundles were mainly intended to build light boats made of vegetal stems (Vandier 1969: 446–510; Leach-Tait 2000: 235) and to make mats and furniture. Information on the manufacturing stages of papyrus itself was never given, probably because such highly skillful craftsmanship remained a trade secret until the Ptolemaic and Roman periods, when it became a state monopoly. Most of the papyrus was for exportation to Mediterranean countries. The etymology of the word papyrus – *Pa-per-aa* "That-of-the-Palace" – indicates that under the native dynasties its manufacturing process was under the control of the Pharaonic state, which saw to its distribution according to the needs of the administration and clergy. The *Zeno papyrus* (third century BCE) indicates that the papyrus fields were systematically set on fire before and after flooding to stimulate the regeneration of young shoots (Lewis 1974; Brink and Achgan-Dako 2012: 129–34). Greek papyri indicate that papyrus was processed in farmhouses in producing areas. Pliny (*Nat. Hist.* XIII 69–89), after Theophrastus, gives the best description of the manufacturing process of papyrus. Their rough edges having been removed, the fibers of the papyrus were used to make ropes, mats, nets, and seats.

The Egyptians understood how to take advantage of these properties of papyrus to manufacture a good writing surface. After the harvest, the process started by peeling the triangular-shaped stems, chopping them into pieces of equal length, cutting them into thin slices, placing the slices in two layers (one horizontal and the other vertical), pounding the crossed slices with a mallet to make the fibers adhere, and then smoothing the surface. The papyrus manufactories needed to be close to the production areas because the material degraded quickly (it dried within forty-eight hours), losing the adhesive properties of its sap. For this reason, the size of papyrus formats was limited. The various formats obtained, as mentioned by Pliny, made it possible to recognize various qualities of papyrus, from the highest (the *hieratic* paper, intended for the administration and religious texts) to the lowest (the *emporitic*, wrapping paper; Leach and Tait 2000: 236–8).

The documentation available – texts written on papyrus or architectural and archaeological remains – makes it possible to affirm that the major temples and the funerary temples were places of storage and of production of different products to meet the daily requirements of the people and of the clergy for liturgical celebrations. In the New Kingdom, the vaulted mud stores found in the Temples of Millions of Years – royal funerary temples on the west bank of Thebes – testified to their storage capacity. These enclosures also contained butcheries, bakeries, and various other workshops.

The supply of these products was ensured by the Pharaonic state, as shown by donations listed in the thirtieth year of the reign of King Ramses III (1186–1154 BCE; see Chapter 6, p. 164). These lists appear in the *Great Harris Papyrus* (British Museum, inv. no. EA9999, 43). For contemporaries, this document evoked royal gifts made to major temples (Thebes, Heliopolis, and Memphis) and to several minor temples in Upper Egypt. The different uses of precious and basic metals, specifying their quantities in units of weight (*deben* or *qite*), of precious minerals with information on their origin, of different species of wood (Baum 1988), and of many types of fabrics were given. There were lists of utilities, cereals, meat and poultry, varieties of bread, beverages, oil, honey, fruits, and vegetables. Aromatics and chemicals (pitch, bitumen, natron, salt) and a multitude of small objects (beetles, seals) and tools, imported from different geographical areas – Egypt, Punt, the so-called God's Land, Kush, and the Oasis – were also listed.

All these products were stored in places originally called "White-House" (𓉐𓏤 *per-hedj*; i.e. treasury). From the Middle Kingdom onwards, on the basis of the primacy of silver over gold in ancient times, they were called "Double Houses-of-Silver-and-Gold" (𓉐𓏤𓉐𓏤𓋞). Some of these treasuries were made of solid stone constructions – material that connoted their religious role – such as that of Thutmosis I (1504–1492 BCE), discovered in Karnak-North (Jacquet 1994), which met both utilitarian and religious needs. Around it there were workshops managed by a staff supervised by the clergy. The excavations of this building show that baking was one of its activities, meaning that the word "treasury" was used in the broad sense of "store." Conical bivalve ovens (100 cm in diameter) opening at the top, found in excavations in the vicinity of this structure, were used to bake flat, round loaves placed on the heated walls. Rectangular-shaped ovens to preheat ceramic bread pans were also found. The furnaces were fed with vegetable waste. The temperature required in the ovens was kept even by placing ceramic shards on the opening. The silos were not far from the ovens (Jacquet 1994: 141–4).

There were few workshops outside the temples. A raw brick structure adjoining the temple of Dendara, previously identified as a sanatorium due to the presence of tuns made of waterproof cement, has recently been reinterpreted as being a

FIGURE 3.1 Procession of metals and mineral bearers. Treasure D'. Dendara temple. © Sydney H. Aufrère.

FIGURE 3.2 *Tinctorium* (dyeing workshop), South-East Dendara temple. © Sydney H. Aufrère.

dyeing workshop (*tinctorium*; Cauville 2005; Cauville and Ali 2015: 264–5). It was a production unit where dyes were made. The linen and wool came from outside, probably from the agricultural fields around the temple, and were dyed to make sacred fabrics for the goddess – fabrics that had to be protected from natural light. It was a place where the dyers worked using techniques described in extracts of texts found in the so-called New Year House of this temple.

Information on dyeing is, however, rather scarce, especially in the case of woad, where the leaves undergo a transformation process of grinding, drying, and fermentation until the moment when the dye material, called agranate (in French *agranat*; the agranate is made of blackish aggregates), is obtained (Vogelsang-Eastwood 2000: 278). The dye is immersed in water and produces a greenish–yellow alkaline medium. The linen is immersed in this bath and, after being removed from it, it oxidizes in the air and turns blue. A similar process is used for wool. This process is much longer, using urine for biting, and it also requires time and heat (Hurry 1930; Ailliaud 1990).

The Egyptian corpus gives details on this dyeing process in a text: "The blue color of lapis lazuli fabric of the goddess Lapis-lazuli is obtained with the help of woad (𓂧𓈖𓎡𓏛 *der-neken*; *Isatis tinctoria*), diluted in the agitated water of the river until the process, which the ancestors mention yields the (same) color (as that of) flax flowers" (*Dendara* IV: 109 ult.-110, 2; Goyon 1980: 33).

Indeed, the blue of the flax flower is similar to that of woad. The dyeing process was considered a trade secret and details of the entire dyeing process were not available. Other texts mentioned the chemical reactions taking place during the fermentation of the froth resulting from the bath of the red cloth (*ines*), which, paradoxically speaking, allowed the dyers to obtain the green color (*wadj*) under certain specific conditions (*Dendara* IV 109,10–11, *Edfou* I 388,5–6; Goyon 1980: 26). The root of the alkanet (𓏏𓈖𓋴𓏏𓏥 *nesty*; *Alkanna tinctoria*) – a Mediterranean plant (attested near Alexandria) – was used in antiquity to dye fat. Egypt used it to give the sacred *medjet* ointment (𓅓𓂧𓏏) a red color for a symbolic reason (Loret 1930: 23–8).

Madder dye (𓇋𓊪𓄿𓏥 *ipa*; *Rubia tinctorum*), extracted from the roots and rhizomes of the plant (Vogelsang-Eastwood 2000: 279), was used by tanners (Loret 1930: 28–32) and also to dye wool. Its dyeing properties were described in the *Satire of Trades* (IV 5–7) because this dye had the color of blood. But the text also connected it to another product (𓃀𓎛𓏌𓏥 *behu*), giving it a pungent smell. *Behu* was probably used as a mordant. The same kind of smell is produced when vinegar is poured onto iron (see Chapter 2, p. 61) to obtain madder in modern recipes. That said, the main mordant used by the Egyptians was alum (𓇋𓃀𓈖𓏥 *ibenu*; see Chapter 6, p. 166).

The false safflower (𓎡𓄿𓍿𓏥 *katj*; *Carthamus tinctorum* L., 1753), harvested in Egypt (Loret 1892: 66, 141; Vogelsang-Eastwood 2000: 279) from the Twelfth Dynasty onwards (1991–1785 BCE), makes it possible to obtain, in successive juices, yellow to orange-red by virtue of oxidation. Yellow would be associated with the gold of the goddess Hathor and red with the morning color of the goddess Isis-Sirius. Sometimes the essence of mandrake (a yellow fruit) was used to symbolically add solar light. Archaeological data show that other dyes, such as henna (*Lawsonia inermis*), could be used (Vogelsang-Eastwood 2000: 279). The use of Polish cochineal dye (*Porphyrophora polonica*) is attested to from first century CE onwards, but snail purple (murex) was not known (Vogler 2013). It is possible that the dyeing workshop of Dendara was made of two parts, one dedicated to a *tinctorium* per se, and the other to a *laboratory*, because the two were, so it seems, closely associated.

According to Egyptian texts, it is clear that several Greco-Roman temples of the Nile valley had rooms considered as so-called laboratories (*is*) for the conservation of recipes, techniques, and scenes of offerings. For the priests, these "laboratories" evoked the world of perfumes, another the *tinctorium*, where fabrics were dyed, and yet another the so-called "treasury" where precious metals and minerals were kept safe. The Mansion-of-Gold or goldsmith workshop evoked all that concerned the designing of jewels specially made to adorn the gods according to rituals.

The importance of these places varied with the location of the temple. The most important laboratory, in terms of amount of textual information available, was that of Edfu, on the walls of which very detailed recipes used to satisfy the gods during liturgies are still found. These texts, including lists of aromatics (at the temples of Edfu, Dendara, and Atripe), enabled the safekeeping of the traditional names of all products used by the Egyptians, the majority of which were of foreign origin. The two treasuries of Dendara provided a complete record of the geographical origin of all imported metals and precious minerals from abroad. But as far as the Mansion-of-Gold or goldsmith workshop is concerned, the one in Dendara is quite unique, inasmuch as it describes all the traditions related to silversmithing practiced in Memphis, a place known for its craftsmanship.

MESOPOTAMIA

Cale Johnson

Urbanization and the large-scale storage economies that developed in Mesopotamia in the fourth millennium BCE laid the necessary social groundwork and economic basis both for specialized craftsmen, who no longer needed to provide their own food, and for urban elites that valued their technological innovations. Rather than individual genius, it was the ongoing support of the state and its rulers for small groups of specialists, engaged with the development of new administrative technologies, such as cuneiform writing, or alternatively the creation of new ways of manufacturing and decorating votive and other high-value goods, that led to significant technical advances throughout Mesopotamian history. The most important processes of standardization and mass production took place in the fields of pottery production and cuneiform writing in the fourth millennium BCE. The pottery of the Ubaid period (ca. 6500–3800 BCE), first found in Eridu, but subsequently aligned with the pre-Uruk-period levels in the city of Uruk, for example, was famous for its elaborate decoration, likely produced on a tournette or slow wheel, while the very different Uruk-period (ca. 3800–3000 BCE) pottery was undecorated and produced on a simple fast wheel (the kick wheel only appears much later; see Petrie 2012: 285). Nissen (1989) argues that the increasingly standardized vessels, mass production, and the abandonment of decoration were direct results of this new fast wheel and that these developments followed from the increased specialization and professionalization of Uruk-period ceramic production centers. The ubiquitous beveled-rim bowl was of a far lower quality than other wares, but was mass produced on a previously unheard of scale (Potts 1997: 150–3).

The earliest cuneiform writing, part of a longer developmental sequence involving plain tokens inside of clay bullae and purely numerical tablets, used the new standardized ceramic vessels as prototypes for a number of the earliest cuneiform signs. Iconic images of these vessels were used to represent the vessels themselves as well as a number of different types of both elite and nonelite foodstuffs. The beveled-rim bowl served as the prototype for the cuneiform sign GAR, which could stand for "bread" (corresponding to later Sum. ninda) or, in combination with the sign for "mouth," yield a sign meaning "to consume, to use up" (Sum. gu$_7$); likewise, higher-quality vessels such as UKKIN and SILA$_3$ were used to identify the elite goods that they normally contained, such as beer and dairy fats, as well as the foodstuffs that were typically associated with them, such as meat and fish (see Johnson 2015 for an overview). In the context of these storage-based urban societies, the relationship between rations, the more-or-less standardized vessels that were associated with the rations, and the bookkeeping mechanisms that were used to track these vessels served as a paradigm for the development of many other domains of economic and technical practice in ancient Mesopotamia, such as the administrative techniques for documenting the use of raw materials in technical workshops.

The most important lexical lists from the Late Uruk period (ca. 3300–3000 BCE) were largely concerned with managing the specialists who operated these workshops, as well as the raw materials they required and the finished goods they produced (see Nissen et al. 1993 for a user-friendly introduction). In contrast to Egypt, we have few images of craft or industrial activity from Mesopotamia, but what we lack in visual representations is more than made up for by the most elaborate bookkeeping procedures for craft and industrial production of any society in antiquity. In the Ur III period (ca. 2112–2004 BCE), where the documentation is extensive and extraordinarily detailed, both raw materials and labor were consigned to a supervisor, who was responsible for meeting specific production goals, and the actual labor performed, quantified in terms of fixed ratios of labor to finished products, was then deducted from the overall consignment of worker time (Englund 1991). The minutiae of these calculations, including varying rates of performance for specific materials and end products and allocations of workers' days off for different age and gender categories, represent the epitome of workshop management and, in fact, the activity of many workshops can be reconstructed from the Ur III textual record alone (see generally Paoletti 2016b). Three studies of purely textual remains have played a leading role in understanding technical practice in ancient Mesopotamian society: (a) Neumann's synthesis (1993) of the Ur III records dealing with raw material inputs and the resulting craft products in five Mesopotamian cities; (b) Heimpel's work (1998; 2009) on "industrial parks" in Girsu and Garshana; and (c) Van De Mieroop's study (1987) of the Isin craft archive from the subsequent Isin-Larsa period (ca. 2004–1763 BCE). Of these, Heimpel's focus on an industrial park in Girsu, where rest houses, animal-fattening, and ship-building centers were collocated with prisons, providing a labor source (Heimpel 1998), and a decade later a project involving "the construction of a ring wall, a triple complex of the food processing facilities brewery, kitchen and flour mill, a double complex of *textile mill and craftsmen's house* (Sum. e_2 uš.bar u_3 e_2 gašam.e.ne), and the rebuilding of residences" (Heimpel 2009: 123, my emphasis) provide us with a comprehensive picture of the physical and administrative situation in which palace-funded workshops existed.

Heimpel argues for a parallel set of technical professions at Garshana and at Ur, the most famous craft center in the Ur III period, "located in the city of the principal royal residence and featur[ing] a port where luxury goods from the Persian Gulf and Arabian Sea arrived" (2009: 157). As Moorey (1994: 15), among others, has emphasized, the key text for making sense of centers of craft production such as these is UET 3, 1498, an inventory of both raw materials and finished products kept in either the "storehouse" or "treasury" (Sum. e_2.kišib$_3$.ba) or the "big warehouse" (Sum. ga_2.nun mah). Centralized "workshops" (Sum. e_2.giš.kin.ti) like these typically dealt with the production and decoration of luxury items for temples and the palace, as well as the extensive bookkeeping of raw materials and finished goods, while the mundane production of agricultural implements and the like (and presumably the preparation of processed raw materials such as metals) was carried out elsewhere (Neumann's *Schmeidewerkstätten*). Raw materials, especially metals, were carefully weighed before being distributed to specific craftsmen, and the finished product was also weighed, allowing for the calculation of "its (loss) consumed by fire" (Sum. izi gu_7.bi) and "its waste" (Sum. za_3.bar.bi). These practices, in combination with multiple copies of inspections (Sum. $gurum_2$ ak) and regular tabulations of the workforce, allowed a full overview of the raw materials and the labor involved. At Ur, eight distinct workshops are mentioned, each beginning with "house of" (Sum. e_2): the "sculptor" (Sum. tibira), the "goldsmith" (Sum. ku_3.dim_2), the "stonecutter" (Sum. zadim), the "carpenter" (Sum. nagar), the "metalworker" (Sum. simug), the "leatherworker" (Sum. ašgab), the "felt-maker" (Sum. tug_2.du_8), and the "reed-worker" (Sum. ad.kub_4). These different workshops were supervised by a "chief administrator" (Sum. šabra) and a small team of scribes. At Garshana, a shorter list of workshops is found (Heimpel 2009: 161), including that of the "carpenter" (Sum. nagar), the "metalworker" (Sum. simug), the "leatherworker" (Sum. ašgab), the "felt-maker" (Sum. tug_2.du_8), and the "reed-worker" (Sum. ad.kub_4).

Unlike the metalworkers, who also required pyrotechnic installations of course, potters are not included in these lists of different craftsmen, and the "potter's house" (Sum. e_2 $bahar_2$.ra) in both Ur and Garshana seems to have been located at a different site. As Heimpel (2009: 162) reiterates, this lines up nicely with archaeological evidence from sites like Larsa and Maškan-šapir, where the pottery kilns were located outside of the city center. In spite of its seemingly anomalous position within Ur III craft production, since it is not included in these production centers, like most other dependent laborers (Sum. guruš), potters typically worked year-round for the state in specialized workshops, and their rate of production was measured in terms of fixed equivalences of workdays: the well-known Umma pottery workshop, for example, produced in a single year more than 60,000 one-liter vessels, functionally equivalent to the Late Uruk beveled-rim bowl, each valued at 0.066 workdays. This suggests that a single worker was expected to produce fifteen of these vessels in a single day (Dahl 2010, *pace* Steinkeller 1996).

In spite of extensive textual records and numerous excavations, relatively few technical workshops can be securely identified in the Mesopotamian archaeological record. One of the biggest problems is definitional: many workshops are hypothesized on the basis of finished objects that share features or the isolated presence of an oven or kiln. It is best, therefore, to briefly reiterate Tosi's diagnostic features for identifying workshops: (a) fixed installations for processing raw materials (e.g. kilns and furnaces); (b) specific working tools; (c) residues or wasters; (d) raw material in a convenient form; (e) concentrations of finished commodities; and (f) materials for recycling (1984). The application of these criteria, especially the co-occurrence of pyrotechnic installations and debris from the manufacturing process, has resulted in many "workshops" described in the secondary literature being removed from the classification. The most famous example is the workshop associated with the Larsa Goldsmith's Hoard, discredited by Bjorkman in 1993, as well as the reevaluations of numerous sites in Moorey's extensive work (Moorey 1985: 36–7).

Recognized key sites are surveyed and briefly described in the standard handbooks (Moorey 1994; Potts 1997) and more recently in Morandi Boncossi's (2016) entry on "Werkstatt – Archäologisch" in the *Reallexikon der Assyriologie*. Important pottery workshops have been identified in Tell Abada and Yarim Tepe (Ubaid period), Abu Salabikh (Uruk and Early Dynastic periods), Umm al-Hafriyat, Qatna, and Tell Sabi Abyad (in the second millennium BCE). Duistermaat's *The Pots and Potters of Assyria* (2008) offers a particularly well-considered study of the potter's workshop in Middle Assyrian Tell Sabi Abyad and represents an important point of departure for future work, answering Moorey's lament (1994: 146) that "no coherently published potter's workplace in Mesopotamia" had been published at the time he was writing in the mid-1990s. Metalworking workshops have been identified in Arslantepe and Degirmentepe in prehistoric eastern Anatolia and at numerous sites in the Iran plateau (see Weeks 2012: 301–3 for an overview), Tell edh-Dhiba'i in the early second millennium BCE (Al-Gailani 1965; Davey 1983; see below), and more recently in Late Bronze Age Qatna (Iamori 2015). Glassmaking workshops, which obviously appeared later, are reviewed in detail in these same publications, with particular focus on the glassmaking workshops at the "Mitannian Palace" at Tell Brak and the workshop that Mallowan identified "on the south side of room 47 [of the Burnt Palace]," where he found "traces of kilns and glassmaker's kit, including one specimen of sealing wax red glass, probably from a crucible" (Mallowan 1966: 209–10). Even if the literary quality of Mallowan's report cannot be surpassed, the most important identified glassmaking workshop is almost certainly the one identified by Oates and coworkers at Tell Brak. Henderson's (2012) recent monographic treatment of the origins of glass provides an excellent description of the finds of glass themselves (and the scientific analyses carried out on them), but only rarely hints at the presence of workshops. One of the few instances is at Tell Brak, where he refers to "possible direct evidence of glassworking" (Oates et al. 1997: 86). But he states that the evidence "does not, however, in itself constitute evidence of primary glass manufacture from raw materials at Tell Brak" (Henderson 2012: 140). Even if we cannot locate bona fide production sites in Late Bronze Age Syria, it is fairly clear that "the Hurrian Kingdom of Mitanni was responsible for the great leap forward in glass production" through their development of "large furnaces that could reach temperatures of c. 1150–1200° C," which in turn allowed for the development of core-forming in the sixteenth century BCE (Henderson 2012: 144).

The primary terminologies for kilns and furnaces (as well as other types of pyrotechnic installations, both domestic and specialized) in the languages of Mesopotamia, particularly in the lexical list tradition, were surveyed by Armas Salonen in 1964; the key terms for our purposes here are Akk. *kūru* and Akk. *utūnu*. These are, not incidentally, the same two terms that appear in the mid-second-millennium BCE glassmaking texts: the *utūnu* "kiln" figures in the less sophisticated technique involving week-long baking of glass in molds, while the *kūru* "furnace" appears in the more advanced recipes for making artificial precious stones. (This latter term, viz. *kūru*, also serves as an exceedingly rare qualification for these artificially created stones; e.g. *ûqnu kūri* "lapis of, viz. from, the furnace.") The corresponding Sumerian terms, strictly speaking, are Sum. udun = Akk. *utūnu* and Sum. dinig = Akk. *kūru*, but this leaves the more common term for kiln or furnace in older Sumerian texts, namely Sum. gir$_4$ (or gir$_4$.mah), out of the picture. The lexical tradition equates Sum. gir$_4$ with Akk. *kīru* (not to be confused with *kūru*), and clearly both *utūnu* and *kīru* are Sumerian loanwords in Akkadian, but it is tempting, nonetheless, to suggest that Akk. *kūru* derives from Sum. gir$_4$ as well (suggested by Salonen 1964: 118), even though it is not directly supported by the lexical tradition or the standard dictionaries. In texts from the third millennium BCE, Sum. gir$_4$ simply means "oven" and is used to bake bread and cook meat, but particularly in second- and first-millennium BCE sources, after the introduction of Akk. *tinūru* (equivalent to Arabic *tannur*) and similar terminology, Sum. gir$_4$ generally refers to a "furnace" used in the production of metal, glass, and similar materials.

The overall correctness of Salonen's identification of Sum. udun with "kiln" and Sum. gir$_4$ with "furnace" is largely confirmed by the qualifications and components associated with these two terms in the lexical tradition, although Salonen consistently translates Sum. udun with "Ofen" and gir$_4$ with "Brenn-/Schmelzöfen" (both contrasted with "Backöfen" for Akk. *tinūru*). Kilns are often qualified by professional designations such as the "kiln" (Sum. udun) of the "potter" (Sum. bahar) or of the "brewer" (Sum. lu$_2$ ^{kaš}kurun$_2$.na), or the roasting of the particular type of material such as "sourdough" (for beer-making; Sum. bappir) or "dried fermented mash" (Sum. titab). In contrast, Sum. gir$_4$, particularly in second- and first-millennium BCE sources, serves as the point of reference for a number of terms that describe parts of the furnace such as the "chamber" (Sum. daggan = Akk. *takkannu*), the "peephole" (Akk. *ḫayyāṭu*), and/or the "vent" (Sum. igi = Akk. *īnu*) and the bellows (Sum. bun$_{1/2}$ = Akk. *nappāḫu*), all of which occur almost exclusively with Sum. gir$_4$, rather than Sum. udun, in the lexical lists. Only some of these terms from the lexical tradition reappear in the late glassmaking texts, and, in particular, a more complex Akkadian terminology for crucibles and related elements for holding or enclosing the crucible such as the saggar is found in the late glassmaking texts, including terms like Akk. *imgurru, dabtu, ḫaragu,* and *maṣādu* (Oppenheim 1970: 69–74).

The linguistic contrast between "kiln" (Sum. udun) and "furnace" (Sum. gir$_4$) does not line up in any simple way with the archaeological and technological record; each of these Sumerian logograms enters the writing system at different points in history, and there is no one-to-one relationship between these Sumerian logograms and the proto-cuneiform signs that depict ovens or kilns. The southern Mesopotamian alluvium may not have been at the forefront of copper smelting technology in the Uruk period (ca. 3800–3300 BCE), and so we also have to factor in the possibility that iconic depictions of metallurgy in proto-cuneiform writing may not correspond to contemporary metallurgical practice in fourth-millennium BCE Iran. Furnaces begin to be used at the end of the fourth millennium BCE in Proto-Elamite-period Iran, but the crucible may still have played a central role in metallurgy in Late Uruk-period Mesopotamia (see the complementary overviews of metallurgical processes and archaeological contexts in Weeks 2012; Weeks 2013), so we will be largely concerned here with pottery kilns. The iconic forms of two proto-cuneiform signs (ca. 3300 BCE), namely MAH$_a$ and AD$_a$, were probably modeled on two distinct types of kiln.

Although any attempt to correlate proto-cuneiform signs with fourth-millennium BCE kiln designs is necessarily fraught, the vertical and curved profile of the MAH$_a$ sign can be equated with the type of two-chambered domed design attributed to the "protoliterate pottery kiln from Chogha Mish" on the basis of a dozen or so parallels from both Mesopotamia and Iran (Figures 3.2 and 3.3; Alizadeh 1985; see the recent survey in Streily 2000 as well). In contrast, the horizontal configuration of the AD$_a$ sign corresponds well to the type of horizontal kiln that became increasingly common in the Early Dynastic period. As Moorey explains:

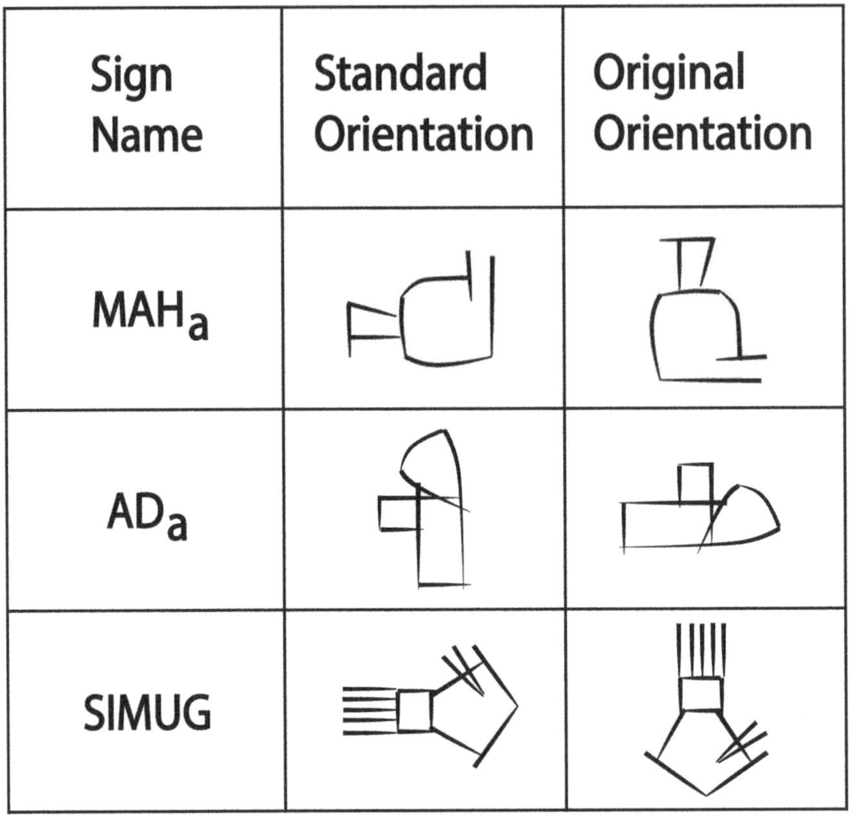

Sign Name	Standard Orientation	Original Orientation
MAH$_a$		
AD$_a$		
SIMUG		

FIGURE 3.3 The proto-cuneiform signs representing ovens and kilns: MAH$_a$, AD$_a$, and SIMUG. Drawing by the author, after signs drawn by R.K. Englund.

In a horizontal kiln the firing chamber and the fuel pit are at opposite ends of the rectangular space. The heat produced near the entrance is drawn by one or more chimneys on top of the firing chamber; in such kilns there is no refractory grid, heavy supports or vaulted roof as found in vertical kilns.

(Moorey 1994: 157)

When both signs are rotated 90° to the right, into their original orientation, the opening to both kilns is on the right and both also have a flue on top, but MAH$_a$ has a tall, rounded chamber, while AD$_a$ has a boxlike, horizontal chamber. Thus, in all likelihood, the proto-cuneiform signs MAH$_a$ and AD$_a$ represent two different types of kiln – vertical and horizontal, respectively. In the subsequent Early Dynastic period (ca. 2900–2400 BCE), contemporary with the increasing use of horizontal kilns, the AD sign is combined with ŠU$_2$ or U to form GIR$_4$, which is first attested in the Archaic Ur corpus (ca. 2900 BCE),

FIGURE 3.4 Photograph of a two-chambered oven from Abu Salabikh. Courtesy of J.N. Postgate.

although centuries later, in the ED III period, AD often still functions alone as the logogram for "oven," presumably gir_x(AD).

The first attestations of Sum. udun only occur at the end of the third millennium BCE, and the new orthography of Sum. udun (namely U.MUHALDIM) is presumably modeled on the orthography of Sum. gir_4 (U.AD), with MUHALDIM (meaning "cook" in Sumerian) replacing the sign AD within the new logogram, presumably in order to differentiate an oven for cooking food (Sum. udun) from a kiln for baking pottery or other nonculinary practices (Sum. gir_4).

Although Moorey identifies a few early examples of crucibles, he goes on to say that "furnace development is a subject for which there is very little hard evidence from the ancient Near East" (Moorey 1994: 243). Craddock provides a clear developmental sequence for furnaces and crucibles on the basis of archaeological work in the Levant, which has served as the primary context for defining metallurgical developments until recently. Central to Craddock's account is the use of crucibles, directly heated by piled-on charcoal, and blowpipes (without a distinct "furnace" in the earliest phases). As Craddock puts it:

> ... in common with other Bronze Age metalworking centres, the absence of recognizable furnace fragments or of the clay bellow pipes, the tuyeres, with the crucibles, does imply that there were no specific furnace structures or bellows at this stage in pyrotechnic development.
>
> (Craddock 2000: 157)

Likewise in the fifth- and fourth-millennium BCE smelting sites in Iran (see Thornton 2009 and Weeks 2013 for illuminating surveys of this material as well as a critique of the "Levantine paradigm"), the earliest evidence for copper smelting involved crucibles rather than furnaces. The proto-cuneiform sign SIMUG, which comes to mean "metalworker" later on in Sumerian, "appears to show the plan of a smelting furnace with attached blow pipes or tuyère" according to Moorey (1994: 243), but in light of the fact that the sign SIMUG much more closely resembles a crucible than a furnace (alongside the likelihood that southern Mesopotamian metallurgy was probably focused on crucible rather than furnace technologies), we should now amend Moorey's statement: SIMUG is probably a representation of a crucible with its blowpipes, although the lid or flue in the middle of the sign (with fire coming out of the top) remains problematic.

Other than the frequently discussed shift from a "slow wheel" (or "tournette") to a "fast wheel" in the Uruk period and its role in the emergence of mass-produced pottery (Nissen 1988: 46–7; Potts 1997: 161), one of the most important archaeological finds in the ancient Near East is the set of copper-working tools that were recovered from the Isin-Larsa-period workshop at Tell edh-Dhiba'i, just outside Baghdad (Al-Gailani 1965; Davey 1983). Davey (1983) describes and offers illustrations of (a) baked clay pot bellows, (b) crucibles, (c) a mold for casting a pin, (d) a baked clay model ax-head, (e) a baked clay ladle, (f) a fragment of tuyère, and (g) small round dishes (Moorey 1994: 268). The most important aspect of this set of tools, as Davey demonstrates, is that the form of the crucibles from Isin-Larsa-period Tell edh-Dhiba'i precisely matches the form (and likely the production process as well) in the depiction of smelting from the tomb of Mereruka in Saqqara, Egypt. Davey describes the process as follows:

> Adopting the practice illustrated in the Egyptian Old Kingdom tomb reliefs, a plug is placed at the entrance of the crucible and it is kept in place until the material is ready to pour. When that occurs, the charcoal is quickly pushed aside and the plug removed so that the metal can quickly flow into the preheated mould.
>
> (Davey 1983: 182)

The key difference between the depiction from the Egyptian tomb and the set of equipment found at Tell edh-Dhiba'i is the presence of pot bellows and a tuyère in the latter. These items suggest that, even if direct work on the crucible was still practiced, it was carried out in a much more carefully appointed pyrotechnic installation, with pot bellows replacing a team of workers equipped with blowpipes.

We have focused here on workshops, installations, and tools that can be included in specific technical processes with certainty, either because they are specified as such in textual sources or because the preponderance of Tosi's

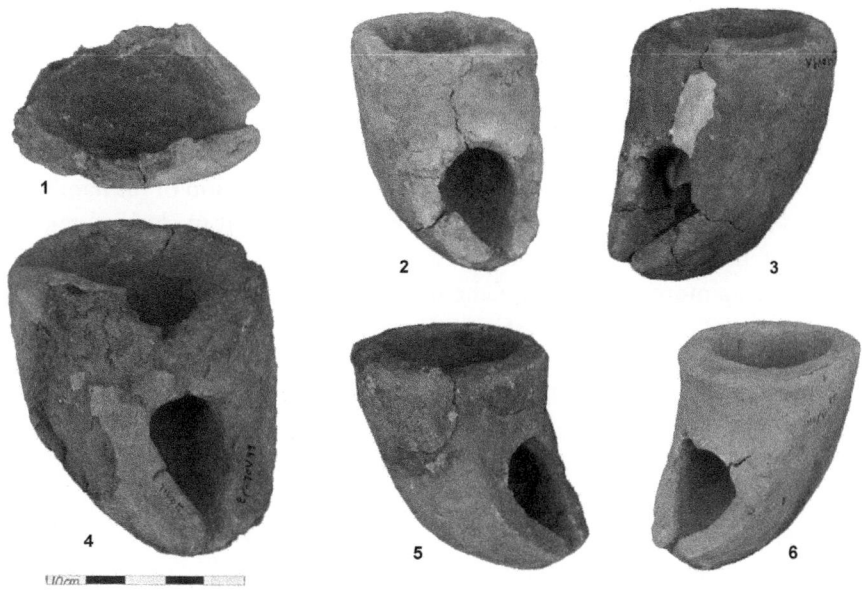

FIGURE 3.5 Tools from Tell edh-Dhiba'i. Photograph © C.J. Davey.

criteria makes it clear that they were involved in specific technical processes. Many other tools could easily have had their primary use in food preparation, for example, and only occasionally found their way into some kind of chemical practice: the various bowls, funnels, steamers, sieves, strainers, and the like, carefully surveyed and described by Ellison (1984), were always primarily used for food preparation, but could of course be repurposed for other uses (see Faivre 2009 and Michel 2012 for recent surveys of vessel use). This idea was the basis for Levey's erroneous suggestion (1960) that distillation was already being carried out at Tepe Gawra in the middle of the fourth millennium BCE. In the absence of clear contextual evidence of some kind, arguments based exclusively on the "possible" use of relatively simple decontextualized vessels must be ruled out. The counterexamples, even if their operation and significance still eludes us, are the so-called "Parthian galvanic cells" consisting of a "clay jar, a cylinder made from copper sheet and an iron rod" (Eggert 1995), which König (1938) suggested were used for electroplating (Keyser 1993 surveys the different theories, ranging from electroplating – actually invented in Birmingham in 1839 – to medicinal uses), but whatever their function, the internal complexity of these mechanisms precludes the possibility that they are repurposed kitchenware. And as Eggert reassures us, "as is always the case in experimental archaeology, successful experiments can only show a supposed ancient technique to be possible, but never by themselves that it was, in fact, applied" (1995: 14).

GRECO-ROMAN WORLD

Matteo Martelli

There is no ancient term for either a chemical or an alchemical laboratory. The word *chymeion*, which one can read in early modern treatises on alchemy, was probably introduced by Andreas Libavius in the second edition of his handbook *Alchymia* (1601). The term *laboratorium* too is late: absent in classical texts, it started to be used in medieval sources, and only by the mid-sixteenth century had it assumed a more specifically alchemical nuance. On the other hand, we do have the ancient Greek term *ergastērion* (ἐργαστήριον), which could refer either to a workshop or to a shop of different craftsmen, from butchers to perfumers, from bakers to smiths (Martelli 2011).

Despite the lack of a specifically alchemical connotation, the general term *ergastērion* could also apply to working spaces where ancient chemical arts were practiced. Indeed, it could refer – along with more specific names such as *bapheion* "dyer's workshop" (or *porphyreion*, if specialized in purple dyeing), *hyelourgeion* "glassmakers' workshop," or *chrysochoeion* "goldsmith's workshop" – to the workshops of various craftsmen active in those technical areas that attracted the attention of the earliest alchemical authors. These craftsmen carried out their activities by using customized sets of tools and devices, which necessarily varied in accordance with the specific needs of their areas of expertise.

Dyers' workshops were usually equipped with dyeing vats, furnaces, water supplies (e.g. basins or fountains), and various types of vessels, which were necessary to prepare the dyestuff and to use it to treat fabrics and cloths. These workshops could be either independent buildings or part of the house, as it is possible to infer from some contracts for leasing or selling *ergastēria* preserved in Geek papyri. For instance, a sale contract of dyers' workshops (*baphika ergastēria*) equipped with a leaden pot and an earthenware cask is preserved by *P.Oxy.* XIV 1648 (second century CE). Archaeological evidence confirms the presence of dyeing workshops in Late Roman Egypt. In this period, various rooms of the Egyptian temple of Repit at Athribis were reused as working spaces (Müller 2015: 188). Here, a dyers' workshop was excavated by Flinders Petrie, who discovered a cistern and many vats in a raised bench: "these vats," Petrie writes (1908: 11), "are lined with cement and deeply stained. Most of them are black blue with indigo, and some are red." Six dyeing workshops have been excavated in Pompeii as well. Usually parts of larger houses, they are recognizable by high furnaces over which large lead cauldrons have been installed (Flohr 2013b: 60–2; Lowe 2016). In four workshops, sets of cauldrons of different sizes were discovered. The largest cauldron was probably used for the pretreatment of wool with alum and other mordanting substances, which allowed dyes (such as madder) to be fixed to the fibers. Indeed, lists of mordanting substances capable of fixing colors are included in the chemical papyrus of Leiden (e.g. § 92 in Halleux 1981: 106), and they are very similar to the lists of astringent drugs provided by medical texts, such as Galen's *On the Capacities of Simple Drugs* (I 34 = XI 440–1 Kühn) or *On the Method of Healing* (III 5 = X 199 Kühn). The smaller cauldrons were then used for dyeing. Cauldrons consisted of lead kettles slotted into a surrounding structure made of bricks and lime mortar. Experimental archaeology has shown that lead represented an unusual choice due to its physical properties. It is, in fact, heavy (and kettles were usually filled with ca. 90 litres of dyeing liquid), malleable, and with a low melting point. When heated, it could change its shape, thus making the apparatus somewhat fragile: evidence of "lead creep" has been detected in the kettles unearthed in Pompeii. However, unlike other metals that affected the different stages of the process, lead appeared from modern replications to act as an inert material that had no effect on the results of the mordanting and dyeing procedures. Moreover, it certainly helped to keep the temperature constant during the processes (Hopkins 2008; Puybaret et al. 2008; Kania et al. 2018).

Archaeological sources provide us with little precise information on early perfume technology, especially for the centuries that precede the end of the Hellenistic period. Recent excavations in Cyprus have unearthed a large installation from the mid-second millennium BCE for the production of perfumes in the area of Pyrgos. Here new evidence has been examined that might point to a very early use of distillation. A rich set of vessels has been discovered, which experimental archaeologists have reconstructed in the shape of various distillation devices for the production of essential oils and perfumes (Belgiorno 2017). In the framework of these experimental reconstructions, the controversial interpretation that Martin Levey (1955; 1960) proposed for the Tepe Gawra channel-rimmed pot found in Iraq (fourth millennium BCE) as an apparatus for distillation was reconsidered. Analogous vessels, dating to the mid-second millennium BCE, were found in Cyprus and Spišský Stvrtok (Slovakia), and replicas of these devices have been used in experiments for distilling scented waters (Belgiorno 2018).

These findings would antedate by ca. 1,500 years the discovery of alembics, whose earliest descriptions in Greco-Roman textual sources date to the first centuries CE (see below). Indeed, Greco-Roman classical texts devoted to the production of perfumes (primarily Theophrastus' *On Odors* and Pliny's *Natural History*, XIII 1–26) never mention distillation among the technologies employed in the field. Aromatics were simply added to different kinds of oil (mainly olive oil) and boiled together, a procedure confirmed by archaeology. For instance, in the Hellenistic perfume shops excavated in Delos and Paestum, archaeologists discovered furnaces, large marble mortars, and stone press beds. Each press bed has neatly carved circular channels converging on an outlet groove: this facilitated the collection of the oil produced by a vertical edge press that originally rested on the stone bed (Brun 2000). Vertical edge presses are actually depicted in frescos of Pompeii (House of the Vettii; see Figure 3.6) and Herculaneum (Casa dei Cervi); they consisted of a wooden structure with different rows of wedges set one upon the other, which were used to compress the olive paste (or other fruits and seeds) that was produced after crushing olives in mortars (Mattingly 1990; Brun 2000). In his *Mechanics*, only extant in Arabic translation (Nix and Schmidt 1900: 102–3), Hero of Alexandria (mid-first century CE?) explains that these edge presses were particularly suitable for extracting fine oils for perfumes.

Furnaces were also critical tools in many fields of ancient craftsmanship. For instance, in reconstructing the history of ancient glassmaking, scholars have stressed how the introduction of a new kind of furnace (along with other tools, such as iron blowing pipes) played a vital role in the development of glassblowing (Stern 1999). Probably discovered along the Syro-Palestinian coast, this new technique of shaping molten glass by blowing into it through iron pipes was perfected in Italy. Here, during the Roman period, glassmakers introduced a novel glassblowing furnace equipped with a closed heat chamber into which the pipe entered horizontally. In this way, glass was not simply heated in the side turned to the fire, but hot air surrounded it, thus allowing the glass to expand evenly (Stern 1999: 446).

FIGURE 3.6 House of the Vettii, Pompeii. Frieze depicting cupids working in a perfumery. Photograph by De Agostini/Getty Images.

Like glass, various metallic ores can be melted. Metallic ores were often treated in workshops next to mines, in installments that hosted different kinds of furnaces. Litharge, for instance, was a by-product of the cupellation of silver–lead ores (or argentiferous galena; see Chapter 2, p. 78). At Laurium, archaeologists found *tubuli* (cones) of litharge that had been lifted with bars (Halleux 1975: 75; Healy 1999: 320–2).

Another interesting case study is the mineral medicine called *pompholyx*, which was produced by burning copper ores. Galen (*On the Capacities of Simple Drugs*, IX 3.25 = XII 234,3–12 Kühn) specifies that in Cyprus some furnaces were designed to transform cadmia into *pompholyx*. A more detailed description of these furnaces is provided in a long passage by Dioscorides, which is worth quoting in full (V 75):

> It (i.e. *pompholyx*) is made this way: in a building of two chambers a flue is built and at the upper chamber a hole is made of equivalent dimensions as the flue opening from the parts above. The wall of the chamber close to the flue is bore through with a small hole level to the melting pot to receive the bellows. The chamber has also a door of proper size built by the craftsman to get in and out.
>
> Attached to this building is another room wherein are the bellows and where the bellows blower works. So, coals are placed in the furnace and lit, then the attending craftsman sprinkles the calamine finely crushed from stations above the top of the furnace and the helper does the same and at the same time throws continuously coal until all the charge is consumed.
>
> For as it burns, the part that is thin and that is composed of light particles is borne to the upper story and settles on its walls and ceiling; then it solidifies and becomes at first like bubbles that rise from rushing waters; then, as more particles are added, it becomes like fleeces of wool (transl. by Beck 2011: 363–4).

Along with this complex structure described by Dioscorides, other furnaces were used to collect the fumes of mineral ores. Vitruvius (*On Architecture*, VII 8.2) explains that, in a workshop (*in officina*) next to a mine for cinnabar, chunks of the minerals were thrown into furnaces and dried, to get rid of their moisture: the fumes that rose from them (i.e. mercury) precipitated on the floor of the furnaces. A special device for extracting mercury from cinnabar is also described by Pliny the Elder (*NH* XXXIII 123) and Dioscorides (V 95.1), who used a different terminology referring to its parts. An iron spoon containing cinnabar was put in a clay vessel (called *patina* by Pliny and *lopas* by Dioscorides), which was covered by a convex lid (a *calix* according to Pliny) or a upside-down vessel that Dioscorides calls *ambix* (ἄμβιξ): this term was then rendered in Arabic as *al-'inbīq* (with *'inbīq* as simple transcription of the Greek *ambix*), from which our "alembic" derives. The device was constantly heated, and the moisture (*sudor/hymor* in Pliny's words) or soot (*aithalē* in Dioscorides' words) that condensed on the lid became mercury when scraped off.

Scholars usually agree in interpreting these passages as one of the earliest accounts of a sublimation technique that exploits the low boiling point of mercury (356°C): cinnabar reacts with the iron in the spoon, thus liberating mercury that evaporates and condenses on the colder surface of the upper vessel. As mentioned by Healy (1999: 343), the remains of condensers dating to the Greco-Roman period have been found in Ladik, an area containing many cinnabar mines in Anatolia. Various scholars argue that similar rudimental apparatus were developed and improved by Greco-Egyptian alchemists: they (a) separated the lower container and the upper pot (or condenser) by a pipe and (b) added to the upper pot a tube (or, in some cases, two or three tubes) with a digestion vessel (Taylor 1945: 186–7; Forbes 1970: 16–24).

According to the Egyptian alchemist Zosimos of Panopolis (*Authentic Memoires*, VII 2 in Mertens 1995: 23), Maria the Jewess described how to assemble various alchemical devices, such as alembics for the production of "sulfur waters," the *kērotakis*, and various kinds of furnaces. Moreover, the cooking apparatus called bain-marie (*bagnomaria* or *Marienbad*; i.e. a water bath) is usually associated with her name (Lippmann 1913: 185–200; Forbes 1970: 24); the expression is common in Latin medieval sources (*balneum Mariae*), but never used in Greek alchemical texts, although a similar device for cooking food (with different vessels slotted into one another) is already described in classical writings, such as Hippocrates's *On Diseases* (III 17.17 = VII 160 Littré = Potter 1980: 98). This tool was then called *diplōma* ("double vessel") in the works of later medical authors, such as Dioscorides' *De materia medica* (II 77), Galen's *On the Composition of Medicines According to Places* (XIII 23 and 36–7 Kühn), and the sixth-century CE medical encyclopedia by Aetius of Amida (books I 122,2 and 123,2; IV 196,73 Olivieri). The use of cooking tools in alchemical practices, indeed, is well documented in the works of Greco-Egyptian alchemists. For instance, Zosimos saw a particular device used to steam poultry in the kitchen of his wealthy pupil Theosebeia (*Authentic Memoires*, VIII 1 in Mertens 1995: 26–7). After discussing this equipment with the chef in charge of the kitchen, Zosimos decided to take the Jewish alchemical books from Theosebeia's library and look for the description of a similar device designed to treat arsenic ores with sulfur's vapors (Mertens 1995: clxii–clxiii; Dufault 2019: 119–22).

A treatise *On Furnaces* is also attributed to Maria the Jewess in alchemical sources (Festugière 1944: 365), but it has not been preserved in Byzantine manuscripts. Various passages from her works, however, are quoted by Zosimos, who records Maria's description of a three-arm still (or *tribikos* in Greek; *Authentic Memoires*, III 1 in Mertens 1995: 14–5):

> I shall describe to you the *tribikos*. For so is named the apparatus constructed from copper and described by Maria, the transmitter of the art. For she says as follows: "Make three tubes of ductile copper a little thicker than that of a

pastry-cook's copper frying pan: their length should be about a cubit and a half. Make three such tubes and also make a wide still-head (copper vessel, *chalkeion*) of a handbreadth width and an opening proportioned to the still-head. The three tubes should have their openings adapted like a nail to the neck of a light receiver … . Towards the bottom of the still-head are three holes adjusted to the tubes, and when these are fitted, they are soldered in place, the one above receiving the vapour in a different fashion. Then setting the still-head upon the earthen pan (*lopas*) containing the sulphur, and luting the joints with flour paste, place at the ends of the tubes glass flasks, large and strong so that they may not break with the heat of the water, heat that comes with the distillate." Here is the figure.

(transl. by Taylor 1945: 190, slightly modified)

This detailed description of a still – which interestingly includes a comparison with a cooking tool (a frying pan) – is followed by a second description of an alembic only equipped with one tube, a device that, in all likelihood, should be ascribed to Maria the Jewess as well. Both descriptions refer to drawings of the devices, which are likely to have complemented Zosimos' original text. The Byzantine manuscripts do include some images of these stills, which also include captions referring to their different parts. As one can infer from the images in Figure 3.7, the terminology used only partially matches Zosimos' descriptions of the devices.

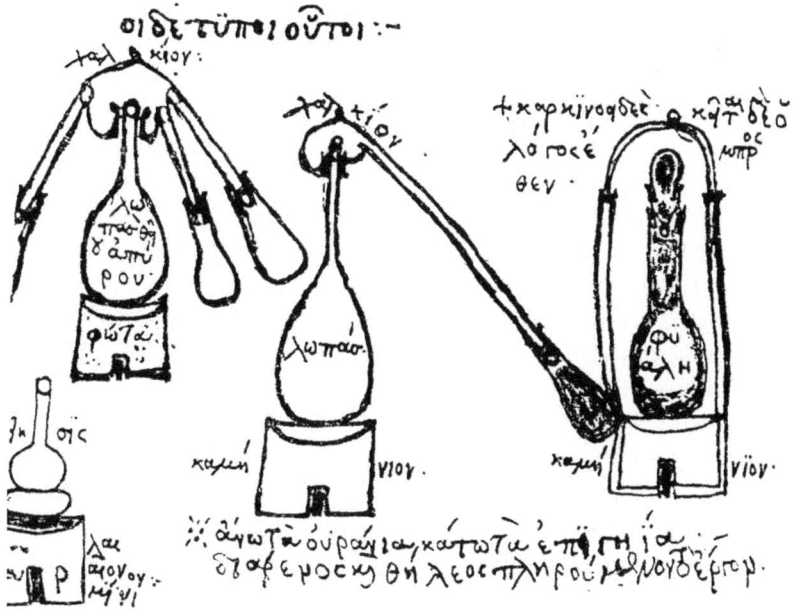

FIGURE 3.7 Distillation equipment in the Byzantine MS *Parisinus* gr. 2327 (fol. 81v) – reproduced in *CAAG* I 161. Wikicommons.

The actual use of similar alembics is not evident in early alchemical texts, which often refer to the production of sulfur water, whose recipe, however, has no mention of distillation or distillatory devices, at least in the version transmitted by the Leiden Papyrus (§ 87; see Chapter 2, p. 80). In other passages, Zosimos describes how to distill eggs in order to produce alchemical dyeing waters (*Authentic Memoires*, IX in Mertens 1995: 30–3). Moreover, scholars have argued that these stills could have been used to distill alcohol during the first centuries CE; that is, long before the period in which the earliest medieval recipes describing similar procedures were compiled (twelfth century CE). Hermann Diels (1913) had the above-discussed alchemical alembics in mind when he tried to interpret a reference to flammable wine he found in the early third-century CE treatise *Refutation of all Heresies* attributed to Hippolytus of Rome (IV 33.2). He speculated that alcohol could have been produced if wine was slowly heated in devices such as the alembics described in Greco-Egyptian alchemical literature. A tantalizing passage from Pseudo-Hippolytus prescribes the boiling of "seafoam" (salt?) with sweet wine (simply referred to as *glyky*, lit. "sweet," in the recipe) to produce an easily flammable liquid that, if poured upon the head (in the context of a gnostic baptism), does not burn. Diels' hypothesis was firmly criticized by Lippmann in a series of papers (collected in Lippmann 1923), where he emphasized, among other points, that sophisticated cooling methods were necessary to isolate alcohol. Even though more recent laboratory tests seem to confirm that distillation of aqueous ethanol can be performed with Hellenistic stills (Butler and Needham 1980), the scattered information provided by ancient sources – recently collected and analyzed by Anne Wilson (1984: 46–9, 56–64), who supports Diels' hypothesis – makes it difficult to solve the problem.

On the other hand, ancient alembics were certainly used to distill liquid substances, as one can infer from Pseudo-Democritus' description of a still as recorded in the fourth-century CE commentary on Pseudo-Democritus' work by the alchemist Synesius (Martelli 2013: 128–31; see also Taylor 1930: 195–7; Martelli 2011: 301–5). Various "bodies" are mixed with mercury and distilled in an apparatus, whose parts are referred in a quite different terminology to Maria's alembic: in particular, the head of the still is described as a glass vessel having a breast-shaped protuberance – a *mastarion* in Greek. This word, a diminutive form of *mastos* ("breast"), represents a technical term implying an analogy between the female body and the shape of the instrument.

An increasing specialization of alchemical vocabulary is also recognizable in the use of the word *kērotakis*, a term related to a second category of alchemical instruments. A similar device was probably described by Maria the Jewess (see Zosimos, *Authentic Memoires*, VII 2 in Mertens 1995: 23), and scholars usually agree in tracing back the *kērotakis* to the palette of ancient painters (*CAAG* II 250,4s.). They melted wax colors on a small metal shovel, which was heated over a vessel containing hot charcoal (*CAAG* I 144; Forbes 1970: 25–6; Mertens 1995: cxxx). In the alchemical texts, the same shovel (actually a metallic leaf) was probably heated and treated with dyeing substances in order to change its color: the word *kērotakis*, in fact, seems to refer both to the leaf itself (*CAAG* II 102,20, 146,13, 169,12–13) and to the specific instrument used to treat it.

On the basis of Zosimos' writings (*Authentic Memoires*, VII 4–6 in Mertens 1995: 24–5) and of the images preserved by the Byzantine manuscripts (Taylor 1930: 132–4; Mertens 1995: 246–51), scholars have tried to reconstruct a specific device composed by different parts slotted together (see Figure 3.8): the instrument could probably have either a cylindrical or a spherical shape. A lower vessel contained the source of heat (Figure 3.8A), while a volatile substance was put into a second vessel (Figure 3.8B) – very often made of glass – fitted to the first one. On the top of the second vessel alchemists placed the metallic leaf (Figure 3.8C), which was covered by a glass cup (Figure 3.8D). Most scholars agree that the metallic leaf was transformed by the vapor of the

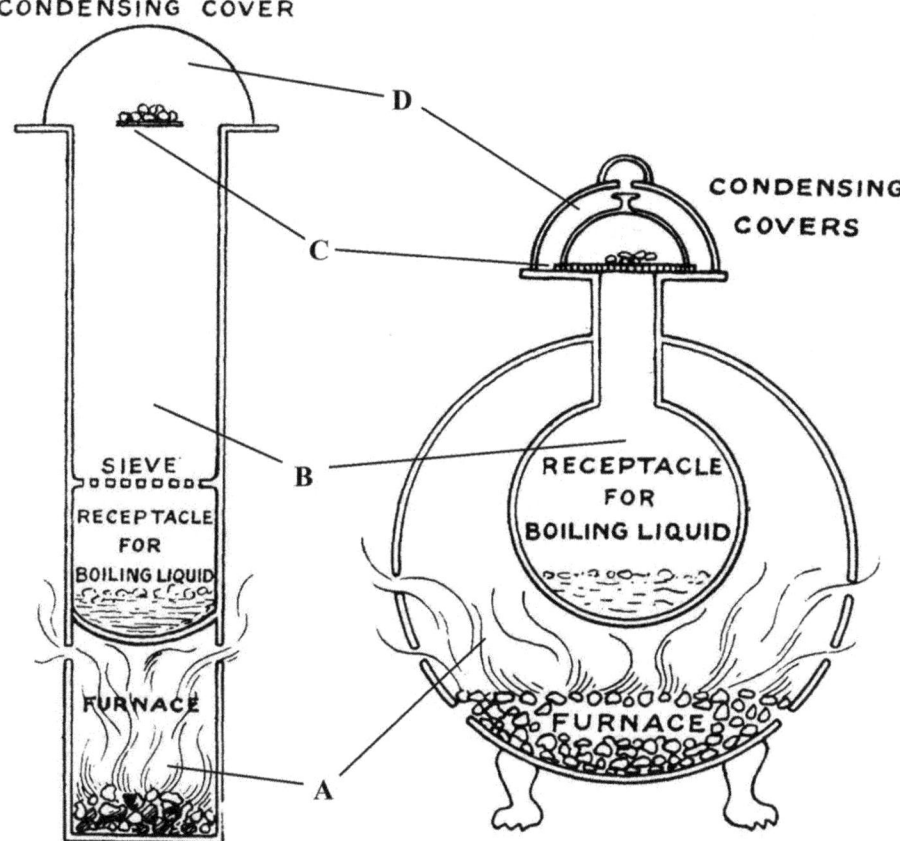

FIGURE 3.8 Two types of *kērotakis* (from Taylor 1930).

volatile substances that were put in the central container (Taylor 1930: 133–7; Mertens 1995: cxxx–clii); in other cases, some reactive chemicals applied to the leaf itself could cause the color changes (*CAAG* II 146,13f.).

CONCLUSIONS

Marco Beretta

The innovations introduced in chemical apparatus and experimental practice by the ancient civilizations created most of the instruments and devices that were used in early modern laboratories. These remarkable achievements were the result of a slow accumulation of improvements that were adapted to stable technical routines of experimentation. This evolution was possible thanks to the importance that the chemical arts acquired in both the Egyptian and Mesopotamian civilizations, where they were under the authority of the political and religious powers. In Egypt, the exploitation of many mineral resources and chemical processes occurred in the areas surrounding the Nile valley. Being under the control of the state, the actual manufacture of chemical commodities was performed by qualified craftsman and was often kept secret. Although archaeological findings are few, the remaining evidence suggests that important chemical workshops were mostly situated in the temples until the Greco-Roman period.

In Mesopotamia too, the state and its rulers supported the emergence of specialists in the chemical manufacture of valuable commodities that, in several cases, were produced in specialized workshops supervised by chief administrators and scribes. Archaeological remains have shown the evolution of kilns, furnaces, crucibles, and other tools.

In the Greco-Roman world, the appearance of several terms to denote workshops devoted to the chemical arts illustrated an unprecedented specialization, especially in dyeing, pharmacology, mining, cosmetics, and glassmaking. Consequently, a variety of new instruments and apparatus was introduced. Archaeological findings have been recently interpreted as evidence of the existence of distillatory techniques. The reference to the first treatise on furnaces, attributed to the alchemist Mary the Jewess, revealed specialized literature on the making of chemical devices.

CHAPTER FOUR

Culture and Science: *Gods, Myths, and Religions*

SYDNEY H. AUFRÈRE, CALE JOHNSON,
MATTEO MARTELLI, AND MARCO BERETTA

EGYPT

Sydney H. Aufrère

It is almost impossible for anybody who has had the opportunity to review what remains of the ancient Egyptian text corpus – over 4,000 years of Pharaonic culture – to claim that they can offer the reader a coherent description of ancient Egyptian culture, from both a geographical and a temporal point of view. What was valid in one time period or in one region was no longer valid later or elsewhere, except by contamination from local theologies more powerful than others.

The basis of Egyptian culture rested on a mytho-scientific perception of the world (see Chapter 1) from the infinitely large to the infinitely small. In this culture, knowledge was always associated with gods or other emblematic characters. In fact, the expression of such a culture and the choice of divine or human beings recognized for having forged it varied over time and according to local traditions. In a very early period, considering the gods and the emblematic characters as the precursors of knowledge, the clergy initiated, perfected, and gradually imposed an association with the divine, which varied with their degree of connection with knowledge kept secret in privileged intellectual spheres. Memphis undoubtedly played a leading role in this domain. From a general point of view, the Greeks considered this traditional knowledge as reflecting the many philosophies peculiar to the Egyptians (Aufrère 2016b).

The origin of this knowledge supposedly transmitted to men was sometimes based either on the intervention of the divine or on a fortuitous discovery. Diodorus Siculus (*Bibliotheca historica*, I 86, 8; see Clement of Alexandria, *Stromata*, VI 4, 36, 1) recorded a legend that, in ancient times, the *Sacred Book* (*hiera biblos*) – the name designating the corpus of religious texts – was brought to humankind by a falcon descending from heaven, which is the reason why the hierogrammats – the most learned men – bore a frontal head band on which two straight falcon feathers were attached on each temple. This legend already figured in hieroglyphic texts, especially in the temple of Edfu, where the *Divine Book* (*medjat netjer*) – the equivalent of the *Sacred Book* of the Greeks – is said to have fallen from the sky (*Edfou* VI, 6, 4; Sauneron 1983: 84–5). Such legends explain the miraculous discovery of renowned medical formulas. According to the London Medical Papyrus (British Museum, inv. no. EA 10059, l. 25), a formula written on a papyrus, illuminated by a moonbeam (Bardinet 1995: 35), appeared during the night at the bottom of a window of the Temple of Coptos.

That the names of the gods, kings, and men associated with the spread or the mastery of this culture are given hereafter is not to suggest that it was a coherent ensemble, but only to tentatively give meaning to it without going into the complexity of a chronological presentation. The elements gathered here should not lead us to believe in a systematic Egyptian distribution of sciences (Sauneron 1967: 111–70; Aufrère 2016b). As it was extremely difficult to trace the origin and know-how of those techniques used, which were strictly regulated (see Chapter 6, p. 166), humans had to invent gods, wise men, preservers of knowledge, or legends to explain their divine origin. Learned gods and wise men involved in techniques of production were asked to intervene only in specialized circumstances. Likewise, according to traditions often referred to in the texts, sovereigns and "emblematic" characters intervened only as mediators of the knowledge that was inherited from the gods. *A priori* it is not necessary to distinguish between the learned gods and these owners of techniques. Generally speaking, if the gods excelling in creation were polymaths, it is nonetheless possible to highlight the specialized contexts in which they intervened. Beside the abovementioned cultural approaches, the Egyptians also had concepts involving cognitive functions (memory, *phasis*, praxis, *gnosis*, executive functions). These deified concepts underlay creators' thought, such as Hu (for the creative verb), Sia (for intuition or perception), and Heka (for magic power). The mastery of knowledge, in the Egyptian sense of the word, rested on these three deified concepts. They are at the heart of the theological systems built around creator gods such as Atum-Rēʿ in Heliopolis, Ptah in Memphis, Thoth in Hermopolis, and other creator gods whose clergy had elaborated as many complex philosophico-theological concepts as their predecessors: Amun-Rēʿ of Karnak, Khnum of Esna, Horus of Edfu, and many others.

Like the Greek gods, Egyptian gods were associated with activities of craftsmanship influenced and favored by the theologies in competition with one another. Creative thought was conceptualized under the guise of skilled activities the gods were supposed to have mastered. Craftsmanship, which requires intrinsic qualities and stimulates the intervention of the senses and a knowledge of the interactions between materials and substances, is set up as a model of intellectual activity (Traunecker 2004). Like Ptah and Khnum, other gods dealt with concepts involving biological transformations or alterations. Their role was to control the interaction between the human body and certain substances.

A Memphite tradition specializing in metalwork mentions that Ptah and Sokaris, who the Greeks considered to be the equivalents of the divine blacksmith Hephaestus, were both experts in the manufacture of military equipment (Sauneron 1954) and in goldsmithing. Khnum, worshipped in Elephantine and in Middle Egypt, was known as a land surveyor (Barguet 1953) and a divine potter. As a craftsman god, creating human beings on his potter's wheel, he was a connoisseur of shapes and became a god of sculptors. A local aspect of this god – Khnum of Shashotep – specialized in the creation of animals (Sauneron 1964: 33–7).

The god Min, lord of the Eastern Desert, worshipped in two cities – Panopolis and Coptos – was regarded as an expert on precious metals and minerals. The texts describe him as crossing the deep desert, searching for mines and quarries. In the Greco-Roman period, he became merged with Pan of the Desert, to whom some of his prerogatives were attributed (Cuvigny 1997; Aufrère 1998d).

Anubis, "head of embalmers," oversaw all operations relating to the preparation of the deceased before burial, an activity that demanded mastery of many chemical processes. His main role was to prevent the putrefaction of the body through the use of products described at length in the *Ritual of Embalming* (Sauneron 1962b; Goyon 1972: 17–84). A recently published papyrus (*P.Louvre*, inv. no. E 32847; Bardinet 2017: 221–2) testifies to the care given to the preparation of the body by the embalmer, who was seen as a manifestation of Anubis, as this quotation explains:

I am Anubis, the one who heals (in) the Place of Embalming, the divine embalmer in the Secret Place. I came to cover [with strips] the parts of the body at risk of decomposition, to cure the burned parts and to take care of the necrotic parts which occur after the death.

(Bardinet 2017: 213)

Persons bitten by poisonous snakes or stung by venomous spiders or scorpions were placed under the protection of the goddess Serket. The professional title of "Conjurors of Serket" (Känel 1988) given to these physician–magicians implies that the goddess inspired the treatment they administered to patients to lessen the toxicity of the venoms; she put at their disposal an important pharmacopoeia along with the recourse to magic evoked in the second part of the ophiological Papyrus Brooklyn Museum, inv. no. 47.218.48 and 85 (Sauneron 1989), as well as in the Papyrus Brooklyn, inv. no. 47.218.138 (Goyon 2012). This latter papyrus also describes how to counteract the potentially lethal toxicity of the venomous bites and stings of serpents, arachnids, and insects.

According to five classical authors, Apis, the sacred bull of Memphis, was considered to be the god of the origin of medicine, thus bringing this art back into the Memphite sector. One of the reasons for this choice was that the sacred animal, venerated in the Serapeum of Saqqara, pronounced oracles in favor of the healing of patients (Aufrère 2001c: 98–9). However, it is worth mentioning that the deceased patient's corpse also required the specific care of the embalmers, thus suggesting that this may have had some influence on the efficiency of the pharmacopoeia. Among the names of its gods or sacred animals (Ptah, Apis) mentioned in connection with medicine, Memphis particularly highlighted men having reached the rank of deity (Imhotep-Asclepios, considered as the son of Apis), and even kings, as will be discussed below. Memphis thus became by tradition the cradle of science and technology.

Several deities related to production are attested as well, among which are Sekhet and Heb, two divine concepts related to field products and those of hunting and fishing, activities that required the implementation of a conservation process (Meeks 1971: 27). Two other deities, Tayt and Hedjhotep (Meeks 1971: 27–8), were responsible for fabrics and weaving, as well as dyes. Two leonine deities, Shesmu and Bastet, were responsible for the manufacture and conservation of perfumes. It may also be that the association of Bastet (𓎬𓂝𓏏) with perfume was arbitrary, based on the fact that a cylindrical vessel (𓎬 phonetic value *bas*) features at the beginning of her name, a vessel that, since the most ancient times, was used to prepare oils and perfumed ointments. Shesmu, god of the wine press and of grape-treading (𓍱𓀀𓏏𓏏) – depicted by the initial hieroglyph that suggests a traditional device to extract oil or must – also played a role in all activities requiring a treading operation, whence his importance in wine production and in the preparation of oils, ointments, and perfumes (Meeks 1971: 28–30; Cicarello 1976). This explains why the name of this god was often mentioned in the sacred laboratories. The goddess Menket was in charge of the manufacture of beer and of fermented beverages in association with another goddess, Tenmet (Meeks 1971: 30), while crops and harvesting (especially grapes) were under the responsibility of Thermouthis and Neper (Broekhuis 1971; Meeks 1971: 30–3).

The model of the divine savant on whom the other gods depended is unquestionably Thoth, a Greek transcription of the Egyptian name *Djehouty*. As

FIGURE 4.1 Tayt, goddess of the dyeing process. Eastern staircase of the temple of Dendara. © Sydney H. Aufrère.

a scholar, he was the heir of Rē', the original holder of knowledge, when the latter decided to leave earth to flee from humankind and reach the sky. Hence, Thoth, in terms of divine administration, became the vizier of the Sun (Yoyotte 1977; see Diodorus Siculus, *Bibliotheca historica*, I 17, 3; Guilhou 1989: 14; Osing 1998: 173, 174, n. d; Aufrère 2007: 234). That was why he sometimes merged with the Sun in the form of Thoth-Rē'. The inventor of the divisions of time and of the calendar as time master, and consequently of calculation and writing, he was the scholar par excellence. In a country where priests used theological wordplay (Sauneron 1967: 123–7; Aufrère 2018), a fanciful Egyptian etymology given in the texts of the temple of Esna about the name of Thoth associates him with bitterness (Sauneron 1961: 234–5; Sauneron 1968: no. 206, 11, p. 33, 2). It is most likely that Thoth was associated with the idea of separating and recomposing the elements, which he did as master of the divisions of time and the calculation of fractions of bushels of cereals, represented by the different elements composing the eye of Horus (𓂀) (Aufrère 2007: 235–9). Thoth, with his daughter Isis – who then transmitted the knowledge to her son Horus – regulated the natural processes and ensured their coherence and harmony. This was contrary to Seth-Typhon, who caused emotional chaos among the gods

FIGURE 4.2 Text associated with the goddess Tayt. Eastern staircase of the temple of Dendara. © Sydney H. Aufrère.

resulting from the scattering of divine body fluids; these were substances that played an indispensable role in Egyptian thought and showed a parallel with atomism. These divine fluids dispersed by this god of chaos had to be reunited by the upholders of science for the benefit of the other gods (see Chapter 1).

In the Late Period, Thoth was credited with all human knowledge and as the inventor of the properties of plants (*pharmaca*; Aufrère 2007: 246–7). Indeed, Thoth was considered a master in the art of medicine, although Egypt also accepted gods skilled in different types of care, such as Duau and Khentyenirty. Khonsu specialized in the treatment of tumors (Bardinet 2017: 99–209). The animals dedicated to Thoth, the African sacred ibis (*Threskiornis aethiopicus*) and the hamadryas baboon (*Papio hamadryas*), presided over purity, regularity, calculation, and intelligence. Redefining the nature of the sharing of knowledge from one generation to the next, Greek thought provided a model that was more in line with philosophy. Thoth, assimilated by the Greeks to Hermes, was soon regarded as the one who passed on his knowledge to his daughter Isis through apprenticeship – following the Greek educational model (*paideia*) – replacing the traditional Egyptian model in which the goddess stole the knowledge of her father by violence and cunning (Aufrère 2016a). According to Plutarch's *De Iside et Osiride* (2, 351E; Froidefond 1988: 179), Isis became a goddess philosopher (i.e. a goddess associated with knowledge in general, including philosophy). She presided over *physis* (Latin *natura*; Hadot 2004) after reaching the same level of knowledge as her father, according to the hymn devoted to her. The Greek traditions associating Thoth with writing and science were prefigured in an Egyptian tradition in which Thoth played the role of "ritualist magician" in the divine government or "the assembly of the gods." This role was that of a "hierogrammat" for the Greeks, a function considered as that of a man of great erudition, well beyond the simple framework of religious knowledge but always under its control (Gardiner 1938).

Scenes depicting offerings made during the Ptolemaic and Roman periods show that in priestly thought the pharaohs were considered as providers of products for cult purposes and acted like the gods they represented. Thus kings, representing prospectors or operators of mines and quarries, acted as if they were the god Min of Coptos, and they represented oleoresins and perfumes as if they were Shesmu.

Nevertheless, some kings were associated with a tradition of important epistemological rupture points. Indeed, in the epitome of one of his works known as *Ægyptiaca*, Manetho of Sebennytos, an emblematic figure who became a broker of knowledge, presented the history of the Egyptian kings (Waddell 1980), mentioning the names of some relevant sovereigns. King Athothis, of the First Dynasty (3150–2925 BCE), is said to have practiced medicine and written a book on anatomy (Fragments 6–7a–b; Waddell 1980: 30–3). King Tosorthros of the Third Dynasty (2700–2625 BCE; i.e. Djoser, for whom the Step Pyramid in Saqqara was built) was considered to be the inventor of the art of hand-hewed stone building, to have practiced medicine, and to have been skilled in writing.

However, this is a late interpretation, since Egyptian tradition attributed the knowledge of stone architecture to Imhotep, considered in the New Kingdom as "the inaugurator of the stone" (*up iner*, literally "The one who opened the stone"; Lauer 1985). Imhotep was also associated with writing and the practice of medicine, which is why the Greeks, who transcribed his name as Imouthes, compared him to Asclepius. Moreover, by Ptolemaic times the names of Imhotep – omniscient scholar, able to read the books of the past and foresee the future – and Djoser were already associated by an age-old tradition. Indeed, in the story written in Ptolemaic times by the clergy of Elephantine and reported in the *Famine Stela* located on Sehel Island, they were cited as witnesses of a famine caused by the droughts of the Nile (Aufrère 2003).

From the Saite era (Twenty-Sixth Dynasty, 663–525 BCE) onwards, and reconnecting with the distant past, the same Imhotep was revered on several festive days (Gauthier 1918). He was represented wearing the skullcap of the ancient Memphite architects, sitting with dignity on a chair equipped with a splash back, writing on papyrus. This representation gave him an intellectual power related to writing and thinking, like his father, Ptah. According to the Royal Canon of Turin (fragment 40; Gardiner 1959: pl. ix), this character, as evidenced from the Nineteenth Dynasty (1295–1188 BCE) on, was the main craftsman-god of Memphis. Hymns chanted his character as a healing god – often associated with Amenhotep son of Hapu, or, according to the Greeks, Amenothes, another holy revered figure (Sauneron 1963; Kerchove 2012: 50–1; Quack 2014). Imhotep even figured in demotic Tebtynite documents (second century CE) debating with the king on subjects of common interest, possibly intellectual.

Manetho also identified King Suphis – also known as Khufu, whom Herodotus named Cheops – as, if not the author, at least the one who ordered the compilation of the so-called *Sacred Book* (*hiera biblos*; Jacoby 1923: 3, no. 609, F 3b; Aufrère 2012a: 332), which contained all the ancestral, scientific, and religious knowledge available to the Egyptian priesthood. Thus, next to a tradition recalling the marvelous by describing a book fallen from the sky (cf. *supra*, p. 114), the example of Suphis showed that the acquisition of knowledge resulted from a long process consisting of the compilation of cultural data across many centuries and preserved up to Greek and Roman times. According to Clement of Alexandria, who read the writings of Chairemon of Alexandria (first century CE), an expert on the customs of Egyptian priests, this compilation embraced both religious and technical data. According to Clement and other classical authors, Thoth was considered to be the author of forty-two books. Thirty-six of them dealt with priestly knowledge, under the authority of high-ranking priests. The other six dealt with medicine under the supervision of pastophors, the only ones who could connect with the public. These books fell into the category of the so-called *techne* (Latin *ars*). Moreover, it was not uncommon for traditions to turn to Cheops, claiming that certain cultural facts or legendary events had taken place under his reign (Chassinat 1928; Daumas

1953; Daumas 1973: 7–20), giving him great notoriety. Cheops also appeared repeatedly in the late Hellenistic alchemical literature. The historiographical tradition about Suphis/Cheops in Manetho's work was undoubtedly at the origin of the emergence of the *True Book of Sophe the Egyptian*, sometimes attributed to Zosimos of Panopolis (Mertens 1995: lxvii–lxix), a work that would have dealt with science and wisdom.

We have shown the importance of holy personalities such as Imhotep and Amenhotep son of Hapu, archetypes of the scholar (Wildung cf. *supra*, p. 120). Although they originated from Memphis and Thebes, respectively, they became healing gods of all Egypt. These two intellectual personalities became leading figures of Egyptian culture in the Greek period (Bataille 1952: 98–102). Apart from his status as a writer, Manetho of Sebennytos was one of those emblematic characters related to Egyptian culture who, for the Greeks, transcended previous models. He acted as an intermediary of Egyptian priestly culture by translating the *Sacred Book* into Greek (Aufrère 2012a; Aufrère 2014). Among the works attributed to him, there is a book entitled *About the Kyphis*, the kyphi being a complex perfume, the recipes for which were known, at least in fragments, by Plutarch and other classical authors (Aufrère 2005a: 246–53). The use of the plural in the title of this work of Manetho seemed to imply that Plutarch, mentioning both solar and lunar kyphis (Plutarch, *De Is. et Os.* 80 383C–4C; Froidefond 1988: 249–51), said to be made of different components, probably referred to this author, and thus indicated that this perfume could have been available in different varieties.

In the Late Period, King Nechepsos or Nechepso, etymologically speaking "Necho-the-scientist" (*Nekau-pa-shesa*), emerged as a "scientist" because his model, Necho II (610–595 BCE), had taken part in the observation of an astronomical event. If Nechepsos was knowledgeable about the astronomical sciences (Fournet 2000; Heilen 2011; Ryholt 2011), it is possible that he may also have mastered other scientific domains. This was true of Petosiris – related to Nechepso (Heilen 2011) – an important personality mentioned in ancient texts without being associated with any technology. Similarly, Cleopatra (VII) (50–30 BCE), recognized for her great learning, was also traditionally known for her mastery of the art of poisons.

Egypt has historically been considered as the land of chemistry and of alchemy (Aufrère 2001d). This tradition stems from the important role played by the abovementioned personalities (Aufrère 2001d). Some believe that the Greek word from which "chemistry" derives, *chēmeia*, may have come from the original Egyptian term designating Egypt, *Kemet*, "the black country." This phrase refers to the color of the arable soil after the annual flood of the Nile due to the silt deposited by it. This land was constantly an object of astonishment for the authors of antiquity who, favoring the wonderful facts, observed that this country changed throughout the year and was subjected to extreme heat, to which the virtues of plants were attributed. But the country also provided many magicians and physicians, not to mention its funeral practices, its ointments, and its complex perfumes that, according to immemorial tradition, it perfected in the laboratories of its temples, or the privilege of providing useful substances. It is therefore not surprising that this country was considered par excellence as the region where alchemy – an aspect of technical hermeticism (Aufrère 2021a: 118) – was born.

MESOPOTAMIA

Cale Johnson

All major forms of technical or specialized knowledge in early Mesopotamia were routinely traced back to the gods. This is the starting point for many discussions of the origin of science and technology in early Mesopotamia, including Tikva Frymer-Kensky's still groundbreaking work *In the Wake of the Goddess: Women, Culture and the Biblical Transformation of Pagan Myth* (1992). Where Frymer-Kinsky differs from others is in her recognition and emphasis on the earliest history of domestic technologies (including a wide variety of kitchen chemistry and domestic forms of production) and their relationship to an array of female goddesses in early Mesopotamian thought. Whether weaving, brewing, pottery-making, bread-baking, or midwifery, each of these domestic technologies was paired with a goddess and given a fixed place in the universe of Mesopotamian technical knowledge.

Among the most important of the Sumerian literary texts in Frymer-Kinsky's account is a decidedly curious conglomerate of mythological vignettes known today as *(The Myth of) Enki and Ninhursag*. Enki is the male deity of fresh water, semen, and technical knowledge, while Ninhursag is the most important of the Mesopotamian mother-goddesses, so little wonder that one of the central preoccupations of the text is birth as a generative device and the womb as the primary site for transformational processes. Enki's ever-fecund semen is the nondescript protagonist that runs through the different vignettes that make up the text. In the key episode Enki impregnates Ninhursag, who easily gives birth in nine days, rather than nine months; Enki then impregnates each subsequent daughter so as to form a matrilineal genealogy. However horrifying this little tale is to us today, it had its purposes in the mythology and reaches its denouement with Uttu, the spider-woman goddess, who is also the goddess of weaving and the domestic loom (not to be confused with the sun-god Utu). Each goddess in the matrilineal line has a smaller "womb" (Sum. ša_3) than the one before, and part of the mythology is rooted in a pun or orthographic decomposition of the word for "(agricultural) field" (Sum. a.ša_3), which combines the words for "water, semen" (Sum. a) and "womb" (Sum. ša_3). This pun is combined with a series of ever-smaller geographical units that are embedded within the names of the female deities – from "mountain range" (Sum. hur.sag) in Ninhursag down to the logogram for "textile" (Sum. tug$_2$) in Uttu's name – so that the size of each female deity (and consequently her womb) decreases as we move down the genealogy to Uttu.

Uttu is special, however: she is the goddess of weaving, which is both the epitome of civilization and domestic production and, more importantly (in the logic of the myth), she is the last of the matrilineal line to be impregnated by Enki. With the pregnancy of Uttu, her womb is too small to allow for the easy delivery of her female progenitors and Ninhursag has to extract Enki's semen from her womb and cast it onto the ground. From this repossessed semen of the god of technical knowledge and subterranean waters, a series of medicinal plants grow. Enki and his vizier Isimud go about examining the plants: Isimud cuts a sample of each plant, Enki tastes it, and, in the end, Enki confirms that various medicinal properties now reside "within" (Sum. ša$_3$ yet again), and in doing so he "determinates the fate" (Sum. nam.tar) of each plant. Some of these plants (or drugs) remain part of the Mesopotamian medical tradition down into the first millennium BCE.

Besides an etiology of disease and the treatment of illness through medicinal plants, which follows in the next section of the myth (and recapitulates several themes from the Enochic myth of the fallen angels; see Martelli's discussion in the next section), the real importance of *Enki and Ninhursag* lies in its depiction of Enki and his vizier working their way through a list of medicinal plants. The dialogue goes something like this:

(Enki says to his vizier): "Which is this one, which is that one?"
(Isimud, the vizier, answers): "Master, the wood-plant," as he plucks it and Enki eats it.
(Isimud): "Master, the honey-plant," as he pulls it out of the ground and Enki eats it.
(Isimud): "Master, the … plant," as he plucks it and Enki eats it.
(Isimud): "Master, the a-numun plant," as he plucks it and Enki eats it.

And on it goes in a perfect rendition, literally *avant la lettre*, of a junior scribe reading a technical vocabulary to his teacher. Lexical lists like this were the central epistemological device (and preoccupation) of Mesopotamian scribes, scholars, and savants in the cuneiform record, and the list of medicinal plants that appears in *Enki and Ninhursag*, lines 190–7, provides the organizing matrix for Enki's assignment of a destiny to each plant, a destiny that represents the function or pharmaceutical effectiveness of each plant/drug. This list also sets in motion, as it were, a series of alignments between medicinal plants, the parts of Enki's body that become diseased as a side effect of ingesting his own semen, and a series of minor deities that Ninhursag "births" in the process of removing these diseases from Enki's body. Keith Dickson, building on a model that Bruce Lincoln developed for Indo-European materials, argues that these list-mediated parallels "establish ... what Lincoln calls 'homologous alloforms' involving specific things – plant, body part, ailment, divinity – whose correspondences implicitly underwrite a magic-medical pharmacopeia" (2005: 501).

If *Enki and Ninhursag* presents us with a justification or etiology for the pharmaceutical effectiveness of plants (and the prescription recipes that include them), the final section of a mythological text known variously as *Lugal-e* or *The Exploits of Ninurta* offers a quite different picture of the "destiny, fate, or function" (Sum. nam.tar) of stones and minerals. Growing out of a long-standing tradition of dragon-slayer myths in Mesopotamia (Karahashi 2004; Gilan 2013), *The Exploits of Ninurta* pits the male warrior-god Ninurta against a monstrous opponent known as Asag. With the defeat of Asag, his army of stones is cast into a heap and forms the Zagros Mountains, described as a range of mountains (and not incidentally using the same Sumerian term [Sum. hur.sag] that forms Ninhursag's name – the text is in part an etiology of her name as well). These are the foothills and mountains that run along the northeastern side of the Mesopotamian alluvium; it was through this mountain range, in one way or another, that nearly all stones and minerals reached Mesopotamia. At the end of the text, each of the important types of stone are presented in a litany in which the characteristic features of each stone are described in poetic terms and their fate determined, largely on the basis of whether or not they fought against Ninurta or remained neutral in the fight.

Whereas the powers of medicinal plants were ascribed to earth/netherworld ("chthonic") or astral deities or both, the stones and minerals participate in a somewhat different apotropaic tradition of preventing ill effects and attacks of all kinds by placing the statues of defeated enemies along the exterior of a temple – the same tradition that locates the famous genii at the entrances of Neo-Assyrian palaces. Within both medical and magical practice, however, these apotropaic stones were gathered into a pouch and placed around the patient's neck in the hope of preventing illnesses that did not easily respond to pharmaceutical treatment. These descriptions represent one of the earliest catalogs of stones and minerals in human history, but it is equally clear that the "fate" of each stone is also based on its mineralogical properties. The king of the stones, interestingly enough, was the "plant" stone (Sum. $^{na4}u_2$), and we can now be quite certain that the "plant" stone is emery, made king of the stones because of its ability, in combination with a metal carrier such as lead or a copper drill bit, to grind away all other forms of stones, particularly in the fashioning of cylinder seals (Gwinnett and Gorelick 1987; Heimpel et al. 1988; Simko 2015).

These two well-known myths – *Enki and Ninhursag* and *The Exploits of Ninurta* – seem to complement one another, providing us with a mythic overview of the two most important domains of chemical practice in ancient Mesopotamia. Unfortunately, there is no evidence that these two contemporaneous pieces of mythology were in any kind of intertextual dialogue. Both were likely composed in the Ur III period (ca. 2112–2004 BCE), although elements of *Enki and Ninhursag* are far older. The key observation of Heimpel et al. (1988), now updated and extended back to the Ur III period in a recent paper from Simko (2015), is that the depiction of the "plant" stone in *The Exploits of Ninurta* aligns perfectly with the growing use of emery as an abrasive in the Ur III and Old Babylonian periods (ca. 2100–1600 BCE). Unlike the other stones that figure in *The Exploits of Ninurta*, the "plant" stone operates in much the same way as the plants that appear in medical recipes: ground up in various ways, plants/drugs (Sum. u_2 = Akk. *šammu*) were typically soaked in a liquid carrier under the stars to release their pharmaceutical powers and ingested by the patient in one way or another. Likewise, the power of the "plant" stone (viz. emery) is released in its ground form, when it is capable of drilling or polishing harder stones such as hematite. And this intrusion of a medical model of effectiveness into the world of the lapidary also rebounds in the other direction as well, for, like us, "calculi" such as kidney or bladder stones within the human body were also known to the ancient Mesopotamians as "stones" (Sum. na_4 = Akk. *abnu*), and emery is also found as the decisive ingredient in medical recipes for the destruction of "stones" within the human body.

If the mythological models that we looked at a moment ago were meant to explain the effectiveness of plants, stones, and minerals against disease and/or misfortune, the Mesopotamian mythological tradition also carved out distinct domains of technical practice and assigned each of these domains to a specific deity. Older links between domestic forms of technical production and mother-goddesses, as outlined by Frymer-Kensky, may have been the inspiration for this categorization of the technical universe, but there are no direct continuities. The new categorization comes into existence in the Ur III and Old Babylonian periods and is represented by the Sumerian term *me*. The term itself is the stem of the verb "to be" in Sumerian (although it also seems to function as a sign for priestly offices in early cuneiform), but it is reified as a specific domain of knowledge (or even as an emblem standing for such a domain) in Sumerian mythological texts like *(The Myth of) Inanna and Enki*. This is the same Enki who figured centrally earlier in this section, but now paired with Inanna, the goddess of love, war, and chaos, who was left out of an earlier cosmogony of crafts and skills in *Enki and the World Order*. The mythological narrative *Inanna and Enki* tells how Inanna, goddess of the great cosmopolitan center of Uruk, talks a hundred or so of these *me* out of Enki, who represents the even more ancient urban center of Eridu, while they were both deeply inebriated. When Enki wakes from his drunken stupor, he tries to get the *me* back, but to no avail.

The list of cultural and technical skills represented by the *me*, now shifted to Uruk, provides us with a unique native picture of how the Mesopotamians conceptualized technical domains at the end of the second millennium BCE. Most of the *me* deal with religious and/or political offices and the insignia and chief concepts associated with these offices, but items 65–72, in Farber-Flügge's enumeration (1973: 56–9, 112–13), should be familiar from our discussion of the workshops in Chapter 3:

Items 65–72 in the list of *me* in *Inanna and Enki*
65. nam.nagar "carpentry"
66. nam.tibira "sculpture, fine metalworking"
67. nam.dub.sar "tablet-writing, scribalism"
68. nam.simug "metalworking, smelting"
69. nam.ašgab "leatherworking"
70. nam.azlag$_x$(LU$_2$.TUG$_2$) "laundering"
71. nam.šidim "construction work"
72. nam.ad.kub$_4$ "reed work"

All eight of these technical domains are represented by the name of a profession preceded by the *nam- prefix, which marks abstract nouns in Sumerian. Thus, the *me* can be conceptualized as demarcated bodies of technical knowledge that originate with the gods, but over time were devolved to groups of human technical specialists.

The most intriguing way in which Mesopotamians, particularly the scholars and teachers in the Tablet House (viz. the primary locus for training students in cuneiform and Sumerian literature), conceptualized domains of skill or activity, technical or otherwise, in the Old Babylonian period was through the composition of dialogues between two protagonists that subdivide a particular domain: *Hoe and Plough* (agricultural tools), *Summer and Winter* (seasons), *Ewe and Wheat* and *Shepherd and Farmer* (both concerned with agricultural production), and, among a number of others, *Copper and Silver* (metals; see Vanstiphout 1991: 25–6). But when we actually look at the content of these debates, they never delve into technical language or process. The hoe addresses the plough in what is a competition between a simple tool that can do many things (the hoe) and a complex tool that can do only one (the plough), and it is clear that these texts are oriented to the schoolboys in the Tablet House rather than a technically knowledgeable audience.

Hoe and Plough 12–15
You cannot heap up the earth in the basket,
You cannot press clay or make bricks,
You cannot lay foundations or build a house,

You cannot strengthen the base of an old wall.
>
> (Mittermayer 2009: 110–11; transl. after Vanstiphout 1984: 421)

And even in *Copper and Silver*, where we might expect a flourish of technical terminology and description, the texts, which are admittedly difficult to construe, do not go in that direction. Overall, then, in the Old Babylonian period at least, it is fairly clear that the technical crafts, which were assigned a *me* in the mythological texts, were not given a place in the nonspecialist surveys made available in the debates. There was, in other words, a clear contrast between specialist crafts that appear in mythological texts (and were devolved from the gods) and ordinary, nonspecialist knowledge, which is not traced back to the gods.

When we turn to the later phases of Mesopotamian history, many of the traditional means of transmitting and categorizing knowledge, which had shown substantial continuity from the end of the Early Dynastic period down into the Old Babylonian period (ca. 2400–1600 BCE), ceased or were fundamentally reorganized at the end of the Old Babylonian period. The dependence of first-millennium BCE scholars on far older intellectual traditions was conceptualized, for the most part, in terms of material "from before the Flood." References to this idea occasionally occur in the midst of an otherwise unexceptional medical text, such as AMT 105/1, line 22 (Lambert 1957: 8), or in connection with the materials that describe the antediluvian ruler Enmeduranki's initial receipt of several forms of divination from Šamaš and Adad, techniques that were then passed on to learned families in Sippar (Lambert 1967: 127). It was Assurbanipal himself, however, who made the most famous such claim:

> I learnt the craft of Adapa the sage, the hidden mystery (*ni-ṣir-tú ka-tim-tú*) of the scribal art (*kul-lat ṭup-šar-ru-tú*). I used to watch signs of heaven and earth and to study them in the assembly of the scholars. Together with the able experts in oil-divination, I deliberated upon (the tablet) "If the liver is a mirror of heaven" .... I looked at cuneiform signs on stones from before the flood (gù.sum *ab-ni šá la-am a-bu-bi*).
>
> (Frahm 2004: 45)

Here again we have the Akkadian word for "stone," but no longer standing for actual stone or the artificial stones from the glassmaking recipes, but rather inscribed clay tablets referred to as stones as well.

The type of scholarship that Assurbanipal claims to have mastered was largely the product of editors and redactors who lived half a millennium earlier in the last few centuries of the second millennium BCE, above all the tradition that culminates in the figure of Esagil-kin-apli, an Akk. *ummânu* "expert" or perhaps "professor" who lived in the eleventh century BCE. Famed for his bipartite editions of technical, scientific, and magical materials such as *The Diagnostic*

Handbook and *The Physiognomic Corpus*, a format inspired by the organizational structure of the Old Babylonian Anatomical Compendium known as Ugumu, the compendial format that Esagil-kin-apli pioneered also likely served as the model for the medical compendia assembled by the scholars of Assurbanipal's Library (Geller 2018: 52). For over half a century, since W.G. Lambert's publication of *The Catalogue of Texts of Authors* in 1962, the following passage has been used to argue that first-millennium BCE Mesopotamian scholars attributed the authorship of a number of key compendia, including *The Diagnostic Handbook* and *The Physiognomic Corpus* (listed below as "physiognomic omens" and "diagnostic omens"), to the god Ea (the Akkadian equivalent of Enki).

> Exorcism, liturgy, astrology,
> Physiognomic omens, anomalous births, diagnostic omens (symptoms),
> Cledomancy, Lugal-e, Angim.
>
> ---
>
> [These are] the authorship (lit. "from the mouth") of Ea.
>
> (Geller 2018: 44)

Two decades after Lambert's publication of this text, Rochberg-Halton's classic paper on canonicity emphasized that "divine authorship, placed as it is in the literary catalog in the context of legendary authors, human authors of great antiquity, and descendants of ancestral scribes, fits into a broader pattern of antiquity of authorship. The antiquity rather than the divinity of authorship clearly emerges as the important criterion for a text's authoritative status" (1984: 136). And while Rochberg was already downplaying the divine aspect of Ea's supposed authorship in 1984, Geller's recent paper on Esagil-kin-apli (2018) provides us with a new possibility: with Ea seeming to author many of the texts for which Esagil-kin-apli is most famous, Geller suggests that Ea here is actually a cryptic rendering of Esagil-kin-apli's name. The name itself means "the one who established (my) son (in) the Esagil-temple," and if the son here corresponds to "Marduk," as it almost certainly does, then the bearer of such a name could have been joking referred to as "Ea," Marduk's father. The centrality of the Marduk-Ea Formula (a standard framework for incantations in which Ea provides an incantation and ritual to Marduk) would have quickly turned a playful in-group reference like this into a logical reality, but the unmediated transmission of texts from the gods to human beings was not the norm in Mesopotamia. Time will tell whether Geller's interpretation proves correct, but if so, it would only reconfirm the preeminent status of Esagil-kin-apli as the most important scholarly figure in all of Mesopotamian history.

Here at the end of this section, a few lines should be devoted to counterbalancing the obsession that some nonspecialists have had, over the past century, with the role of "premature or stillborn children" (Akk. *kūbū*) in the rituals attached to certain first-millennium BCE glassmaking texts. Discussions of how Mesopotamian myths might tell us something about the history of chemistry have focused on Mircea Eliade's *The Forge and Crucible* (1956) and its precursors, several short papers from Robert Eisler in the 1920s. This line of research is entirely the creation of nonspecialists and has been quite severely criticized from its first moments (Darmstaedler 1925; Zimmern 1925; Darmstaedler 1926; Zimmern 1926). There are a number of reasons for doubting Eliade's approach: the Sumerogram that would normally be used to write the Akkadian word *kūbu*, namely Sum. $nigin_3$, is never used in these texts and, in all likelihood, these supposed occurrences of Akk. *kūbu*, at least in the late second-millennium BCE sources, were actually read as Sum. $^{d}ku_3.su_{13}$, a deity associated with torch and censer (Michalowki 1993), which makes much more sense, especially in a ritual for the preparation of an oven or kiln. Variants demonstrate that some scribes in the first millennium BCE believed that $^{d}ku_{1/3}$-*bu* was the correct reading, but these mistakes should simply be discounted (note as well that the reading BU = su_{13} largely disappears after the Old Babylonian period, which may explain why the term was misunderstood in the first millennium BCE). Eliade and his followers used the supposed occurrence of "embryos" in the ritual performed before the use of a kiln, in conjunction with similar ideas from other parts of the world, to suggest that non-gold or non-precious metals "grow" in the "womb" of the earth and that alchemical practices were meant to hasten this maturation of base metals into gold (for thoughtful reconsiderations of theories of metal formation since Aristotle, see Norris 2006 and Martelli 2014b). Needless to say, the entire research tradition that culminates in Eliade's book imposes on the decidedly nonmythological Mesopotamian glassmaking texts an anachronistic model of sulfur- and mercury-driven processes that only came to dominate alchemical thought in the Hellenistic and Islamic periods (see Principe 2013: 35–7 and Martelli's contribution below).

GRECO-ROMAN WORLD

Matteo Martelli

Arts and gods are strongly associated in Greco-Roman mythology. Gods were often portrayed as fabulous craftsmen and generous benefactors, who passed their technical skills on to humans. Among the "chemical arts," this is particularly evident for metallurgy. Hephaestus was not only a skillful blacksmith and fine artist, but he also donated his craft to humankind. The Homeric hymn dedicated to him (*Hymn* 20, vv. 2–3) praises the god and Athena who "taught glorious works to earthbound humans, / who used to live like animals in mountain caves" (Rayor 2004: 95).

The gift of the arts is clearly related to the origins of civilization, a theme further developed in ancient myths explaining the technological achievements of humankind. Well known is the figure of Prometheus, the titan who gave fire to humans after stealing it from the gods – Hephaestus and Athena were robbed of their "technical wisdom" (*entechnos sophia*) according to Plato's *Protagoras* (321d1). For his theft, Prometheus was punished and chained to a mountain in the Caucasus. In the so-called "catalog of the arts" included in Aeschylus's tragedy *Prometheus Bound* (mid-fifth century BCE), the titan recalls the arts he invented for the sake of human beings: how to mix soothing remedies (vv. 478–83) and how to mine useful metals such as copper, iron, silver, and gold (vv. 500–3).

In the fifth century BCE, the divine origins of the arts were questioned by various philosophers. The sophist Prodicus of Keos postulated that Olympian gods were an invention of primitive men, who deified useful goods and their inventors; wine and the discoverer of fermentation were worshipped as Dionysius, bread and the inventor of leavening as Athena, fire as Hephaestus (84 B5 DK; see Mayhew 2011: 175–94). The origins of religion are explained along similar lines in the first book of Diodorus Siculus' *Historical Library* (I 8–29; first century BCE), which probably draws on the writings of Hecataeus of Miletus (550–476 BCE; see Henrichs 1984: 147–8). Early men in Egypt granted a divine status to natural elements as well as to ancient Egyptian kings, who were worshipped as the first inventors of various sciences and technologies. Egyptian priests identified the first king with Hephaestus (I 13.3); under the reigns of Osiris and Isis, special honors were attributed to those who discovered how to work the metals extracted from copper and gold mines in the Thebaid (I 15.5). Isis was also depicted as greatly versed in the art of medicine, having discovered many health-giving drugs and, in particular, "the drug of immortality," which brought her son Horus back to life (I 25.2–3).

Deified culture heroes overlap with the figures of the first discovers of the arts (*prōtoi heuretai*), a theme discussed in a specific literary genre often referred to as heurematography (Kleingünther 1933; Zhmud 2006: 23–44). Most writings on this topic were lost. However, a rich catalog of first inventors is included in the seventh book of the *Natural History* (VII 191–213; see Beagon 2005), where Pliny attributes, for instance, the discovery of woven fabrics to the Egyptians, and of wool dyeing to the Lydians in Sardis (VII 196). According to Ovid's *Metamorphosis* (VI 6–11), Arachne (a skillful weaver transformed into a spider after competing with the goddess Minerva) was the daughter of two Lydian dyers from Colophon. Then, Pliny devotes an extended section of his catalog to metallurgy (VII 197–8). Interestingly, each metal is considered separately: for instance, tin is linked to the islands of Cassiterides, while the invention of ironworking is ascribed to the Cyclopes. As for goldsmithing, various inventors are mentioned, such as *Sol* (lit. "the sun"), son of Oceanus, who also discovered how to produce "medicines" from mined ores (*medicinae ex metallis*).

Similar catalogs are relics of a more complex discourse on the origins of civilization and the role played by technology in the process. Accounts on this topic are included in Vitruvius' *On Architecture* (II 1), Lucretius' *De rerum natura* (V 925–1457), or Seneca's ninetieth letter, sources that follow similar patterns (Cole 1967). Technological discoveries allowed early men to overcome their primitive way of living and marked the different passages of their cultural history. The first men were driven by necessity (*chreia* in Greek) to their invention of the arts, whose products were praised by ancient authors for their usefulness and beauty. According to an *Aristotelian Diaresis* (n. 34), the sciences (*epistemai*) were distinguished according to their objects: sciences of useful things included the making of tools and house-building, while sciences of beautiful things encompassed statue-making and purple-dyeing. A third category included sciences of exact things that lead to demonstrations, such as arithmetic and philosophy (new edition of the *Aristotelian Diaireseis* in Dorandi 2016).

Arts belonging to the first category were especially valued by Lucretius, who emphasized how the discovery of metallurgy made it possible to produce tools essential for weaving and farming (Cole 1967: 17–18); on the other hand, he condemned his contemporaries for being too eager for gold and purple (*DRN* V 1423–4). Seneca dealt with similar topics in his ninetieth letter, where he criticized the middle-stoic philosopher Posidonius (second to first centuries BCE), who, in framing his own *Kulturgeschichte*, attributed technological achievements to ancient wise men and philosophers. Particularly relevant is the case of Democritus, who was credited with the discovery of the arch as well as of technologies for softening ivory and dyeing pebbles, transforming them into emeralds (*Ep.* XC 33).

This transformation of pebbles is strongly related to the emergence of early alchemical literature, which dealt, among other topics, with the making of precious stones. Alchemical authors provided different accounts of the origins of their art. Democritus, for instance, was said to have learned alchemy from the Persian magus Ostanes in the Egyptian temple of Memphis (Martelli 2013: 69–73). Important cultural influences came from the Jewish tradition as well. In particular, the Enochic myth of the fallen angels left a clear mark on early alchemical writings. The myth is fully developed in the first section of the *Book of Enoch* (*1Enoch*, chap. 1–36), an apocryphal text related to *Genesis* (6:4). This section (third to second centuries BCE) is also referred to as *The Book of Watchers*, from the name attributed to a group of angels who, attracted by the beauty of women, left the heavens and taught them secret arts. The revelation ascribed to the fallen angel Azael (chap. 8) is particularly relevant, as we can read in the Greek version preserved by the Cairo Papyrus 10759 (*Codex Panopolitanus*; Black 1970: 22; on the Aramaic and Ethiopic versions, see Bhayro 2005): "Azael taught men to make swords, weapons, shields, and breast-plates –

instructions of the angels – and showed them metals and how to work them, armlets, adornments, (powdered) antimony, paint for the eyelids, all kinds of precious stones, and dyes (*baphyka*)."

This list of crafts – mentioning metalworking, precious stones, and dyes – shows clear similarities with the areas of expertise covered by Pseudo-Democritus' four alchemical books and the Leiden and Stockholm papyri. In addition, more explicit Enochic elements were incorporated by other early alchemical authors. A short recipe book entitled *Isis the Prophet to Her Son Horus* opens with a long account that explains how the Egyptian goddess received the revelation on "the preparation (*skeuasia*) of gold and silver" from two angels who descended to the earth (*CAAG* II 29–30; Mertens 1984: 128–34). Isis, moreover, was asked to swear a sacred oath, thus solemnly promising to reveal the alchemical science only to her son Horus. The goddess swore to heaven, earth, light, and darkness – a formula also used in the Greek magical papyri (*PGM* IV 1705) – to the four elements, and to Hermes and chthonian deities such as the Egyptian Anubis and Cerberus (Mertens 1988).

In his own account of the revelation of alchemy, Zosimos of Panopolis explicitly mentions his sources, namely the "holy scriptures" – that is, the *Book of Enoch* – and Hermes' (lost) treatise *Physika* (Bull 2018). The account is extensively quoted by the Byzantine chronicler Syncellus (Mosshammer 1984: 14) and fully preserved in Syriac translation (*CMA* II 238–9; Martelli 2014a: 11–20). The fallen angels – identified by Zosimos with demons – fell in love with women and, after leaving heaven, taught them the alchemical crafts, which were encapsulated in a book called *Chemeu*, from which alchemy took its name (*chēmeia*). This book, which originally included twenty-four treatises, was then summarized and hidden by the Egyptian priests, who played a key role in Zosimos' alchemical discourse. In his *First Book on the Final Quittance* (Festugière 1944: 363–8 = *CAAG* II 239–46), Egyptian priests are presented as the ministers of those demons who wanted to hide the dyeing techniques they had disclosed, thus trying to regain control over them. Through the agency of their priests, they introduced the "auspicious dyes" (*kairikai baphai*), whose success as dyes depended on the right astrological configurations. However, Zosimos strongly recommended his pupils to avoid these practices and free themselves from the control of demons through a deep study of the treatises of ancient alchemists, who explained the natural dyes (*physikai baphai*; Dufault 2019: 127–31).

Zosimos' alchemical treatises are strongly influenced by gnosticism and hermeticism, which in turn adopted alchemical images and ideas. In the hermetic treatise *Korē Kosmou*, for instance, the Egyptian goddess Isis uses an alchemical vocabulary to describe how God created the substance of souls (Festugière 1967). A model for this account was provided by the Platonic dialogue *Timaeus*, in which a metallurgical lexicon was used to describe how the demiurge forged the human and the world souls (Brisson 1998: 36–8; see Chapter 1). Likewise, the Coptic treatise *Paraphrase of Shem*, found in the gnostic library of Nag Hammadi (*codex* VII), includes a complex cosmogony that employed metaphors drawn from the metallurgical practices of Greco-Egyptian alchemy (Burns 2015). Scholars have long noted the use of alchemical images in gnostic texts by emphasizing similarities between the gnostic baptism and the alchemical practices of dipping metals, stones, or textiles in dyeing "waters" (*baptein* in Greek; Charron 2005). These dyeing waters are referred to as the waters of resurrection in an alchemical dialogue ascribed to Cleopatra (*CAAG* II 292–7), while the Coptic *Gospel of Philip* depicts God and the Son of Man as dyers (Charron and Painchaud 2001). Similar references to alchemy occur more broadly in early Christian writings beyond gnostic literature, in order to explain how God can perfect matter and resurrect dead bodies. Gregory of Nyssa refers to the agglomeration of drops of mercury (*On the Making of Man* in *Patrologia Graeca* XLIV 228), while Aeneas of Gaza alludes to glassmaking and to the alchemical making of gold (*Theophrastus* in Colonna 1958: 62–3; Dufault 2019: 102–3). On the other hand, Zosimos himself used vivid imagery to describe alchemical processes in his famous text usually referred to as the *Visions* (Mertens 1995: 34–47). In a series of five dreams, metals and ingredients become living beings – priests, men of copper with lead tablets in their hands, a barber, and a white-haired man – who are skinned, cut into pieces, dismembered, and boiled in altars having the shape of an alembic. During the dreams, Zosimos talks with these characters, who give a preliminary explanation of what he is witnessing; after awakening, the alchemist himself interprets the dreams as referring to specific alchemical operations.

Already in classical antiquity, various "chemical" activities provided philosophers and physicians with various analogies, which they used to elucidate both their epistemologies and their explanations of natural phenomena. If we go back to the Hippocratic treatises of the fifth and fourth centuries BCE and to the *Problemata* attributed to Aristotle, it is possible to find different attempts to explain the effects of drugs on human bodies and their diseases (Stannard 1961; Scarborough 1983). In some cases, the drug acted as a solvent, as a hot substance able to liquefy or melt the pathological material and thus facilitate its expulsion (*Problemata*, I 40, 48; Touwaide 1998b: 260–1). The ideas of heat melting and dissolving appear to be rooted in artisanal practices, also evoked by the author of the Hippocratic treatise *On Regimen*, who describes various crafts (*technai*) to clarify invisible physiological processes (Bartoš 2015: 138–64). Digestion is explained by means of an analogy with the work of goldsmiths, which implies a comparison between vital fire and the fire used in metallurgy and cookery (*On Regimen*, I 20 = VI 494 Littré):

Men work on gold, beat it, wash it and melt it. With gentle, not strong, fire it is compacted. When they have wrought it, they use it for all [purposes]. So a man beats corn, washes it, grinds it, applies fire and then uses it. With strong fire it is not digested in the body, but with gentle (fire).

(Bartoš 2015: 156)

Moreover, analogies between nature and crafts were used in the explanation of geological phenomena. Early attempts to explain the formation of minerals can be already located in the Peripatetic school; according to Aristotle, for instance, metals and ores are the solidifications of watery or smoked exhalations trapped in the underground (*Meteorology*, III 6, 378a15–87b6; Wilson 2013: 271–7). Similar ideas also emerge in the *Historical Library* by Diodorus Siculus (II 52.1–4), who refers to procedures for dyeing stones dipped in coloring baths (Halleux 1981: 50–1):

For the rock-crystals, so we are informed, are composed of pure water which has been hardened, not by the action of cold, but by the influence of a divine fire, and for this reason they are never subject to corruption and take on many hues when they are breathed upon. For instance, *smaragdi* and *beryllia*, as they are called, which are found in the shafts of the copper mines, receive their colour in a bath of sulphur, and the chrysoliths, they say, which are produced by smoky exhalation due to the heat of the sun, thereby get the colour they have.

(Oldfather 1935: 55)

The contribution of various chemical arts to medicine is evoked by Galen of Pergamon in various passages of his immense *oeuvre*. His *Commentary on Epidemics* (V 1 = XVIIb 299 Kühn = Wenkebach 1956: 257), for instance, lists a number of experts who act as "servants" of medicine. Among them we find cooks and all the people who provide physicians with instruments and substances: blacksmiths who forge metallic surgical instruments, "root-cutters" (*rhizotomoi*), "herbalists" (*botanikoi*), and "those who process minerals and other similar products" (Boudon-Millot 2003: 114–16). Moreover, in a long passage of the sixth book of his work *On the Composition of Drugs According to Places* (VI 2 = XII 905–7 Kühn), Galen explains how, after returning home from his studies in Alexandria, he could infer the medical properties of walnuts by looking at their uses in dyeing procedures. Dyers prepared a juice made out of walnut shells, which could produce a lasting color on wool (see also Plin. *NH* XV 87). Galen is astonished that physicians, even after observing this practice, did not understand the strong astringent "power" (*dynamis*) of the thick liquid. Conversely, he refers to the meager medical talents of dyers, blacksmiths, fullers, shoemakers, and cooks in his harsh critique of the medical school of Methodism: they were all the students of Thessalos, one of the leading figures of this school, who pledged to teach medicine in six months (*On the Method of Healing*, I 1 and I 3 = X 5 and 19 Kühn). However, behind Galen's sarcastic picture of a medical system that achieved wide popularity between Hellenistic and Roman times, it is possible to recognize the exchange between experts of different arts and Greco-Roman physicians.

Dyers and pharmacists certainly shared habits in the ways they manipulated natural substances. For instance, both Galen and Dioscorides often mention the dyeing properties of the simple drugs they describe in their pharmaceutical handbooks. In the second part of the alchemical papyrus of Leiden, its compiler included the description of eleven mineral substances taken from Dioscorides' pharmaceutical treatise *De materia medica* (Halleux 1981: 109). Artisans such as dyers, metallurgists, and fullers were certainly experienced in handling specific sets of "chemicals" employed both in their own arts and in medicine. The term *pharmakon*, indeed, had different meanings and could refer to a medicine, a dye, or a poison. The same semantic spectrum is maintained in its derivatives, such as the word *pharmaxis*. This term is used by Plato with reference to a medical treatment (*Philebus*, 46a9) and by Plutarch with reference to the treatment of metallic alloys by means of *pharmaka*. The beautiful metal statues of Delphi displayed a patina very different from verdigris or rust. Shining with a deep blue tinge, the color of their metallic alloy (bronze) was the result of a chemical treatment of their surface, a *pharmaxis* in Plutarch's words (*De Pyth. Orac.* 395b; Jouanna 1975).

An increasing interest toward the effects of toxic and poisonous substances allowed for the further exploration of the properties of *pharmaka*. As observed by the fourth-century BCE physician Diocles of Caristus (fr. 177 in Eijk 2000: vol. 1, 286–9), the bites of very little insects (e.g. scorpions, spiders, and vipers) had a lethal effect on the whole body, despite the small quantity of the injected venom (Touwaide 1998a: 196–8; Eijk 2001: vol. 2, 334–41). A similar idea also emerges in the earliest alchemical literature. In his book *On the Making of Gold*, Pseudo-Democritus emphasized how a small amount of some natural drugs could bring about great transformations: "a drop of oil can remove much purple and a pinch of sulfur can burn many species" (*CAAG* II 48,1–2 = Martelli 2013: 96–9). These powerful, almost divine, properties of *pharmaka*, however, cannot be properly assessed by those alchemists who do not spend their time in testing the drugs as physicians do (Martelli 2013: 95–9). Already in the Hellenistic period, the physician Herophilus – who defined drugs as the hands of the gods (fr. 248b in Von Staden 1989: 417) – and his followers greatly contributed to the development of pharmacology and to the study of toxic substances. Pseudo-Dioscorides' treatise *On Deleterious Substances and Their Prevention* (Sprengel 1830: 1–41), for instance, inherited this scientific legacy, showing a certain awareness of the toxicity of various drugs (Touwaide 1983; Touwaide 1992). Of the minerals, the author devotes specific chapters to white lead (§ 22), gypsum (§ 24), litharge (§ 27), mercury (§ 28), and lime, along with arsenic ores (§ 29). These observations certainly relied on the experience of physicians, but also on that of various experts dealing with these substances in different fields (see Chapter 5, pp. 154–5). The ancients, for instance, were aware that metallic vessels could alter the taste of the cooked food. Pliny (*NH* XXXIV 160) recommends coating copper cooking pots with tin, since they could spoil the taste of the meals because of their "poisonous verdigris" (*virus aeruginis*; presumably the taste of copper comes directly from the reaction between the metal and acid in wine or food; Halleux 1977: 563–8; Healy 1999: 319). Columella (*On Agriculture*, XII 19.1 and 20.1) and Pliny (*NH* XIV 136) advocate the use of lead vessels for boiling must and producing grape syrups, such as *sapa*, often used to sweeten wine. According to Waldron (1973: 393), lead has the property of inhibiting enzyme activity, "so it is not surprising that the Romans and Greeks found that *sapa* prevented fruit souring and fermenting and used it extensively as a preservative." Trimalchio preferred serving wine in glass cups rather than in cups made of the precious Corinthian bronze, since glass does not smell (Petronius, *Satyricon*, 50). Vitruvius, on the other hand, specifies that the taste of water conveyed in earthen pipes is better, while warning that lead pipes – widely used in Rome's impressive aqueducts – could poison the water they carried (*On Architecture*, VIII 6.10–11). Pseudo-Dioscorides describes the toxic effects produced by these mineral substances only when ingested or drunk. In fact, most of these minerals did have medical uses, often being applied to the body externally, or they could serve as cosmetics. For instance, the author claims that it is not possible to hide the ingestion of white lead since it whitens the palate, tongue, and teeth (Sprengel 1830: 32). Indeed, in antiquity, white lead was largely used by women to whiten their faces. On the other hand, each chapter of Pseudo-Dioscorides' treatise lists various substances that could counteract the effects of each intoxicating drug. This feature reflects an important tendency in Hellenistic and Roman pharmacology, often devoted to the search for powerful antidotes, which led to the growth of the so-called polypharmacy (Totelin 2004: 7). Complex antidotes, including more than forty ingredients, are recorded in ancient sources. The *Mithridatium* was named after Mithridates VI, King of Pontus (120–63 BCE), who was credited with its discovery: he both immunized himself by taking small quantities of poison on a daily basis and developed the *pharmakon* named after him. This medicine was then modified by the physician Andromakos the Elder (first century CE), who thus created the famous theriac, a kind of universal antidote and panacea (Boudon-Millot 2002; Totelin 2004).

CONCLUSIONS

Marco Beretta

Over the 4,000 years of history covered by this volume, the interrelations between different cultures and chemical arts have naturally changed considerably, and it would be impossible to characterize them within a homogeneous frame. In spite of these differences, there are a few essential elements that have remained constant. In ancient Egypt, culture and science were both associated with gods, and this common ground projected onto gods the role of precursors or founders of specific branches of knowledge. This characteristic, shared by the Mesopotamians and, later, by the Greeks, identified specific gods, such as Ptah, Kunum, Thoth, and Anubis, as the founders and the superintendents of arts and crafts. In Mesopotamia, these roles were principally played by Enki, the male deity of fresh water, and Ninhursag, the mother-goddess, thus placing the generation of things at the center of theoretical speculation. Another important myth was that of the war opposing the god Ninurta against the monstrous opponent Asag, which resulted in the victory of the former and the formation of mountains, stones,

FIGURE 4.3 Scene of filling the Udjat-Eye. Eastern Osirian chapels, inner courtyard, southern wall. Detail of the procession. Temple of Dendara. © Sydney H. Aufrère.

and minerals. In Greece, the role of mythology in the justification of the chemical arts was equally important, and it was influenced in no little degree by Egyptian culture. Hermes, Hephaestus, and other Olympian gods were mere translations of Egyptian deities. The mythological origins of the arts bridged the gap between nature and artifice. Technical achievements and inventions were celebrated by the Greeks by resorting to intermediate figures between gods and men and to a literary genre – the heurmatography – in which the lists of the first inventors were given. By a gradation of roles, gods were followed by heroes and by inventors. This hierarchy legitimized a progressive autonomy of the artisans. Conversely, the central importance of the *technai* in Greco-Roman culture was a source of inspiration in philosophy, medicine, and other branches of natural philosophy. It is therefore not surprising to see that many chemical terms played a significant role also in other disciplines.

CHAPTER FIVE

Society and Environment: *The Alteration of the Ancient Landscape*

SYDNEY H. AUFRÈRE, CALE JOHNSON,
MATTEO MARTELLI, AND MARCO BERETTA

EGYPT

Sydney H. Aufrère

In order to avoid the recurrence of famines caused by the whims of the Nile (flooding or droughts), the Pharaonic kinship group had to establish an efficient organization. The objective was to improve the standard of living of a gradually increasing number of people, to meet their essential needs (e.g. food and consumer goods), and to respond to the demands of the aristocracy for imported products and artifacts for funeral uses or liturgical purposes.

Through the creation of departments, each responsible for a specific economic sector, the hierarchy of working methods grew. New techniques of craftsmanship for the manufacture of goods were developed, impacting to a greater or lesser extent on the environment, in accordance with the growth of the population and the improving standard of living. These new techniques considerably increased the production of artifacts (metal, mineral, glass), with several peaks in the importation of substances during the Middle and New Kingdoms, witnessing the domination of Egypt over the Near East and Lower Nubia. The availability of a great diversity of materials allowed more sophisticated technologies to develop, for armaments in particular, which mobilized techniques in a number of different fields, including metallurgy, quarrying, and woodwork. The study of the techniques used for weapon-making and shipbuilding showed that materials were selected because of their resistance or flexibility, the end point being improved performance.

The manufacture of chariots, bows, arrows, and javelins brought these complex technologies to their peak. Their manufacturing processes not only enhanced the proliferation of craft experiments and the control of the chemical processes involved, but also led to the overexploitation of both nonrenewable resources like ores and renewable resources like wood. Indeed, the scarcity of wood in the lower Nile valley and in Mesopotamia compelled the Egyptians to obtain wood from the forests of Lebanon – oak, cedar, juniper, and Cilician pine – whose exploitation largely contributed to the deforestation of this region.

As the information available in texts does not cover the wide range of trades involved in the transformation of materials and substances used in chemical processes, a brief overview will be presented here. Archaeological and textual data indicate both the short- and long-term impacts of this activity on the environment in Egypt and abroad. Among the methods mastered to build monuments, ranging from civil architecture to pyramids, the making of mud-bricks was the most important, since it was inexpensive and did not require skilled staff (Kemp 2000: 83–4). Mud-bricks were made from a mixture of mud and straw as well as mineral and vegetable residues that ensured their cohesion. The wet mud was packed in wooden shape-molds then dried to obtain bricks to build walls. Some of these bricks were stamped with the cartouches of builder-pharaoh. Gaps between bricks were filled with a mortar made of sand and mud (Kemp 2000: 92–3). The outside of mud-brick structures had to be covered with roughcast, then covered with a thin layer of plaster to ensure additional protection against the elements and to prevent the upward movement of salt.

Made of natural elements, the mud-brick was not reusable, but it did not pollute the environment. Remnants of old monuments were exploited by the *sabakhin* (in the nineteenth and twentieth centuries BCE) who collected the *sebakh* (i.e. a decomposed dry organic material rich in nitrates) in order to fertilize cultivable soils. In ancient times, the mud foundations of old buildings were not evacuated – other levels were simply rebuilt on them, creating kom-shaped (Greek *kome*, "village") structures. When reddened by fire, the remains of mud-brick structures polluted the soil. The use of clay-bricks, which also polluted the soil, appeared in the New Kingdom.

It should be pointed out here that thanks to a process invented in the Third Dynasty (2700–2625 BCE; see Chapter 4), the hand-hewn stone used for divine and funerary architecture was sometimes reusable. Many monuments that were partially built with such materials in Egypt still exist. Monuments of the great burial sites could be stripped. For example, in the reign of Ramses II and owing to the unavailability of cut stone – limestone, sandstone, or granite – the engineers had recourse to the stripping of ancient monuments at Giza for new architectural projects. Such a detail calls into question the myth of a political will to maintain the structures of the past. However, Khaemouaset, the fourth son of Ramses II and high priest of Ptah in Memphis, distinguished himself in this regard, because he restored several old monuments (Aufrère 1998e), thus gaining posterity in Egyptian literature under the name of Setne (Charron and Barbotin 2016).

Mining and stone-quarrying operations showed that, when necessary, the administration was capable of mobilizing a large workforce both in the Nile valley and in the surrounding deserts. This makes us conscious of the impact of human activities in this desert environment. As shown in documents like the *onomastica* (see Chapter 2), the Egyptians tried to control this impact by creating new functions such as those of the steward of the treasury of gold and silver, the steward of the desert in the foreign country of Syria, and the steward of the deserts of the Land of Kush. But the texts also mention that among the Theban priestly titles there was that of the "Great of the Gebels of the Gold of Amon," testifying to the omnipotence of the clergy in the New Kingdom (Ziegler 1981). From this one can infer that Egyptian political and religious systems were based on the plunder of the vegetable, animal, and mineral (especially auriferous) resources of a subjugated Nubia, mainly those of Wadi Allaqi. Thus, the Viceroys of Kush ensured that the royal capitals (Thebes, Memphis, etc.) received important yearly tributes (see Chapter 6), which probably required bonded labor in gold mines as attested in the classical texts. The interpretation of the written texts as well as the observation of the mining and quarrying sites exploited in antiquity show that, even though the first miners did not intend to ravage divine resources in order to carry out their necessary rituals, they nonetheless modified the environment by overexploitation of these resources in the long term. Thus, to facilitate the settlement of people in those deserted areas, if only episodically, the availability of vegetal resources was more important than the transformation of these materials.

Massively exploited without any control until the materials were exhausted – not only by the Egyptians themselves but also by the late Greek and Roman conquerors of Egypt – the desert became a place of fear and irrationality (Keimer 1944). In spite of the shortage of wood and water resources, men and animals could be settled to labor if a network of fortified wells or water tanks along with new means of transporting heavy loads were made available. Over time, most operating sites contemporaneous with the Pharaonic and Greco-Roman periods laid bare whole sections of the mountains and emptied veins to the smallest nuggets or mineral pellets, while mounds of accumulated waste rocks and erratic blocks of rock were discarded in the quarries. Moreover, close to the gates of Roman forts, there were discharges of all forms of waste related to daily life (Reddé 2018).

Ceramic production was a highly polluting activity resulting first from the combustion of kilns and then from the processing of the material itself. Indeed, once crushed, the material had little chance of being recycled. However, some shards of ceramics could be used to spread the mud-bricks of arched vaults, as well as employed as ostraca (surfaces for writing). However, this reuse was very limited and did not solve the problem of waste. Since ceramics were indispensable in everyday life, detritus was thrown on isolated mounds outside villages, making the wasteland infertile. As food consisted mainly of bread, tooth enamel was prematurely abraded by the fragments of silica detached from the querns and grinding wheels during the milling of flour. Dental attrition was observed even in the ruling classes.

In the Nile valley, the copper-melting industry, which mobilized significant numbers of workers, polluted the air with toxic fumes released from open or closed furnaces. The use of toxic substances and cow urine in tanneries and the workshops of curriers produced a stench and polluted the waters of the Nile. Tanned leather was widely used for the manufacture of cloth and weapons. The use of vegetable tanning limited pollution to a certain extent. Dyeing tanks also emitted unpleasant smells, as reported for madder in the *Trades Satire* (IV 5–7), but this also applied to many other dyes like woad (see Chapter 2). Moreover, according to the classical authors, embalmer workshops around Thebes, located on the west bank of the Nile and far from inhabited areas, were known to release appalling smells caused by the preparation of bodies for embalming after evisceration and desiccation (Riggs 2014). This industrialized postmortem activity required the involvement of a large number of specialized personnel and the use of large quantities of chemical substances such as natron, widespread in the environment. The preparation of bodies to safeguard their passage to eternity required that linen strips be woven, that coffins be made to standards required by the nobility, and that floral ornaments be prepared (David 2000). This industry around death applied broadly to sacred animals as well, as is revealed from animal necropolises in Saqqara and elsewhere.

If Egypt was considered by tradition to be a country in which simple drugs were prepared to relieve pain, it was also famous for making poisons (Cumont 1937: 174–7). During the Roman period, this reputation was mainly attached to the legend of Cleopatra (Marasco 1995). However, plants containing alkaloids – the opium poppy (*Papaver somniferum* L., 1753) or common henbane (*Hyoscyamus niger* L., 1753) – though not very toxic, and mineral poisons such as arsenic sulfides (orpiment and realgar), mercury sulfides (cinnabar), and copper sulfates were well known as poisons. In ancient times, these poisons were generally unstable, and it is not until the Renaissance that they were stabilized and really became highly toxic (Collar 2007). Nonetheless, Alexandrian Egypt was considered the land of poisoners (*pharmakoi*), also known for its love potions and aphrodisiacs. During this period, poisoning was part of daily life. For rulers, the risk of death by poisoned foods could be prevented by using the services of pretasters (Kaufman 1932: 106; Cumont 1937: 174–6).

MESOPOTAMIA

Cale Johnson

Unlike today, when raw materials are bought, containerized, and shipped around the world, the acquisition of rare or precious raw materials in antiquity was a difficult proposition. And while it can be easy to apply an anachronistic model to the ancient Near East, in which peaceful trade predominated, two particularly important modalities for acquiring raw materials are often left out of such accounts: the direct acquisition of metals through mining carried out by an expeditionary force and the acquisition of materials through military force, whether as tribute or plunder (Englund 2006). While the acquisition of raw materials through trade is deferred to the next chapter, this section first looks at these two methods of acquiring raw materials through direct action, and then turns to both archaeological and literary evidence for waste, pollution, and recycling.

Because we do not expect ores and other raw materials to be readily found in an alluvium like Mesopotamia, it is little wonder that nearly all discussion has focused on the extraction of ores and metals from well-known mining centers to the northwest and to the east of Mesopotamia. Copper and other raw materials from or via the Gulf, especially Oman, play a role later on, and in the Amarna Age in the second half of the second millennium BCE trade becomes central. By contrast, in the earliest phases, the Mesopotamians looked to eastern Anatolia or the Iranian plateau (see generally Potts 2007; Weeks 2013), and only at the end of the third millennium BCE to sources beyond Iran, perhaps as far as the Bactria Margiana Archaeological Complex (BMAC; see Steinkeller 2016). There is extensive evidence for ancient mining and metallurgical work in eastern Anatolia, prior to the intervention of southern Mesopotamian states during the Uruk Expansion (Lehner and Yener 2014: 536); as Lehner and Yener point out, "important developments in extractive metallurgy occur ... at the beginning of the fifth millennium BC ... in Cilicia (Garstang 1953), [including] the development of casting technologies and the possible smelting of ores into metal," but the best early evidence for extractive metallurgy is from the site of Degirmentepe at the beginning of the fourth millennium BCE (539). Of course, mining activities were not limited to this site: "Amongst Anatolia's 91 recorded copper ore deposits, 36 reveal evidence of prehistoric mining" (Wagner and Öztunalı 2000: 31; Kassianidou and Knapp 2005: 218).

Likewise on the Iranian plateau, certainly in greater proximity to Mesopotamia, there is a long history of mining and metallurgy that extends back to the fifth millennium BCE (see generally Thornton 2009; Weeks 2012; Weeks 2013). Much of the literature focuses on Talmessi/Meskani near Anarak, but the centrality of this area for the early extraction of copper ores is now disputed (Pernicka 2004; Weeks 2013: 278). The earliest evidence for crucible smelting of copper is from the fifth-millennium BCE site of Tal-i-Iblis in the southeast. Over the course of the fourth millennium BCE, this technological tradition continues to develop, as evidenced in the archaeological sequence at Arisman, culminating in copper-smelting furnaces in the Proto-Elamite period at the end of the fourth millennium BCE (Helwing 2011; Steiniger 2011; Weeks 2013: 279). Although the proto-cuneiform sources (ca. 3300–3000 BCE) include administrative documents dealing with small amounts of metals, especially copper, only in the third millennium BCE sources do we find evidence for the movement of metals between major urban centers (Englund 2006). Despite the efforts of much later literati to claim technological priority for Uruk, we can be fairly certain that in the Uruk period (ca. 3800–3000 BCE), the Mesopotamians were on the receiving end of this long metallurgical tradition.

One of the most important innovations, at the beginning of the third millennium BCE, was the development of tin bronzes, and it is telling that scholarly opinion on the nature and sourcing of materials for tin bronzes in Mesopotamia has changed dramatically in the last decade. It was long argued that tin bronzes, in contrast to arsenic bronzes, were intentional alloys and that tin sources must have been sought far off in Central Asia, but both of these ideas have been challenged in the last decade. There is "some evidence for the intentional (if uncontrolled) production of arsenic-rich materials for deliberate alloying purposes" such as arsenide-rich slags from Arisman and iron–arsenic speiss slags in Hissar (see Weeks 2013: 280 for an overview). Moreover, recent work in the central Zagros mountains, at the Deh Hosein mine in Luristan, suggests that tin-copper ores were mined in this region, in between Mesopotamia and the Iranian plateau, on the eastern edge of the Zagros in Iran, in the early second millennium BCE (Nezafati et al. 2009; Weeks 2013: 280), and consequently "a naturally mixed source of copper and tin [like] Deh Hosein" may have been the source of tin bronze artifacts in the third millennium BCE (Nezafati et al. 2009: 225). If so, the production processes for arsenic bronzes and tin bronzes may have been more alike than previously suspected. This possibility tends to undercut the idea of taking the Old Assyrian trade at the beginning of the second millennium BCE as a model for what was happening in the third millennium BCE (even if recent finds of a Bronze Age palace at Kültepe/Kaneš offer further support for this; see Ezer 2014), namely that tin was moving from Central Asia, via the Iranian plateau and the Mesopotamian alluvium, into Anatolia, while gold and silver from Anatolia were moving in the opposite direction.

Steinkeller (2016) has emphasized the enormous amounts of gold and silver that were given by Ebla, on the Euphrates in western Syria, as tribute to Mari on the middle Euphrates, and he goes on to postulate that during this period of massive political centralization in the Old Akkadian and Ur III periods, it was an influx of gold and especially silver into southern Mesopotamia that served as the distinctive economic medium of that imperial age. But some elements of Steinkeller's overall model seem to be contradicted by new findings such as the discovery of tin-copper ores in the Zagros and Archi's argument that Durgurasu, which Steinkeller, following Biga, identifies with Egypt, is actually a place in western Iran (Archi 2016).

The one well-documented instance of mining in the early cuneiform textual record, from the Ur III period (ca. 2112–2004 BCE), involves neither trade nor local specialists, but rather a team of "metalworkers (who) dig (in) the mountains" or "miners" (Sum. simug hur.sag ba.al), presumably sent from the southern Mesopotamian city of Lagash to extract ore, on site, in the Iranian foothills to the east of the alluvium (Lafont 1996). This is probably in the same region and also roughly contemporary with the Deh Hosein mine studied by Nezafati and coworkers. The text itself is a flour distribution record from a single day, in which a team of ostensibly 100 miners each receive 6 liters of flour while on their way to the city of Adamdun, or better Adamshah (we cannot assume that the team consisted of 100 individuals, since Ur III bureaucrats did occasionally use simplified summaries like this for smaller groups of workers over a number of days). Lafont reviews two other allusions, in literary texts from the nearly contemporary Gudea inscriptions, to "digging (Sum. ba.al) copper (Sum. uruda)" in the cities of Abullat and Kimaš (Statue B vi 21–3 and Cylinder A xvi 17), and on the basis of this data argued for a route from Lagash, through Adamshah, as a transit point rather than source, and on to Kimash in the Iranian plateau. This prompted a substantial literature on the locations of both Adamshah and Kimash, which has been admirably synthesized and evaluated by Potts (2010). Alongside a renewed investigation of a wide variety of cuneiform textual sources as well as Layard's mid-nineteenth-century investigation of the Tiyari mountains in Iraqi Kurdistan, Potts also points to the archaeological evidence for mines in the northwestern Zagros (2010: 251), and ultimately favors locating both Adamshah and Kimash in this region. Of the three regions briefly discussed here (Anatolia, the Iranian Plateau, and the northwestern Zagros), it is presumably no accident that the one cuneiform tablet that directly reports on mining activities refers to Adamshah, which is probably located in the northwestern Zagros, the closest of these three regions to southern Mesopotamia.

Although we speak of trade networks easily, under the influence of present-day models, Englund has emphasized that the most readily available source of raw materials in ancient Mesopotamia would have been the finished goods kept in the temples, palaces, and storerooms of competing or antagonistic urban centers. The historical record makes it quite clear that temples were one of the major beneficiaries of military campaigns and often received a great deal of the spoils from other conquered cities, and that one of the most widely used techniques for the extraction of raw materials was, as Englund has described it, the "simple plunder" of other people's temples and palaces.

> The violent removal of desired goods from Anatolia, Persia and other Gulf regions such as Bahrain and particularly Oman (ancient Magan), or their removal under threat of annihilation, was a preferred means of Babylonian elites to satisfy their needs for goods not native to Mesopotamia … . This threat of violence stood squarely behind the more benign extortion of taxation of domestic populations and close neighbours, and the demand of tribute from those more distant from Babylonian seats of power.
>
> (Englund 2006: 40)

In recent years, the clearest example of this process (now enshrined in Steinkeller's model of the silver-based economy that emerges at the end of the third millennium BCE) is the conflict between Ebla and Mari at the end of the Early Dynastic period.

Just prior to the emergence of the Old Akkadian empire, we have quite detailed evidence for more than four decades of conflict and diplomacy (ca. 2380–2334 BCE) between Ebla and Mari. Earlier in that period of time, Mari was ascendant and Ebla was forced to pay more than 1,000 kg of silver and 60 kg of gold in tribute (Archi 2017: 165). Mari would eventually fall to Ebla (and this entire zone of conflict would be subsumed within the Old Akkadian empire almost immediately afterward), but the cuneiform textual record makes clear that the acquisition of metals, whether through plunder or tribute or diplomatic "gift," was a central preoccupation of these city states. As Archi puts it:

> War was, therefore, a primary way to obtain valuable goods. The documents prove that the Eblaite army left for war nearly every year (a custom well exemplified in the following centuries by the Hittite and the Assyrian annals), but only a single document (ARET XIV 56) registered the booty collected during these campaigns.
>
> (Archi 2017: 164)

Although we do not have such clear documentation of subsequent Old Akkadian-period extractions, we do see this practice registered in Sumerian literary compositions, likely composed in the Ur III period, such as *The Curse of Agade*. No doubt written by Sumerian literati in southern Mesopotamia in order to shame their colonial overlords, especially Old Akkadian rulers like

Naram-Sîn, *The Curse of Agade* depicts the flourishing new capital of the Akkadian empire, Akkad, with Inanna as the key deity there, in contrast with the traditional religious center of Mesopotamia, the Ekur temple in Nippur. The booty and other wealth, especially in the form of precious metals and stones, which had previously found its way to Nippur was now diverted to Akkad, and Naram-Sîn's efforts to rebuild the Ekur temple in Nippur would even be misrepresented as a kind of unholy recycling of the raw materials held in the Ekur temple. This story is reacting against a central belief in Mesopotamian temple-building ideologies, namely that the raw materials for a new temple ought to be sourced from the full extent of the state (or macrocosm) that the palace represents and should not be recycled from previous incarnations of the temple or palace (Liverani 2001; Johnson 2014), an idea that probably stood in stark contrast to the usual state of affairs.

The most frequent types of waste or pollution in the archaeological record are various forms of nonchemical debris, such as the flakes produced in the production of stone tools, but slags and other waste products from the production of metals and glass are the most prominent debris from chemical processes. Potter's kilns and metal- and glass-making sites are generally swamped with all manner of discards, slag, and other debris from the production process. The best, almost literary, example of this kind of debris field may be found in Woolley's description of a potter's workshop in Pit F at Ur:

> At varying levels there were found, buried in the mass of wasters, the actual kilns in which they had been fired. The kilns had been used each many times and constantly repaired ... [They] were circular and though differing in size were all approximately the same pattern ... . The kiln proper was 1.30 m. in diameter, the ledge 0.20 m. wide round the rim of the furnace-pit serving as a support for the roof of the latter; the walls were of bricks ... these were set in clay mortar and were liberally plastered with fire-clay which had been burnt by the heat to a greenish-white, while the soil round was deep red, shewing that the kiln was in part buried so as to preserve longer the heat of the furnace; the roof of the kiln was rebuilt for each firing and destroyed so as to remove the pots when the firing was complete.
> (Woolley 1956: 65–6: Moorey 1994: 146)

The reality is that the entire site consisted of waste products or one kind or another.

Certain kinds of waste products from the smelting process were distinctive and, in more recent descriptions, have acquired a kind of identity. In his description of the smelting of nonsulfurous copper ore, for example, Forbes describes zinc oxide residues "condensed against the furnace-roof" and known as "Ofenbruch" by early modern German metallurgists. Agricola describes it as "a kind of white liquid [that] flows from the furnace which is noxious to silver because it burns the metal" (Forbes 1950: 281). Other well-known processes such as cupellation (in order to separate silver from lead, for instance) would have led to residues, although evidence of this has only been identified in a couple of sites in the ancient Near East (Kassianidou and Knapp 2005: 226–7 mention only Habuba Kabira South in Syria [Pernicka et al. 1998] and Mahmaltar in Anatolia [Wertime 1973: 883]). By far the most important domain for complex procedures involved in the production of refined metal (and the corresponding precipitation of all manner of residues and slags) is the refining of sulfurous copper ore. In summarizing Roman through early modern metallurgy, Forbes even remarks that "from Roman times to the days of Biringuccio and Agricola the main struggle was the perfection of the method of dry extraction of copper from sulphide ores" (Forbes 1950: 306), yielding as residues primarily sulfur, arsenic, and antimony. Forbes summarizes the five-step procedure outlined in Agricola: (a) roasting the ore, (b) smelting the roasted ore into a mixture of copper and iron known as copper matte, (c) smelting the coarse materials with charcoal (coke) and siliceous fluxes, (d) resmelting the blue metal, and (e) refining the black copper. We do not know how much of this complex process was known to ancient Near Eastern smelters, but if the smelters had to use a simpler process they would have had "to sacrifice efficiency and extract a smaller percentage of the copper present in the ore and leave more copper in [their] slags. In fact, if the earlier processes may seem simpler, they were so at the expense of efficiency" (Forbes 1950: 310). There was, in other words, a continual trade-off between relatively simple procedures that produced large amounts of heterogeneous waste (including still-useable metal) and more complex procedures that yielded more refined metal and discrete types of slag. It has even been suggested that the shift from arsenic to tin bronzes was motivated, in part, by the toxicity of arsenic fumes, but others have found this argument unconvincing (see Charleston 1978: 30 and Pare 2000: 7; Steinkeller 2016, however, cites counterarguments from both the Near Eastern [McKerrell and Tylecote 1972] and Andean metalworking traditions [Lechtman 1979]).

In recent years, increasingly global and scientific approaches to pollution have focused on the *longue durée* evidence for pollution that can be identified at metal-refining sites or even, via atmospheric dispersal, in the Greenland ice sheet. As Grattan et al. frame the question, "the broad conclusions from investigations of lead and copper from cores from the Greenland Ice Sheet, as well as mires at high latitudes or high altitudes in Europe ... indicate that many of the primary contributions of metalliferous pollutants to the global atmosphere from late prehistory to the modern day reflected changes in both the manner and intensity of mining, ore processing and smelting in the ancient Classical World" (2007: 87).

FIGURE 5.1 Pollution visible in a Greenland ice sheet core. Photograph by Jeff Overs/ BBC News & Current Affairs via Getty Images.

Since ancient Near Eastern workshops do not seem to have registered in the global environment in this way, the only method available is to test the chemical composition of on-site residues. Gratten and coworkers have conducted a number of studies of both human bone remains and smelting-site residues in the Feynan valley in western Jordan and found major toxic residues in both (see Gratten et al. 2002 and Gratten et al. 2005 for the skeletal evidence). In a study of toxic residues in three smelting sites, for example, Grattan et al. (2007: 100) identified concentrations of lead (~45,000 parts per million [ppm] at one site, ~14,000 ppm lead at another) and thallium (~90, ~35, and ~145 ppm at three sites) that were "clearly hazardous." For some perspective on these numbers, the US Environmental Protection Agency limits the amount of lead in a playground to 400 ppm, so the amount found in the smelting residues were more than 100 times above this limit.

Present-day concerns with toxicity or pollution can easily lead us to misunderstand how "waste" or "pollution" were viewed in antiquity. Rather than focusing on the deleterious health effects of these "additives," the primary concern in Mesopotamian and early Mediterranean antiquity was the fraudulent debasement of precious metals (or other ingredients). Already in the second half of the third millennium BCE at Ebla, and in the Ur III state at the end of the third millennium BCE, both gold and silver could be purified and classified on the basis of their purity. During the Amarna Age in the second half of the second millennium BCE, when precious goods again circulated among elites throughout the Mediterranean, Syria, and Mesopotamia, questions of purity and artificial creation come to the forefront. This was the same time frame within which recipes for producing artificial lapis lazuli were first written down, and in the famous Amarna Letters (ca. 1360–1330 BCE), we find explicit discussions of the debased gold that the Pharaoh of Egypt would send as gifts to his "brothers" (other kings of major kingdoms), including the rulers of Babylonia, the Hurrians, and the Hittites.

Though only a topic of epistolary conversation among the great kings (and thus limited to the first thirty letters in the corpus), the word "gold" appears more than 500 times in the first thirty letters in Moran's translation; it can be judged without hesitation as one of the chief obsessions of the Amarna Age. The belief of the non-Egyptian rulers, especially in the letters from Burna-buriash of Babylonia (to Amenhotep IV) and the Mitanni ruler Tushratta (to Amenhotep III), was that "in my brother's country [= Egypt], gold is more plentiful than dirt" (EA 19 rev. 20–1; Moran 1992: 44; Rainey 2015: 144–5), although we know from the map of the Wadi Hammamat gold mines and other sources that the gold-mining operations in Egypt were massive undertakings under direct control of the state (see Stierlin 2007: 74–5; Van de Mieroop 2009: 172). In letters from both Babylonia and the Mitanni Empire in Syria we repeatedly hear about debased gold coming from Egypt. In EA 10, Burna-buriash writes to Amenhotep IV:

> As for your messenger whom you sent to me, the 20 minas of gold that were brought here were not all there. When they put it into the kiln, not 5 minas of gold appeared. [The ... th]at did appear, on cooling off, looked like ashes.
> (EA 10 obv 18–24; Moran 1992: 19; cf. Rainey 2015: 96–7)

There is no mention of melting down the Pharaoh's golden gifts in kilns in the correspondence from Tushratta, the ruler of the Mitanni state, but there as well he complains that the gold in question "was not gold and that it was not solid" (EA 29 obv 71; Moran 1992: 95; cf. Rainey 2015: 308–9). Elsewhere, in an inventory of wedding gifts, one total reads "1200 minas of gold" (EA 14 ii 34; Moran 1992: 30; cf. Rainey 2015: 118–19), which is at roughly the same order of magnitude as the huge amounts of precious metal that Ebla gave to Mari as tribute a millennium earlier, but now of gold rather than silver.

The most important literary commonplace for waste products and pollution in ancient Mesopotamian literature is the physical character and environs of the washerman, both in pieces of comedic literature in Akkadian, such as *At the Cleaners*, and in earlier Sumerian and Sumero-Akkadian bilingual lists of humorous professions and embarrassing physical characteristics. We will look

FIGURE 5.2 EA 19 letter from Tushratta. Courtesy of www.BibleLandPictures.com/ Alamy Stock Photo.

more closely at the literary form of *At the Cleaners* in Chapter 7; the crucial point here is that, although Mesopotamian fullers and washermen do seem to have made use of beer and other fermented liquids in their work, there is no evidence that they used stale urine, in particular for the ammonia that it contains, as a cleaning agent, so the horrendous smell associated with washing garments in later Greco-Roman texts should not necessarily be extended back to earlier Mesopotamian sources (as emphasized in the most recent edition, Wasserman 2013: 263–4, where he references a wide variety of possible cleaning agents in the Mishnah: "tasteless spittle, water from boiled grits, urine, nitre, soap, Cimolian earth and lion's leaf" [m. Niddah 61b–2a]). Even if they did not make use of stale urine, fullers and washermen necessarily did use a wide variety of alkalis, animal fats, and dyes, so their chemical stews may already have gained attention. Unlike the well-known lists of professions that extend, in several forms, back to the origin of writing at the end of the fourth millennium BCE, the list known by its incipit as Lu-azlag (viz. Sum. lu$_2$.azlag = Akk. *ašlāku* "fuller, washerman") is first attested in the Old Babylonian period and most exemplars are bilingual (Veldhuis 2014: 161–2). As Veldhuis goes on to say, "terms for professions are exceedingly rare in Lu-azlag, which concentrates on physical characteristics (including illnesses), psychological states, and human activities." But it is probably no accident that lexical elements derived from the Lu-azlag list also served as building blocks for humorous scholastic dialogues such as *The Class Reunion*, so it is likely that the list served as a compendium of humorous motifs in the Old Babylonian period (Johnson and Geller 2015: 10–11).

GRECO-ROMAN WORLD

Matteo Martelli

Deforestation represented a critical result of the urban, economic, and technological development of the Greek and Roman civilizations. Several factors, such as large-scale agriculture, urbanization, and ship- and house-building, certainly contributed to an intense exploitation of wooden areas and forests, which resulted in depletion of these natural resources in various districts and regions of the Mediterranean (Williams 2006: 62–86; Hughes 2014: 68–87; Hughes 2017). Deforestation of Attica is mentioned in a well-known passage of Plato's dialogue *Critias* (111b–c). In its mythical past, the region was unspoiled and full of forests, while in Critias' days, the mountains of Attica could only sustain bees (Harris 2011). The shortage of timber (in particular tall trees for ship- and house-building) was also reported by other classical sources: in Plato's *Laws* (IV 706b), Attica is said to lack shipbuilding wood, which in mainland Greece was only produced in Macedonia and some parts of Thrace, as Theophrastus observes in his *Enquiry into Plants* (IV 5.5).

Metallurgy (especially ore smelting) and metalworking – along with other pyrotechnical activities, such as ceramic production, glassmaking, tanning, and dyeing processes – were often considered among the prime factors in deforestation, even if recent studies tend to reduce the scale of their impact to the local situations (Rehder 2000: 153–9; Williams 2006: 77–80). Wood and charcoal represented the most important fuel for ancient furnaces, and they were in high demand in mining regions. A much-discussed case study is the Athenian silver mines at Laurion, where, according to Hughes' calculations, the need for fuel exceeded the annual forest growth of the whole of Attica (Hughes 2014: 136–42; Hughes 2017: 204). Copper ore smelting must have contributed to the deforestation of Cyprus, according to the Hellenistic multitalented scholar Eratosthenes of Cyrene (third century BCE). Strabo's *Geography* (XIV 6.5) partially linked the destruction of forests on the island to the activities of the local mines, important suppliers of copper sulfate (*chalkanthos*), iron oxide, and other medicinal products.

Intensive mining activities could certainly reshape the landscape of exploited areas. Enormous heaps of spoil and discarded rocks were sometimes piled up next to the mines, such as the gold mines of Thasos, where Herodotus (VI 46–7) saw a mountain turned upside down by the Phoenicians. According to Pliny's account (*NH* XXXIII 70–6), galleries were dug in the mines by crushing the rocks with fire, vinegar, and iron picks, and supported by arches. When the vein was exhausted, the arches were knocked down, so that the fractured mountain collapsed with a tremendous crash. In order to wash this deposit away into the sea, a large supply of water was required. Pliny claims that "the land of Spain owing to these causes has encroached a long way into the sea" (*NH* XXXIII 76; Rackham 1952: 61). Water was collected in large reservoirs set above the mine, often being brought from distant sources through a net of artificial channels that rutted the ground for kilometers (Craddock 2016: 208–11). Ancient authors often embellished their accounts on these environmental transformations with moral judgments. Pliny denounces how the earth was violated by human workers who penetrated its bowels and dug deep tunnels, seeking gold, silver, electrum, copper, iron, pigments, gems, and medicines. "We trace out all the fibres of the earth," Pliny writes, "and live above the hollows we have made in her, marveling that occasionally she gapes open or begins to tremble" (*NH* XXXIII 1; Rackham 1952: 3). Along similar lines, in the seventeenth pseudo-Hippocratic letter (§ 5) Democritus harshly attacks those who run mining and extracting enterprises:

> They have no shame at being called happy for digging gaping holes in the earth using the hands of chained men, some of whom have died from the collapse of porous earth, and others of whom stay on in endless bondage … They search for gold and silver, seeking out tracks and scrapings of dust, gathering sand from here and there and excising earth's veins for profit, even turning mother earth into lumps. But it is one and the same earth that they walk on in wonder.
>
> (Smith 1990: 81)

In the second century CE, the copper mines of Soli, an ancient Greek city in Cyprus, were visited by the Greek physician Galen, who traveled in various regions of the Roman Empire (Palestine, Lycia, the island of Lemnos) in order to collect and purchase the best-quality medicines. Galen was troubled by the working conditions of the miners, chained slaves (probably condemned criminals) who were accustomed to quickly running naked through a passageway that, starting from the building at the entrance of the mine, led down to an underground pond of green, thick, warm water (Galen, *On the Capacities of Simple Drugs*, IX 3.34 = XII 239–40 Kühn). In the tunnel it was almost impossible to breathe because of the dense warm air, which also prevented the lamps from burning for any length of time. The water of the underground pond was brought to the surface and poured into square ceramic basins, where it evaporated and left copper concretions settled in the receptacles (Mattern 2013: 101–3). All the mining work was carried out under the strict control of a procurator of equestrian rank, who allowed Galen to collect as much copper ore as he wanted from the multilayered deposit in the mine (*On the Capacities of Simple Drugs*, IX 3.21 = XII 226–7 Kühn).

Ancient miners often worked in dreadful conditions, of which ancient authors have left vivid accounts. The Greek historian and geographer Agatharchides (second century BCE) described the dangerous working conditions in the gold mines of Nubia. Only excerpts of his work on the Red Sea have been preserved by later authors, in particular Diodorus Siculus (III 12–18) and the Byzantine scholar Photius (*Bibliotheca*, codex 250). Photius' excerpts were also copied in the oldest alchemical manuscript, the *Marcianus* gr. 299 (tenth century CE), which also includes a list of gold mines in Egypt (*CAAG* II 26–7; Letrouit 1995: 66–8). Greco-Egyptian alchemists, indeed, were interested in mining activities. Zosimos of Panopolis, in particular, says that in Egypt the mining and working of precious metals were under the strict control of the kings (*CAAG* II 239–40; Festugière 1944: 363–4); according to the commentary by the late antique alchemist Olympiodorus (*CAAG* II 69–73; Viano 2018), Zosimos himself would have discussed how to wash and treat auriferous sand in his lost work *On the Action*.

Miners were certainly exposed to toxic fumes, as recorded by the Latin poet Lucretius (*DRN* VI 808–15): the noisome stench of sulfur and bitumen was heavy in the underground, and after breathing these vapors the pallid workers in the silver and gold mines in Scaptensula, a Thracian town, invariably died. Pliny observes that the exhalations from the silver mines were dangerous to all animals, but especially to dogs (*NH* XXXIII 98). Because of the toxic vapor of the ores extracted in the mines of Mount Sandarakourgion in Asia Minor – probably a mine for arsenic-based ores, from *sandarachē*, "realgar" – the 200 workmen (mostly enslaved criminals) often fell sick and died (Strabo XII 3.40). In order to reduce these risks, ventilating shafts were opened in the mines, which dispelled the toxic fumes (Plin. *NH* XXXI 49).

Poisonous exhalations were also produced in the workshops where the mined ores were processed. Dioscorides (V 94), for instance, warned against the noxious properties of cinnabar, which he identified with the Spanish *minion* (from the

Latin *minium*, usually referring to red lead) and distinguished from *kinnabari* or "dragon's blood," namely the exudation from the oriental plant *Dracaena* (see also Plin. *NH* XXXIII 116). While the vegetal substance had many medical applications, cinnabar was toxic, as was evident from the suffocating smell it produces when heated in the furnaces (Trinquier 2013; Martelli 2014b: 39–42). Pliny was skeptical about the recipes that prescribed cinnabar for medical uses and urged those who treated cinnabar in the workshops for the production of red pigments to protect their faces "with masks of bladder skin in order to avoid inhaling the dust which is highly pernicious" (*NH* XXXIII 122; Caley 1928: 424). Likewise, workers were advised to avoid breathing the toxic fumes of lead that was processed for medical purposes: in fact, when the lead was smelted in earthen vessels with sulfur, deadly vapors leaked from the furnaces (Plin. *NH* XXXIV 167). Ancient alchemists tried to seal their devices (e.g. alembics and the *kērotakis*) and make them airtight in order to prevent dangerous fumes from escaping. Lutes of different compositions (*lutum sapientiae* or "lute of wisdom"), in fact, were used to coat alchemical devices. In his commentary on Zosimos' alchemical work, Olympiodorus mentions a glass cup used to burn arsenic ores, which Julius Africanus called an *asympton* (F72 in Wallraff et al. 2012: 185): Olympiodorus specifies that another vessel was set on the top of the *asympton* so that the arsenic could not dissipate as a vapor (*CAAG* II 75,19–22; Beretta 2009: 116). Tubes or funnels could also be used to draw noxious fumes out of the furnaces. Strabo (III 2.8), after praising the unique variety of metals mined in Spain, mentions the local silver-smelting furnaces that were equipped with high chimneys "so that the gas from the ore may be carried high into the air; for it is heavy and deadly" (Jones 1921: 43).

The emission of gases and fumes had consequences both for human health and for the environment. Various lakes near mining areas were contaminated in antiquity. During the Roman period, especially the silver–lead deposits in southeastern Spain were exploited (e.g. the Rio Tinto region; Richardson 1976). Laguna de Río Seco in southern Spain, for instance, records a marked peak in lead content between 2100 and 1700 years ago, which can be associated with Roman metallurgy (García-Alix et al. 2013: 454). Northwestern Spain was especially abundant in gold deposits; analysis of the water in Laguna Roya, a small glacial lake 35 km south of Las Médulas mines, has shown high concentrations of lead, antimony, bismuth, and arsenic, probably due to the extraction and processing of gold ores (Hillman et al. 2017). The significant atmospheric effects produced by ancient metallurgical industries can be still detected in the ice layers of Greenland (see Figure 5.1), which represent "unique frozen archives of past changes in large scale atmospheric fluxes of metals" (Hong et al. 1996b: 191). Recent studies have shown that the concentrations of lead and copper in Greenland ice cores were significantly higher during the Greco-Roman era than in the previous and following centuries (Hong et al.

1994; Makra 2015: 31–3). Atmospheric copper emissions increased between 200 BCE and 350 CE, with a peak at the height of the Roman Empire (Hong et al. 1996a; Hong et al. 1996b). High levels of lead emissions are registered especially between the mid-fourth century BCE and the second–third centuries CE, as shown by Greenland's ice sheet that dates back to between 2500 and 1700 years ago: lead is present at concentrations four times greater than natural values (Hong et al. 1994; Capasso 1995). Fluctuations are recorded in conjunction with plagues, wars, or periods of political instability (especially during the Roman Republic), which probably caused temporary declines in mining activities (McConnel et al. 2018).

During wars, water could be intentionally poisoned, an underhand strategy particularly effective in siege craft. During the First Sacred War (ca. 595–585 BCE), for instance, in order to defeat the city of Kirrha under siege for ten years, Athenians used hellebore to poison a canal that supplied water to the city (Grmek 1979: 146–7; Mayor 2003: 100–4). According to a pseudo-Hippocratic treatise (*Embassy*, § 4 in Smith 1990: 114), it was the doctor Nebros (lit. "Fawn"), ancestor of Hippocrates, who planned to pollute the water with unspecified poisons (*pharmaka*). Poisonous fumes were also used in sieges, when pitch, sulfur, and bitumen were employed both as incendiary accelerants and to produce noxious smoke (Mayor 2003: 207–50). For instance, the Spartans defeated Plataia (427 BCE) by fueling a massive fire in front of the city's walls with pitch and sulfur (Thucydides, II 77). The same substances could also be used underground, as has been revealed from the archaeological evidence in the eastern Roman city of Dura Europos, besieged by the Persians in 256 CE. Here, a net of tunnels was dug under the city walls by both Romans and Persians; the latter, however, burned sulfur and bitumen to kill their enemies trapped in the galleries (James 2011). Indeed, in the East, petroleum products had military applications: in Samosata (on the Euphrates), a local inflammable mud (called *maltha* by Pliny; perhaps containing naphtha) was used as an incendiary weapon against Lucullus' army (Plin. *NH* II 235; Partington 1999: 3–5). Sulfur, quicklime, bitumen, and other ingredients are combined in the recipe for a "self-lighting fire" given by Julius Africanus in the *Cesti* (fr. D25 in Wallraff et al. 2012: 116–19), a multivolume "encyclopedia" that included, among other topics, military literature and alchemical recipes (Partington 1999: 5–10). These incendiary weapons anticipated the infamous "Greek fire," namely a "fluid fire" of crude inflammable oils (naphtha or petroleum) that was projected by a sophisticated device with siphons, pipes, cauldrons, and pumps, which could be accommodated in Byzantine ships (Haldon 2006).

Along with the extraction and processing of metals, other activities, such as brick and glass production and textile industries (which included both fulling and dyeing processes), polluted the air and water in urban areas as well.

Polluting smoke was produced by Roman glassworkers who often blew molten glass: the process required a sophisticated pyro-technology with furnaces that could reach constant high temperatures (ca. 1050–1150°C; Stern 1999: 450–4). Sometimes recycled material was remelted, as can be inferred from scattered references to the barter of broken pieces of glass by various Flavian poets, in particular Martial (*Epigrams*, X 3,3–4), Statius (*Silvae*, I 6,73–4), and Juvenal (*Satires*, V 47–8). It is difficult to locate glassmaking workshops in Rome: a *Vicus Vitrarius* (an area that took its name from glassmakers) in nearby Porta Capena is recorded by the fourth-century *Regionary Catalogues* (Holleran 2018: 461), and Martial mentions a Transtiberine street peddler who exchanges sulfur for broken glass (*Epigrams*, I 41,3–5). Scholars have sometimes interpreted this passage by Martial (along with the related passages by Statius and Juvenal that are referred to above) as a reference to the use of sulfur to remelt or glue broken glass (see, for instance, Post 1908: 21; Kardos 2002: 122). However, this interpretation, already questioned by several scholars (Leon 1941; Harrison 1987; Santorelli 2013: 93–5), does not seem to be confirmed by ancient sources.

Already in the second century CE, glassmaking was probably moved to the suburbs of Rome in order to remove the nuisance of its smoke from the city center (Douglas and Frank 1972: 4). Similar restrictions were applied in Palestine, at least according to the short sixth-century CE treatise on urban planning by Julian of Ascalon (Saliou 1996), who prescribed setting up workshops that produce polluting fumes – particularly the shops of glassmakers, blacksmiths, statue-makers, dyers, and fullers – on the outskirts of cities (Dell'Acqua 2004: 138; Saliou 2012: 44–5). Likewise, the *Mishnah* (Jewish exegetical oral traditions that began to be collected in the first century CE) includes restrictions on the location of workshops: a dyeing shop could not be located under another person's storehouse, while tanneries were to be relocated outside the town (Mamane 1987; Saliou 2012: 47).

Tanning was among the crafts that the ancients considered responsible for water pollution; the use of urine along with various astringent mordants, both vegetal (e.g. oak gall, pomegranate, or sumac) and mineral (especially alum), could contaminate the rivers where the treated leather was soaked. A fifth-century BCE inscription (*IG* I^3 1.257; 430 BCE) found in Athens reports a decree that prohibits using the waters of the river Ilissos, upstream from the temple of Heracles, to rinse skins or wash away the waste produced by tanners (Rossetti 2002; Beazot 2017: 55–6). The application of this decree had beneficial effects, if we rely on the testimony of Plato, who some decades afterwards praised the fresh waters and the scented air of the region (*Phaedrus*, 230b–c). Fulling and dyeing workshops often raised similar issues. *Fullonicae* were commercial laundries that washed used cloths and, less often, finished new textiles. Even though we find some fulleries at the margins of the city (e.g. the large *fullonica* excavated in Casal Bertone; Musco et al. 2008), they were more often situated in the centers of towns, as archeological evidence in Pompeii and Ostia clearly shows (Flohr 2013a: 229–36). The use of urine and sulfur in the processes could harm the workers (especially because of ammonia; Bresson 2017: 186–7) and produce unpleasant smells, ubiquitously mentioned in literary sources (Flohr 2003). The impact on health and the environment was probably more serious in the case of dyeing workshops, where chemicals were processed in heated cauldrons (Flohr 2013a: 184–6). Strabo (XVI 2.23), for instance, refers to the polluting dyeing shops on the island of Tyre:

> The city was also unfortunate when it was taken by siege by Alexander; but it overcame such misfortunes and restored itself ... by means of their dye-houses for purple; for the Tyrian purple has proved itself by far the most beautiful of all; and the shell-fish are caught near the coast; and the other things requisite for dyeing are easily got; and although the great number of dye-works makes the city unpleasant to live in, yet it makes the city rich through the superior skill of its inhabitants.
>
> (Jones 1930: 269)

Martial (*Epigrams*, I 49,32; II 16,3; IX 62) often mentions the unpleasant smell of purple-dyed garments, probably caused by use of urine in the dyeing procedure (see also Plin. *NH* IX 127). Perhaps in order to solve this problem, a recipe on purple dyeing included in Pseudo-Democritus' alchemical books (Martelli 2013: 80–1) prescribes using an aromatic substance to fumigate the wool that had been left to soak in urine for two days.

The textile industry is often associated with women in literary sources, especially with reference to weaving and spinning. However, the real involvement of female workers in the different steps of textile production is difficult to assess. Fulling and dyeing are especially opaque fields. Latin inscriptions do mention *purpurariae*, women who produced or retailed purple dyestuff or purple products (Larsson Lovén 2013: 115–16; Larsson Lovén 2016: 203–7); we also find two female perfume dealers (*unguentariae*; CIL VI 10006; X 1965), and a woman is mentioned among five jewelers (*gemmarii*; CIL VI 9435: see Larsson Lovén 2016: 204). The frescos in Pompeii's *fullonicae* display some women at work, usually associated with the last phase of the production process and with works physically less demanding (Clarke 2003: 116–17; Flohr 2013a: 282–3). Greco-Egyptian papyri refer to female fullers (*knaphissa* or *gnaphissa*; see, e.g., *P.Mich.* IV 359; *P.Oxy* XXIV 2425; *P.Iand* III 43), and the papyrus *P.Oxy.* XIV 1648 (second century CE) preserves a contract in which three women (a mother with two daughters) sell their dye shop equipped with a leaden pot and an earthenware cask (Wipszycka 1965: 148). Some letters from the archive of

Apollonios, a *strategos* (ca. 113–120 CE) who ran a large weaving enterprise in Upper Egypt, mention his wife Aline and his mother Eudaimonis engaged in purchasing purple dyestuff and assessing its quality (Martelli 2014c: 118–20; see Chapter 6, p. 184).

The recipes on purple dyeing included in the chemical papyri of Leiden and Stockholm do not explicitly mention the possible users of the copied instructions. In general, however, Greek alchemical literature often refers to women as significant actors in the early development of this art in Greco-Roman Egypt. As already seen in Chapter 3, Maria the Jewess was counted among the earliest authorities in the field, especially for her accurate descriptions of various devices, such as alembics, the *kērotakis*, and furnaces. Zosimos of Panopolis wrote his own alchemical writings for Theosebeia, a wealthy upper-class woman perhaps of Roman lineage, who is addressed as "purple-robed lady" and "my mistress" by the alchemist (*CAAG* II 226, 246; Hallum 2008). For a while she joined the Egyptian priest Neilos and his alchemical circle, to which the Egyptian woman Taphnoutia also belonged (*CAAG* II 190–1; Letrouit 1995: 22, n. 49).

Women were, after all, the first to receive the book on *chēmeia* that was revealed by the fallen angels, according to Zosimos' account, and the angel Amnael disclosed to the goddess Isis the secrets of the preparation of silver and gold (*CAAG* II 28–9; see Chapter 4, p. 132). Likewise, *chēmeia* is defined as the preparation of gold and silver in the Byzantine Lexicon *Suda* (χ 280) too, which also refers to the Roman emperor Diocletian (244–311 CE), who destroyed all the books on this subject. The same story was already recorded by John of Antioch in his *Universal History* (fr. 191 Mariev):

> Diocletian full of hatred and grudge against those who rebelled against his power in Egypt, was not satisfied with ruling the country unfairly and cruelly, but traversed it and damaged Egypt with proscriptions and murders of distinguished people. He tracked down and burnt the books on the *chēmeia* of silver and gold, which had been written by their ancestors: so Egyptians would have no longer gained money out of this art and thenceforward they would have no longer had confidence in the abundance of their substances and risen up against Romans.

Diocletian's hostility against alchemy does not seem to have been maintained by early Byzantine emperors. Indeed, the table of contents of the oldest alchemical manuscript (*Marcianus* gr. 299, fol. 2r; Saffrey 1995: 4–5) lists works on alchemy ascribed to the emperors Justinian (ca. 482–565 CE) and Heraclius (ca. 575–641 CE); to Heraclius, moreover, the alchemist Stephanus of Alexandria addressed his own *Lectures* on the sacred art of gold-making (Papathanassiou 2017: 213–8).

CONCLUSIONS

Marco Beretta

The increasing value attached by ancient civilizations to commodities that were the product of chemical manipulation such as metals, minerals, precious stones, and glass impacted on their society and environment in two ways: (a) territorial expansion and (b) environmental changes.

When they were not able to trade valuables, Egyptian and Mesopotamian rulers expanded their territories to where sources of new richness could be found. Tin, which was needed to alloy with copper for bronze, was imported from Anatolia. Even less valuable materials like wood, which were scarce both in Egypt and Mesopotamia and essential to keeping going the batteries of the furnaces implemented by chemical workshops, were imported from Lebanon (particularly cedar wood), Nubia (ebony), and from other areas rich in woods situated in the Near East.

The systematic exploitation of these resources led to deforestation, probably the most visible and extended form of alteration of the landscape. Although the polluting effects of the chemical arts became rapidly apparent, they did not lead to any attempt to mitigate these effects. Moreover, the dreadful conditions of workers such as miners, mostly slaves, were regarded as normal. Pollution and waste were not yet perceived as environmental problems, but the situation began to change with the Greco-Roman authors who, in a few texts, denounced both the greed and the dangers of exploiting the earth without any control. The progress made in medicine raised the awareness of the toxic properties of certain substances and chemical processes: tanning and dying manufactures poisoned the water and toxic fumes exhaled from the metallurgical and glassmaking workshops and noxious airs often filled the mines. The large-scale employment of the chemical arts introduced by the Romans encouraged the authorities to take some measures to mitigate the effects of the pollution. The growing output of waste in the chemical arts also inspired the first techniques enabling reuse by recycling.

CHAPTER SIX

Trade and Industry: *The Circulation of Trade in the Mediterranean*

SYDNEY H. AUFRÈRE, CALE JOHNSON,
MATTEO MARTELLI, AND MARCO BERETTA

EGYPT

Sydney H. Aufrère

Land and maritime trade routes determined the offloading points and locations of the state-owned factories where goods were exploited or transformed. This section gives only a brief review because the variety of materials transported was so great and the nature of their processing was extremely diversified.

The territory of the present-day country of Egypt forms a rough quadrilateral consisting of 95 percent desert land. By contrast, ancient Egypt was a narrow fertile valley watered by the Nile from Elephantine to Memphis (modern Aswan to Cairo). Initially narrow at the first (northernmost) cataract, this valley, like a papyrus stem, ultimately opens into a delta where, according to the descriptions made by the major classical authors (Herodotus, Diodorus Siculus, and Strabo), the waters of the Nile formerly reached the Mediterranean Sea through five to seven branches. Although the Egyptians ventured out on the high alluvial terraces of the Nile valley, until they exploited them temporarily for pastoral or hunting reasons, the deserts – sandy stretches or plateaus –

that border this valley were alternately places of confrontation or interaction between the Pharaonic ethnic group and Bedouin tribes that could be either hostile or cooperative. These areas – the mountainous Eastern Desert and the sandy Western Desert – were crisscrossed by several caravan tracks that transported humans and goods to their destinations. The main track was the gateway to Egypt. Starting northeast of the Delta at Sile (*Tjaru*), it connected Africa to Asia along the northern Sinai. There were other traditional tracks crossing the Isthmus of Suez from Syria-Palestine, which allowed access to different latitudes of the Nile valley by following the Red Sea coast route and adjacent wadis. Moreover, the Egyptians could get to the shores of the Red Sea from different points of the valley to reach favorable port areas (Wadi el-Jarf, Mersa Gawasis), where boats, safely stored in galleries (Wadi el-Jarf; Tallet 2015; Tallet 2017: 15–21), could be loaded pending the next favorable season to set sail for the coasts of the Sinai, Yemen, or Punt in modern-day Somalia (Meeks 2003; Tallet 2009; Espinel 2011).

Foreign vessels coming from the Mediterranean area could use the westernmost branch of the Delta, known as the Canopic or Heracleotic branch, to reach the port of Memphis (Šichan 2011: 96–100), traditionally Egypt's main industrial town. Most of the products used to make weapons imported from foreign lands arrived there. They were then transported by a waterway to *Peru-nefer* (Ezbet Helmy), considered today as the port of the Ramessid capital, Pi-Ramses (Tell el-Dab'a; Bietak 2009; Šichan 2011: 96–100), located on the Pelusiac branch, not far from the Wall of the Ruler. This "wall" was a network of fortresses protecting the track leading from the Delta to Syria-Palestine (Hoffmeier 2006). The main army corps was stationed close to this archaeological site, already attested under the reign of Horemheb (1325–1295 BCE). Thus, during the Ramesside period (1292–1069 BCE), Pi-Ramses met most of Egypt's needs for military equipment. In the west, several tracks led to the Oases and to Cyrenaica, and in the south to the fourth cataract, in the country of Kush, with which Egypt had traditional commercial ties.

From the Eighteenth Dynasty (1552–1314/1295 BCE) onwards, the coasts of the Delta and the Nile branches were exposed to piracy and pillaging by the Acheans – a collective name for the Greeks in Homer's *Iliad* – and to two invasions by the Sea Peoples, confederations of Mediterranean ethnic groups (Oren 2000), during the reigns of Merenptah (end of the twelfth century BCE) and Ramses III (beginning of the thirteenth century BCE). This situation prevailed during the Saite Twenty-Sixth Dynasty (672–525 BCE) onwards. It was essential to open a route to the Mediterranean Sea in order to promote trade with the Greeks in Caria and Ionia. The latter were obliged by the Egyptians to navigate to the Canopic branch of the Nile, to the emporium of Naucratis, where the Greek cities enjoyed the advantages of residence and commerce. Two steles were erected at two locations as statements to curry political favor with the priesthood under the reign of Nectanebo I (380–362 BCE). The first of these was discovered in Naucratis, the second more recently in Thonis-Heracleion (a port located at the outlet of the Canopic branch), which, by the so-called decree of Sais, testified to the vitality of these exchanges by the payment of customs duties on the imported and exported products at the entrance and the exit of Naucratis (Bomhard 2012). This was already clearly indicated in the contemporary customs papyrus written in Aramaic in the reign of Artaxerxes II (404–458 BCE; Briant and Descat 1998).

Since the New Kingdom, Egypt, according to the standard iconography, serviced its economic exchanges with other foreign countries by receiving tributes, whether they be real or fictitious. Following a traditional pattern, the importance of these exchanges was depicted in the tombs of the major economic actors of the state, especially those of viziers, who held a tight control on foreign trade. The remarkable tomb of Vizier Rekhmirê (TT 100; Davies 1943; Anthony 2017; Güell i Rous 2018), a contemporary with the second half of the reign of Thutmose III (1458–1425 BCE) and that of Amenhotep II (1428–1400 BCE; i.e. at a time of unprecedented conquest in history when Egypt dominated Syria-Palestine to the Upper Euphrates region), is the best example of a group of tribute holders. Each ethnic group is shown paying its tribute by bringing different raw products it had exploited or traded, or with objects that displayed the craftsmanship in which the group excelled. Generally speaking, these traditional commercial exchanges comprised all sorts of "tributes" paid for the conduct of the official yearly ceremony.

The Minoans or Keftyu (who originated from Crete and exerted their influence on the Cyclades) were distinguished by their colored loincloths, their face profiles, and their hairstyles, and they brought oxhide-shaped bronze ingots, worked objects (precious vessels), and elephant tusks. The Egyptians did not always distinguish them from the Mycenaeans, whose trade replaced the Minoans in the fourteenth century BCE. The inhabitants of Syria-Palestine (Retjenu) are depicted clothed in white and bringing chariots, bows, quivers, swords, oxhide-shaped ingots, wine amphorae, and other vessels, as well as art objects (vases), silver rings, horses, a bear, and an elephant. The Puntites, identified by their brown skin without being considered as black Africans, carried a variety of bulk oleoresins, oleoresinous trees (frankincense and myrrh) in baskets, as well as ebony, gold, ivory, and exotic animals on leashes. Finally, the Nubians, whose country was under the control of an Egyptian viceroy, paid their tribute with gold in various aspects, along with cattle, exotic animals (baboons, guenons, giraffes, and panthers), and ivory tusks (Davies 1943: pl. LII). Syrians and Nubians, as well as women and children destined to be educated in Egypt, were also depicted (Mathieu 2000). On a smaller scale, these traditional tribes are found in other tombs: the tribute paid by the Nubians is represented in the tomb of the Viceroy of Nubia,

Huy (TT 40), or in the tomb of Menkheperreseneb II (high priest of Amun, superintendent of the gold and silver treasuries; TT 112), where the members of a Syro-Palestinian ethnic group are seen in procession, bringing ivory, containers, and fabrics. All these pseudo-tributes, some more prestigious than others, showed the nature of international trade with Egypt, all reflecting its requirements.

The intention of the traditional cults of ancient Egypt was to appease the Egyptian gods and encourage them to play their role in regulating the cosmic mechanisms and natural cycles, thus ensuring the regular delivery of exotic products of high value. The prerogative of the sovereigns was to constantly meet the demands of the temples. Inscriptions such as those of the so-called *Annals* of Thutmosis III (1458–1425 BCE) in Karnak (Grimal 2008–18), or in the *Great Harris Papyrus* under the reign of Ramses III (1186–1154 BCE) – the last great reign of the Nineteenth Dynasty – showed that huge amounts of products were imported by the Egyptians through either trade or plunder after wars (Grandet 1994).

The pharaohs of the Middle and New Kingdoms exerted economic pressure on Asia and Nubia. Secured by military garrisons, the domination by Egypt of large territories – excluding those of the Phoenicians who were their allies – ensured that levels of products in stock remained sufficiently high to enable state-controlled stores to meet the demands of temples. The loss of Egyptian influence in the Middle East caused an immediate decline in priestly wealth, a situation extensively reported in the literature, especially in the *Report of Wenamun*, a text written during the Twenty-First Dynasty (1069–945 BCE), under the reign of Smendes (1069–1043 BCE; Lefebvre 1949: 204–20; Goedicke 1975; Egberts 1991; Vandersleyen 2013). All this had a negative impact on the Egyptian economy as a whole, even on materials of symbolic or high technological value.

As Egypt had no ores or minerals that could be exploited, except in adjacent desert areas, trading with neighboring foreign countries became difficult. Egypt sometimes had profitable trade agreements and alliances with its neighbors. Observation of the scenes and reading of the texts engraved on the steles of the site of Sarabit el-Khadim, south of Sinai, show that at the end of the Twelfth Dynasty, under the reign of Amenemhat III (1842–1797 BCE), expeditions were organized under the aegis of both the pharaoh and Palestinian princes to access and process mineral resources (Staubli 1991: figs 16–18). There are also testimonies that nomads from Palestine imported galena mined in the Eastern Desert into the Nile valley (Goedicke 1984; Kessler 1987; Staubli 1991: 30–4, figs 15b–c; Kemp 2005: 319, fig. 112). Exchanges between Egypt and the peoples living in the desert were necessary and beneficial for everyone. In those times, collaboration in the desert was essential in order to survive.

In that respect, from the Old Kingdom onwards, the Phoenicians of the city-states of Byblos, Tire, Sidon, and Ugarit always maintained a close relationship with Egypt, which needed to import rot-resistant timber – cedar of Lebanon and Cilician pine and their derivatives (resins, pitch) – for the manufacture of ships, carpentry, and ordinary and funerary furniture (Killen et al. 2000; Hampson 2012: 61–102). Technological know-how, especially in the field of navigation, was also imported. Until the Late Period, these trading traditions, including the sharing of myths (Aufrère 2004b), were maintained with navigators and traders from these foreign countries. In the funerary Egyptian texts (*Coffin Texts*), Hathor, assimilated to the goddess of Byblos, Baalat, appeared as the seafaring goddess and a symbol of traditional commercial and religious links between Phoenicia and Egypt. The Palermo Stone, which is one of the most important Royal Annals of the Old Kingdom, records during the reign of Snefru (2575–2550 BCE) the arrival of forty ships loaded with cedar from Byblos and shows the building of ships and palace gates with Lebanon cedar or Cilician pine. Later, echoing the contemporary economic situation during the Twenty-First Dynasty, the already-cited *Report of Wenamun* (2: 40–43) relates the dispatching of a man named Wenamun by the clergy of Amon of Thebes, with a mission to fetch timbers to rebuild the Sacred Bark of the god Amon of Karnak. This literary text provides realistic information. It describes a sea journey that took Wenamun to Phoenician ports on the Syrian–Palestinian coast – Byblos, Dor, Sidon, and Tire – and even Cyprus. It also relates the many commercial transactions he carried out and the means of payment he used, such as different types of manufactured objects: pieces of high-quality fabrics, linen from Upper Egypt, large numbers of cables (probably made with papyrus), oxhides, sacks of lentils, fish, rolls of high-quality papyrus (Wenamun, 2: 40–43), and possibly also species of woods such as pine (*ʿāsh*), fir, cypress, and Phoenician juniper (*uʿān*; Bardinet 2017: 255–61).

As they had mastered the art of offshore navigation in the Mediterranean basin, the Phoenicians were consequently the most important traders of oriental products transported by caravans from the Far East to the ports of Phoenicia, including metals (Pons Mellado 2005), minerals, oxhides, and processed products in exchange for Egyptian raw or manufactured goods. Papyrus for writing purposes, aromatic gums, and medicines were made known to the wider world by the Phoenicians, who spread them in a largely regionalized economy for centuries. In the Eighteenth Dynasty, the Letters of Amarna testified to the exchange of diplomatic gifts between Pharaoh and their allies, the kings of the Syrian–Palestinian coast. Metals and high-quality manufactured products sent by the Syrian princes were exchanged for gold objects from Nubia sent by the pharaoh (Moran 1992).

Nonetheless, in a world of fierce competition, countries holding technical knowledge on which they had built their reputation prevented the disclosure of their manufacturing secrets outside the circles of their customary users. However, the countries of the Middle East made sure that their economic requirements were satisfied by transferring some of their technical know-how. Skilled Syrian craftsmen were sought, some of them taken as war prisoners during Egyptian military conquests. This applied to craftsmen competent in bronze and iron metallurgy, currying, carpentry, and utilizing animal by-products, who were particularly sought for the manufacture of better weapons. From the Middle Kingdom onwards, Egypt extended its domination in Asia. As shown in the *Annals* of Amenemhat II (1928–1895 BCE) written on the walls of the temple of Memphis, metallurgists were among other war prisoners (Altenmüller 2015). In the Middle Kingdom and especially in the second part of the Twelfth Dynasty, a high technical level of bronze smelting was developed. Glass manufacture, initially developed by the Syrians, became available in Egypt under the reign of Thutmosis III (1458–1425 BCE). Following many military campaigns in Asia and Nubia, specialized workforces and particularly master glassmakers were compelled to migrate to Egypt. Thus, this technology flourished particularly under the reign of Amenhotep III (1411–1352 BCE), when the dominance of Egypt reached its peak (Cline and O'Connor 2006: 314; Nicholson 2006). Textile technology also came from abroad (Cline and O'Connor 2006: 314–15).

Technologies originating from the period of the New Kingdom were perfected in Egypt, thereby increasing its technological capacity in order to maintain a dominant position. Smiths specialized in the smelting of certain metals and the manufacture of weapons – bows, arrows, and javelins, as well as war chariots. An idea of the level of equipment required to carry out a military expedition into Syria (*Kharu*) is described in the *Papyrus Koller*, of Memphite origin. In this document, the author describes wooden chariots, fully equipped with their weaponry: several bows, eighty arrows in the quiver, a spear, a sword, a dagger, a whip and spare strips, and a javelin from Hatti (Gardiner 1911: 36*–8*).

Egypt was the main exporter of natron (soda) to the Mediterranean shores, but also to the Fertile Crescent, as ensured by the contents of the cargos of Greek ships mentioned in the Thonis customs slip discovered in an Elephantine papyrus (see above). These large quantities of natron were either intended for food preservation or for the glass and earthenware industries (Briant and Descat 1998: 95). The word natron/niter (Greek *nitron*, *litron*, Arabic *natrūn*) is of Egyptian origin (*neter*). Another substance, alum (Egyptian *ibenu*), was also highly sought after. Cuneiform documents of the neo-Babylonian period indicate that this chemical substance, used as a mordant for dyeing, came from the Nile valley and was exported in large quantities: "Egyptian alum in sacks (233 mines)" (Briant and Descat 1998: 96–7). It is possible that Egypt exported bitumen from Gebel el-Zeit (*Mons Petrolius* of the Romans) on the western shore of the Red Sea, but most of it came from the Dead Sea basin during antiquity (see Chapter 3). Sulfur (*keperet*) from Gebel Kibrit, also on the Red

Sea coast, was used for the treatment of skin diseases and amulet casting, and it may also have been exported. The technology of casting sulfur was unknown before the Saite Period (Keimer 1932). Egyptian blue, synthesized from the Fourth Dynasty onwards (see Chapter 2) and improved during the Eighteenth Dynasty, was also exported – it is found throughout the Mediterranean basin, including Pompeii (Bower 2005).

Drugs were among the substances exported from Egypt, and Egyptian medical expertise was esteemed abroad. Homer claimed that medicine was the dominant art among the Egyptians. Egyptian physicians had precedence over those of other peoples. According to Diodorus Siculus (*Bibliotheca historica*, I 82), the Greeks believed that the physicians of the Nile valley had to strictly follow the prescriptions of the so-called *Sacred Book* to prevent the death of their patients (Aufrère 2001c). Herodotus (*Hist*. II 84) listed and prioritized the specialties in which Egyptian physicians excelled – eyes, head, teeth, and abdominal diseases – confirmed by the Egyptian texts. It was therefore understandable that these physicians were much in demand throughout the Middle East until eventually Greek physicians began to prevail over their counterparts in the Nile valley. The so-called tale of Princess of Bakhtan evoked the dispatching of an Egyptian magician–physician by Ramesses to treat his sister-in-law in Bakhtan (Bactria; Lefebvre 1949: 221–5; Broze 1989; Dunand 2006). Cambyses II and his successor Darius I had a famous Egyptian physician from Sais, called Udjahorresnet, who reformed the House of Life in his hometown (Posener 1936: 1–26; Lloyd 1982; Verner 1989). The fame of Egyptian physicians was shared by their rich pharmacopoeia used throughout the Mediterranean basin, with some products, such as kyphi, remaining highly regarded in the treatment of asthma (Derchain 1976; Lüchtrath 1999; Aufrère 2005a: 248–51).

According to a tradition that is echoed in Greek novels – especially the *Aethiopica* (3: 16) of Heliodorus of Emesa and *Leucippe and Clitophon* (4: 15–17) of Achilles Tatius – Egypt was a country of thieves, mystery, magicians, and poisoners. But the drugs in the Egyptian pharmacopoeia were considered to be effective, and Egypt was famous for its production of *pharmaka* – simple drugs or poisons – including a great variety of medicinal plants with curative virtues exported to the Mediterranean and the Near East. Magicians had to follow prescribed religious rituals (*Leucippe and Clitophon* 4: 17) both when gathering (Delatte 1936; Aufrère 2001a) and when administering concoctions of plants.

The nature and contents of economic trends even influenced local legends. Homer (*Odyssey* 4: 219–32) relates a legendary fact about Egyptian physicians, especially the remedies available in Egypt and notably from the city of Thebes, where the "drug of forgetfulness" was prepared. According to Aelian (*NA* 9:21), Polydamna, wife of King Thon – an eponymous character of the harbor of Thonis-Heracleion, recently discovered by Franck Goddio (2006) during his underwater explorations – gave Helen a herb normally used to repulse snakes, and sent her to a safe place in Pharos Island in order to protect her from her own husband's sexual desires. He adds that this herb, replanted by Helen, was named *Helenion*. In this etiological legend, King Thon and Queen Polydamna ("she who controls everything") form a couple who represent a disturbing, ambiguous, powerful, and dangerous Egypt, specialized in all kinds of philters, welcoming Menelaus and his wife Helen, who embody an archaic Greece, dependent on the lower Nile valley for the production of medications and drugs. Helen appears, in another context, in association with the fight against a dangerous Egyptian snake, an episode that still links her to drugs and venom. Nicander of Colophon (*Theriaca* 309–19) related that Canopus, the pilot who led Menelaus and Helen to Egypt on their return from Troy, was cruelly bitten by a *Hemorrous* snake at Thonis and died (Amigues 1990; Aufrère 2012b; Marganne and Aufrère 2014: 396–7). In this story, Helen, who we know to be in possession of the Egyptian pharmacopeia of Polydamna according to Homer, knocked a vertebra out of the snake's body with a heel blow, thus explaining the lateral movement of vipers. Through her, Greece not only receives medicines but also the art of treating snakebites and fighting snakes. It is reasonable to assume that the *pharmaka* were the center of an important trade in the Egypto-Mediterranean port of Thonis-Heracleion. Diodorus Siculus (I 97,5) confirmed that Thonis (a port open to the Mediterranean) was very closely associated with Thebes (the historic capital). The Greeks were probably not the only people who inherited this knowledge in a legendary form. The Phoenicians had long been allies of Egypt and transported all the traditional Egyptian products such as papyrus and *pharmaka* (Aufrère 1998a: 74–7). Aromatic products transiting through the Nile valley or arriving directly into Phoenicia from the western coasts of Arabia through the desert of Petra Arabia were called *ponikijo* (Phoenician products) by the Greeks (Lipinski 1992: 43; Aufrère 1998a: 74).

MESOPOTAMIA

Cale Johnson

Even if trade was not the primary engine of economic life in Mesopotamia, it was ever-present; there is clear textual evidence for the peaceful exchange of goods, to the northwest, to the east, and to the south, from the middle of the third millennium BCE. Particularly in the southern Mesopotamian alluvium, there are no significant deposits of metals or minerals or high-quality wood that could be directly harvested and processed, so the Mesopotamians focused on the production of high-quality textiles for trade or exchange. The fiber revolution at the beginning of the third millennium BCE laid the essential groundwork for the development of expertise in the textile arts, including the dyeing, weaving, and finishing of textiles and garments, over the millennia

(McCorriston 1997). The intense labor involved in this type of work prompted the emergence of large-scale workshops in which enslaved women and children toiled in the production of these items, resulting in an administrative framework that calculated the expected production of textiles per worker per day with sometimes harsh incentives to meet quotas. Throughout their history, from the shepherding and shearing of sheep to the painstaking production of elaborate textiles for export, the Mesopotamians traded vast amounts of labor and skill for exotic raw materials (Adams 1974; Yoffee 1981; Moorey 1994: 5).

The primary limiting factor in these trade networks was the difficulty of moving bulky items overland, so throughout Mesopotamian history waterways provided the key avenues for the movement of goods, both within Mesopotamia proper and beyond, while overland routes were limited to high-value goods such as tin, precious metals, or resins. Although there are documented examples of staple grains being shipped outside of the Mesopotamian alluvium, for example to the Persian Gulf (Leemans 1960: 20–2; Edens 1992: 127, references ITT 2 776 and UET III 1666), for the most part staple crops and domesticated livestock were only moved by boat within the dense network of waterways and canals that crisscrossed southern Mesopotamia. We have an extensive set of records from the Ur III period (ca. 2112–2004 BCE) for the movement of bulk and/or low-value goods between different urban centers in southern Mesopotamia, records that even allow the distance between cities to be approximated on the basis of the travel times recorded in the texts.

This internal trade was always a central feature of Mesopotamian life, but the degree to which Mesopotamian traders and merchants went beyond their borders in order to engage in trade varied dramatically in different periods. At one high point, in the Old Akkadian period, literary sources tell us that merchant ships from Meluhha docked at Akkad, presumably at the approximate latitude of present-day Baghdad, even if the precise location of Akkad remains unknown, while a couple of centuries later in the Ur III period, southern Mesopotamian merchants were traveling to Magan and Meluhha, which almost certainly can be identified as Oman and the Indus valley. In periods when the imperial infrastructure was weaker, however, Mesopotamian merchants seem to have been less adventurous: in the Old Babylonian period, for example, they do not seem to have gone any further than Bahrain (for a detailed overview of the circulation of raw materials, see Potts 2007 and Potts 2017).

Alongside the physical infrastructure needed for the production of trade goods and their movement within and beyond Mesopotamia proper, the Mesopotamians also excelled at the development of managerial and bookkeeping techniques for quantifying the raw materials and labor involved in the production of finished goods. Although there is relatively little textual evidence for long-distance trade in the earliest proto-cuneiform records (ca. 3300 BCE), the culture of calculation and oversight that came into existence more than five millennia ago was continually enriched and expanded, so that by the mid-third millennium BCE, in places like Early Dynastic Girsu, in southern Mesopotamia, or Ebla in present-day Syria, thousands of records, documenting the movement of metals and textiles, allow us to quantify trade in a way that is extremely difficult in most other times and places in antiquity (Maekawa 1980; Biga 2010; Biga 2014; Sallaberger 2014; Archi 2017). In the Ur III period institutional oversight reaches its peak: all manner of materials and labor were quantified in terms of their silver equivalencies, and this nearly universal commensurability is best seen in the accounts of institutional trading agents, responsible for exchanging domestic surpluses for exotic materials and products on behalf of the major institutions (Englund 1991; Englund 1992; Englund 2012). A few centuries later, in the early second millennium BCE, however, new legal instruments, such as the *naruqqu* partnership, come into existence, allowing groups of investors to combine their assets and fund long-term trading networks, such as the famous trade colony at Karum Kanesh in Anatolia (Dercksen 1999; Albayrak 2010).

The movement of small amounts of obsidian and lapis lazuli seem to have been central to the establishment of trade networks in the earliest prehistorical periods (Herrmann 1968; Watkins 2005; Watkins 2008), but it is only in the course of the third millennium BCE that the textual record provides us with clear evidence of the mechanisms involved in trade. Here, therefore, it is important to distinguish between the mere presence of foreign materials, such as metals, stones, and resins, and solid evidence for the pathways these materials followed on the way to Mesopotamia. Proto-cuneiform texts include a wide variety of materials and stones that were not native to the Mesopotamian alluvium (Englund 2006), but other than the use of the KUR sign and a sign corresponding to later DILMUN (see below) there are no textual indications of the sources of these materials. The centrality of lapis lazuli, as the prototypical foreign material in Mesopotamian thought, is highlighted by its orthography: the later Sumerian term is za.gin$_3$ (written ZA.KUR), but the same orthography is already found in the proto-cuneiform sources and clearly corresponds to the "stone" (ZA) from the "mountains" (KUR). In contrast, the examples of DILMUN in proto-cuneiform clearly point to material coming through the Persian Gulf (Englund 1983: Potts 1986; Potts 2007: 126–7).

Both archaeological and textual sources, however, make it clear that trade routes in the earliest periods existed to the northwest (largely following the Euphrates), to the east (centering on the Great Khorasan Road), and to the south via the Persian Gulf (Postgate 1994: 206–22 offers a particularly compelling overview of the early periods; Faist 2001: 194–237 does likewise for the later periods). The well-known routes to Arabia and Yemen, through which gold and incense moved, appear to have come into existence only at the beginning of the first millennium BCE, owing to the introduction of the camel as a beast of burden (Potts 2007: 135; for early caravans into Suhu, see Na'aman 2007).

The route to the northwest largely followed the Euphrates up into Syria and the southern border of present-day Turkey, but Syria and Anatolia could also be reached overland through routes that began on the Tigris at places like Assur and Nineveh and continued across the Habur and Balikh plains toward Carchemish. This northwestern route mirrored the southeastern route down into the Persian Gulf, as is seen already in an inscription from Sargon of Akkad that lists the major trade centers that he conquered from the southeast to the northwest: Meluhha (= Indus river valley), Magan (= Oman), Dilmun (= Bahrain), Mari (on the middle Euphrates near Deir ez-Zor), Yarmuti and Ebla in Syria, and onward to the Cedar Forest and the Silver Mountain (Hirsch 1963: 37–8; "Sargon b 2" = Frayne 1993: 27–9; Moorey 1994: 8; Steinkeller 2016: 128).

Whatever the precise geographical locale of the last few sites, this line running from the Persian Gulf to Anatolia was the primary axis along which the Mesopotamian world was ordered, both conceptually and architecturally. Sumerian tales about Gilgamesh speak of retrieving long timbers for the doors of temples and palaces from the Lebanon mountains, and, as noted in Chapter 5, large amounts of gold and silver moved from Anatolia down into southern Mesopotamia in the second half of the third millennium BCE. The other end of this transect – extending from the Euphrates through Bahrain, Oman, and the Indus valley – was much closer and more easily accessed, so it is little wonder that a wider range of goods were traded in and through the gulf; as Edens puts it, "the commodities of the Gulf trade included metals, textiles, semiprecious stones, ivory, woods and reeds, cereals, alliaceous vegetables and other condiments, oils, unguents, resins, shells, and possibly pearls, and a small array of finished products of wood, metal or stone" (1992: 122).

The Syrian Desert was largely impassable until the beginning of the second millennium BCE (and likely not regularly inhabited until the first millennium BCE), so western routes were forced far to the north, running parallel or coalescing with overland routes into Syria and Anatolia. In contrast, the eastern route along what is now known as the Great Khorasan Road starts out near present-day Baghdad, where the Tigris and Euphrates are at their closest, and passes through Kermanshah, Hamadan, Tehran, and Meshed (Moorey 1994: 8). The proximity of the major rivers in this region, including the Diyala, to this overland route meant that this "eastern front" was often militarized at various points in Mesopotamian history. During the Ur III period, for example, dynastic marriages maintained peace with Mari on the Middle Euphrates to the west, while military campaigns and even occasionally fortified outposts were used to control the flow of people and goods out of Iran and northern Iraq and into southern Iraq (Michalowski 2005: 204–6; Michalowski 2011). It was this complex military and colonial situation that served as the background for the Uruk Cycle, which was composed in Sumerian in the Ur III period. One of the central motifs in this series of epic tales is the competition between Uruk, in southern Mesopotamia, and the land of Aratta, presumably on the other side of the Iranian plateau in or around present-day Afghanistan. Uruk and Aratta found themselves in military and technological conflict, and in the subsequent Lugalbanda epics, the eponymous hero proves his legitimacy, as a future king of Uruk, by crossing the seven mountains that separate Uruk and Aratta alone, so as to carry a message back to Inanna in Uruk (Vanstiphout 2003; Mittermayer 2009).

Already in the Gudea Cylinders, which are roughly contemporary or perhaps slightly earlier than the Ur III period, when the Uruk Cycle was composed, we first see a kind of global repertoire of named mountains and the raw materials that could be extracted from them.

FIGURE 6.1 One of the Gudea Cylinders, ca. 2140–2124 BCE. Photograph by DEA/G. DAGLI ORTI/De Agostini via Getty Images.

In contrast to the unacceptability of Naram-Sîn's recycling of building materials in his reconstruction of the Ekur Temple in Nippur, Gudea correctly sources all his materials from virgin sources that extend throughout the length and breadth of the known world (Johnson 2014):

> Gudea ... took cedars (erin) and boxwood (taskarin) from the Amanus and other timber (za.ba.lum, u_3.suh$_5$ = probably fir (abies), dal.bu.um = plane) from the Commagene (region of *Uršu*). Stones were brought from, among other places, the mountains of Amurru, i.e., the mountains in the regions west of the Euphrates. Copper came from the mountains of Kimaš, and from Meluhha (kur me.luh.ha.ta) came *ušû*-wood (probably ebony in this instance). Gold was obtained from a mountain (hur.sag), named Ha.hu.um, and from Meluhha. Haluppu wood (^{giš}ha.lu$_2$.ub$_2$) came from around a place named Gu.pi$_2$.in^{ki}, evidently identical with Ku.pi.in in the Lipšur Litanies. Asphalt and gypsum came by boat from a mountainous region named ma.ad.ga^{ki}, na.lu.a stones were brought by boat from the Bar.sib mountains, and ^{na4}esi "diorite" came from Magan.
>
> (Leemans 1960: 12, Sumerian terms modernized)

The Lipšur Litanies, which date to the early first millennium BCE, and other lexical sources preserve many of these same equations between exotic raw materials and geographical names (Reiner 1956; Bloch and Horowitz 2015). In all likelihood it was in the context of this type of global approach to the sourcing of raw materials – particularly with the temple or throne room seen as a microcosm of the state or the known universe (the classic example is described in Winter 1983) – that collections of refined raw materials were included in the foundation deposits of temples, as described earlier. Crucially, however, Gudea never mentions trade per se and the Uruk Cycle is largely framed in terms of a military or technological conflict for the sake of regional supremacy. There is no notion here of an open circulation of goods that benefits all participants.

With the purification and grading of precious metals like silver and gold, particularly in the second half of the third millennium BCE, its use as a general device for measuring value in the Ur III period was the logical next step. Although copper was used as a standard of value, to a limited extent, in the latter phases of the Early Dynastic period (ca. 2500–2400 BCE), the influx of silver, particularly under the Old Akkadian kings, led to its use as a standard of value throughout the subsequent half-millennium (Ouyang 2013; Steinkeller 2016: 131). The crown was responsible for establishing not only calendrics and units of measure, but also norms for production and payment; these norms were occasionally spelled out in legal codices or edicts (Renger 2002; Englund 2012: 431), but can be seen more generally in the silver (or barley) equivalencies that were used to calculate the relative value of both raw materials and processed goods, as well as the labor that was required in order to produce specific processed materials. Beyond simply establishing conventional rules of thumb for equating different types of materials ("a shekel of silver (ca. 8.33 g) fetch[es] 300 liters of barley, 30 liters of fish oil, 10 liters of clarified butter, or a healthy sheep"; Englund 2012: 427), equivalencies in silver, in particular the amount of a commodity or labor that could be purchased for one shekel of silver, emerged in this period as a general standard of value, and this in spite of the fact that coinage had not yet been invented. Englund has sketched out each step in the abstraction of value in a number of important papers; I will outline just a couple of these steps here (see Englund 1991; Englund 2012).

For any given quantity of raw material, the labor required to convert that material into a finished product or the value of the finished product itself could be converted into a commodity-based currency, and these equivalencies could then be used to set production quotas and evaluate the performance of a team of workers (and its foreman, who would be held responsible for shortfalls). For example:

1($šar_2$) sa gi	3600 reed bundles
še-bi 1(u) 2(aš) gur	its/their barley: 12 gur (= 3600 liters of barley)
	(SNAT 444, obv. 1–2; Englund 2012: 437)

Within a sector that did not involve external trade, the standard of value could be any convenient commodity: here, one liter of barley for each reed bundle. This led to a wide variety of such equivalencies ranging from "its barley" (Sum. še.bi), "its wool" (Sum. sig_2/siki.bi), or "its oil" (Sum. i_3.bi; viz. the primary items rationed to low-level workers) to far more abstract equations like "its silver" (Sum. ku_3.bi), "its weight" (Sum. ki.la_2.bi), or "its labor" (Sum. a_2.bi); Englund (2012: 435) identifies 4,000 barley equivalencies, 2,000 labor equivalencies, and more than 1,500 silver equivalencies in the Ur III textual record. When institutions in southern Mesopotamia wanted to exchange local goods for materials that they did not have, this task was assigned to a "trader" or "trade agent" (Sum. dam.gar_3), and a running account was kept of all local goods that he was provided with as well as the foreign goods that he was able to acquire. All these items, both the starting materials and the acquired goods, were converted into their silver equivalents; the trade agent was held responsible if the value of the incoming materials fell short. Much the same procedure was also applied to foremen in charge of teams of workers in various domains, yielding a wide variety of institutionally determined labor equivalents in terms of expected production: a healthy male worker was expected to produce 3 sieves or 240 bricks, cut 360–1,080 m^2 of camel thorn per day, or make four one-liter pottery vessels per day (Englund 2012: 449–50). We do not have these types of labor calculations for technical activities other than pottery production, but they do give a clear sense of the wide-ranging equivalencies and production forecasts that played an enormous role in Ur III institutional life.

After the Ur III state fragmented into a number of smaller kingdoms and distinct trade networks arose in the early second millennium BCE, the "trade agent" morphed into the "merchant" (Akk. *tamkāru*), now backed financially by new kinds of financial instruments in which private investors put up the capital for these ventures (see generally Dercksen 1999; Dercksen 2000: Stol 2004: 868–99; Jursa 2010). This type of joint venture agreement (only rarely with limitations on risk to certain parties, so not, strictly speaking, analogous to the modern limited liability corporation) emerged in several different trade networks within Mesopotamia in the early second millennium BCE, but the most interesting of these was the *naruqqu* (money)-"sack" partnerships that supported the trade between Assur and Karum Kanesh (ca. 1900 BCE). Most of these agreements were probably held in the "house of the city," in Assur, which provided oversight for these agreements; few have been recovered, but one example (Kayseri 313) reads as follows:

> In all: 30 minas of gold, the *naruqqum* of Amur-Ištar. Reckoned from the eponymate of Susaja he will conduct trade for twelve years. Of the profit he will enjoy (lit. "eat") one-third. He will be responsible (lit. "stand") for one-third. He who receives his money back before the completion of his term must take the silver at the exchange-rate 4:1 for gold and silver. He will not receive any of the profit.
> (transl. by Larsen 1977; Dercksen 1999: 93)

The term used for these investors was *ummiānum* "scholar" or "specialist," the same term that was used for the leading practitioners of any technical discipline from metalwork to scribalism. These merchants (and their investors) were entrepreneurs and businessmen, interested in maximizing their profits. And as Dercksen (2000: 141) has emphasized, even the crown participated alongside other investors in agreements like these.

The Amarna Age, which corresponds to the fourteenth and thirteenth centuries BCE, takes its name from ca. 380 clay tablets found in Amarna in Egypt; the standard editions are Moran (1992), Liverani (1998), and Rainey (2015). Alongside similar correspondence from Hattusha in present-day Turkey, these letters present us with the best evidence for international diplomacy and trade in the Late Bronze Age. Most of the correspondence depicts the interactions between vassal kingdoms and their overlord (Mynarova 2014), but approximately fifty letters from Amarna and a few dozen from Hattusha were sent between the great powers in the Eastern Mediterranean and Greater Mesopotamia in this period, a group that Liverani speaks of as the Club of Great Powers: Egypt, Babylonia, Mitanni (in northern Syria), Hatti (in central Anatolia), and Assyria

(Liverani 2000; Bryce 2003). The rulers of these nations addressed each other as "brother" and often wrote to each other as if they were petty nobles in a small town. But amid their squabbles about gifts and royal marriages, we also see some of the earliest evidence for international law: local rulers were responsible for the well-being of merchants passing through their territories and several features of present-day international law were in place, such as the use of passports and indemnification of losses suffered by traveling merchants (Westbrook 2000).

These rulers requested and often received elaborate gifts, made up of hundreds of objects crafted in gold, lapis lazuli, and other materials (Cochavi-Rainey 1999; Feldman 2006). Many of these gifts were linked to diplomatic marriages between these kingdoms, but others were to reestablish the link between two kingdoms when a new ruler had come to the throne, or simply in response to requests.

FIGURE 6.2 Pomegranate vase from the tomb of Tutankhamun, New Kingdom (silver or electrum). © Boltin Picture Library/Bridgeman Images.

The items in these lists were carefully described in Akkadian, although local names of finished products were also included, and the total amount of gold and silver in a group of objects was occasionally included. Here is a short section from EA 14, a list of goods sent from Egypt to Babylonia (col. 2, lines 41–54):

> 1 *kukkubu*-container, for [..., o]f silver, [al]ong with its cover. 3 s[mal]l measuring-vessels, of silver; *bumeris* (is) its name – 1 *haragabaš*, o[f silv]er – 1 pail, of silver – 1 sieve, of silver – 1 small *tallu*-jar, of silver, for a brazier – 1 "pomegranate," of silver – 1 (female) monkey, with its daughter on its lap, of silver – 1 oblong pot, for a brazier, of silver – 23 *kukkubu*-containers, of silver, full of "sweet-oil"; *namša* is its name – 6 *hubunnu*-containers, [and] 1 large *hubunnu*-container, also of silver. 1 upright chest, of silver, inlaid – 1 ladle, of silver, for an oil-container; *wadha* is its name.
>
> (Moran 1992: 30)

But alongside finished goods such as these, this small group of rulers also exchanged certain kinds of scholars and technical specialists; Zaccagnini's classic paper (1983; see also Edel 1976; Couto-Ferreira 2013) lists numerous physicians, conjurers, a haruspex, and even a sculptor sent between Egypt, Hatti, and Babylon. The best case study of this phenomenon is Heeßel's description of the life of the Babylonian physician Rabâ-ša-Marduk, from whom we have both a handful of administrative documents from the reign of Nazi-Maruttaš (reigned 1302–1277 BCE) and a Babylonian medical text (BAM 11). Rabâ-ša-Marduk was later sent to Hattusha during the reign of Muwattalli II (reigned 1290–1272 BCE), and ends up as a topic of discussion in KBo I 10+, a letter sent from Hattusili III to Kadašman-Enlil ca. 1255–1250 BCE (Heeßel 2009).

GRECO-ROMAN WORLD

Matteo Martelli

The distinction between workshops and shops (both referred to as *ergastēria*) is not clear-cut in ancient sources (see Chapter 2). Ores and minerals, for instance, were often processed in workshops next to the sites of extraction. However, it is difficult to reconstruct the various steps through which they reached ancient markets, let alone all the actors involved in the process (Photos-Jones 2018). In the fifth and early fourth centuries BCE, the mines in the Laurion region (southern Attica) produced silver on an industrial scale, at up to twenty tons a year (Rihll and Tucker 2002). The extraction activities were regulated by Athenian officials (*pōlētai*), who sold fixed-term leases to individual entrepreneurs in front of the Council of 500 (Aperghis 1998). As reported by the historian Xenophon

(*Vect.* 4.14–15), a force of hundreds of slaves was put to work in the mines by wealthy Athenian investors, such as Nicias (with 1,000 slaves), Hipponicus (600 slaves), or Philemonides (300 slaves). The contracts often mention workshops (*ergastēria*) located in the leased areas, in most cases establishments where ores were preliminarily ground, washed, and prepared for smelting. Only in a few cases are smelting furnaces named (Crosby 1953: 195). Smelting and cupellation, in fact, were difficult operations, which required skillful workers and furnaces reaching over 800°C. This specialized work needed consistent investments, and facilities were usually established near the mines to reduce the transport costs. The cost of transporting unprocessed ores, in fact, was certainly much higher than processing the ores *in situ*, thus leaving the dross behind (Acton 2014: 120–4).

In other cases, metallic ores – such as of lead or copper – were directly smelted by the smiths who intended to work the extracted metals. Smelting furnaces have been excavated in the area of the ancient Agora in Athens, next to the *ergastēria* of metalworkers. Classical sources refer to workshops selling specific types of metallic items, such as swords, knives, spears, and shields (Acton 2014: 124–46). In his first speech *Against Aphobus*, Demosthenes (383–322 BCE) mentions a large factory with thirty-two or thirty-three slaves that specialized in sword manufacture (*Or.* XXVII 9). Plutarch claims that Melon, in Thebes, armed those who joined his party by breaking into the workshops of spear-makers and sword-makers and stealing weapons (*Pel.* 12.1).

Mineral ores were extracted locally and sometimes exported; they were imported from abroad when not available in Greece. The Laurion region was rich in silver and lead, whereas Cyprus was a vital source for copper, while tin, for example, probably came from Iberia. Archaic Ionian colonies had been settled in regions rich in metals, such as the Black Sea area (Treister 1998), while in the Western Mediterranean ancient Greeks and Phoenicians were particularly active in the vicinity of sources of metal, such as Italy, Sardinia, and Iberia (Morley 2007: 24). This trade is confirmed by the shipwrecks found by archaeologists, which often contain mixed cargo with metal ingots (Treister 1996: 252–60). Archaeological evidence also documents the commerce of manufactured metallic items, including precious metals and their imitations (Kron 2016: 367–8).

Along with metals, mines provided a wide array of minerals used in different professional fields. The Laurion area produced red ocher, realgar, orpiment, malachite (*chrysocolla*), and litharge, a by-product of the cupellation of silver ores that could be used either as a pigment or as a medicine (Rihll 2001: 128–32). Alum, another important substance in metallurgy, dyeing, and medicine, came from various areas of the Mediterranean (Borgard et al. 2005); the renowned alums from Egypt and the island of Melos are already mentioned in the Hippocratic writings (Totelin 2016a: 154).

If the journey of these products from mines to the workshops of craftsmen is not completely clear, we can safely infer from the sources that many of these commodities were sold in the markets of ancient Greek cities. A long list of sellers is provided by the Athenian philosopher Critias (fifth century BCE), who names (fr. B70 DK = Pollux, *Onomasticon*, 7.196–7), among others, bronze-sellers (*chalkopōlai*), iron-sellers (*sidēropōlai*), emetic-sellers (*syrmaiopōlai*), wool-sellers (*eriopōlai*), frankincense-sellers (*libanōtopōlai*), root-sellers (*rhizopōlai*), and drug-sellers (*pharmakopōlai*). The last ones, often labeled as quacks or charlatans in ancient sources (see Chapter 8), could be either street peddlers or retailers of *pharmaka*, namely medicines and pigments (Samama 2006; Totelin 2016b). According to Aristophanes (*Nub.* 766–8), they also offered transparent stones used as glass lenses for igniting fires. Theophrastus names a few *pharmakopōlai* who sold poisons or antidotes, such as Eudemos from Chios, who invented an antidote made of vinegar and pumice-stone dust (*Hist. Pl.* IX 17.3). A wide array of "chemicals" was probably available in their retail shops, a choice that, in all likelihood, was not very different from what one reads in the later *P.Oxy* 31.2567 (third century CE): the papyrus records various items kept in the stock of the drug-seller Aurelius Neoptolemos in Oxyrhynchus, including alum, shoemakers' black (*melantēria*), and red ocher.

The Athenian marketplace accommodated sellers and bazaars. According to Theophrastus (*Lap.* VIII 53), the Greek painter Cydias was in a bazaar – a *pantopōlion* in Greek, lit. "a place where all sort of things are for sale" – when a fire broke out. He then noticed that red ocher, when burned, turns purple. Along with ocher, the bazaar probably sold a variety of items and foodstuffs, such as oils, dried or smoked fish, and cheese, all products mentioned in later papyri (e.g. *P.Lond.* 3.1159, second century CE). Cydias then used the newly discovered red/purple pigment in his paintings, such as the picture of Argonauts that was sold for 144000 sesterces to the orator Hortensius (Plin. *NH* XXXV 130). According to Theophrastus (*Lap.* VIII 52), the best red ocher came from the island of Kea (Photos-Jones and Hall 2011: 73–7), while other varieties were imported either from Lemnos or from Cappadocia through the port of Sinope (hence the name of *Sinōpis* given to red ocher). Discoveries of pigments are also described by Pliny in a long section focused on painting, in which he writes: "the very celebrated painters Polygnotus and Micon at Athens made black paint from the skins of grapes and called it grape-lees ink. Apelles invented the method of making black from burnt ivory; the Greek name for this is *elephantinon*" (*NH* XXXV 42; transl. by Rackham 1952: 293). In the same passage, Pliny also reveals that a black paint was prepared by dyers (*infectores*).

Textual and archaeological evidence points to the development of workshops specializing in the production of specific pigments under the Roman Empire. Vitruvius (*De arch.* VII 11) mentions a workshop producing Egyptian blue (*caeruleum*) in Pozzuoli, owned by a certain Vestorius, a friend of Cicero.

By following an Alexandrian procedure, he produced a pigment that was called *Vestorianum* after his name. The same information is reported by Pliny (*NH* XXXIII 161–2), who also mentions *Puteolanum* blue, from *Puteoli*, the Latin name of Pozzuoli, clearly an important center for the production of blue pigments. Archaeological excavations have confirmed the presence of workshops in the Phlegraen Fields near Pozzuoli (Cavassa et al. 2010), while pots still containing traces of pigments and colored powders have been found in Pompeii.

Perfume-sellers – *myropōlai* in Greek or *myrepsoi*, a word supposedly introduced by the abovementioned Critias (fr. B70 DK = Pollux, *Onomasticon*, 7.177) – were also involved in the trade of *pharmaka*: plants and minerals along with exotic (and pricy) commodities coming from Egypt, India, Arabia, or Persia (Samama 2006: 17). In classical Athens, perfumes, cosmetics, and aromatics were sold both by individual retailers and in shops where a handful of slaves was employed (Acton 2014: 239–46). The orator Hyperides (fourth century BCE) wrote a speech against Athenogenes, a dealer of Egyptian origins who owned three perfume businesses in the city, which were managed by a slave and his two sons. The trade was certainly lucrative, but competitive. In order to beat the competition, perfume-sellers used to scent wavering costumers with rose oil, a perfume that was supposed to destroy other odors; hence, customers could not appreciate and buy flagrances in shops belonging to other sellers (Theophr. *On Odors*, 45). Wealthy Athenians were used to spending time in these shops, which also served as social meeting spaces; usually they were located in upper rooms that did not face the sun and were shaded as much as possible (Theophr. *On Odors*, 40); perfumes, in fact, quickly deteriorate if kept in hot spaces or exposed to sunlight.

From the end of the first century BCE, the production of perfumes grew, as has been established from two recently excavated workshops in Delos and Paestum, which accommodated a considerable number of wedge presses that point to a small-scale industry (Brun 2000; see Chapter 3, pp. 105–6). Near-industrial levels were reached in Campania (Monteix and Brun 2009), which, according to Pliny, almost equaled Egypt in the production of unguents (*NH* XIII 26). The region produced the famous and expensive Italian rose oil (*rhodinon italikon*); Capua's varied market of aromatics and drugs, called *Seplasia*, gave the name of *seplasiarii* to the makers/traders of these substances in the Roman Empire (Korpela 1995: 102). Aromatic plants and flowers were farmed in Campania (e.g. in the gardens of Pompeii) as well as imported from distant regions; Galen, for instance, mentions perfume-dealers (*myropōlai*) who visited Crete every year to collect primary ingredients and sell them in Roman markets (Boudon-Millot 2003: 113). In Pompeii, archaeologists have discovered the remains of various perfume workshops situated along Via degli Augustali (Brun 2000: 291; Monteix and Brun 2009: 123–8, with further bibliography).

Alexander the Great's expeditions dramatically expanded the range of substances known to the Greeks (Flemming 2003: 457–61; Totelin 2016a). Alexander's conquests, consolidated by his successors the Hellenistic kings, opened new commercial routes, which were further expanded under the Roman Empire. Alexandria and Rome had flourishing "global" markets, where one could purchase commodities (drugs, pigments, dyes, perfumes, precious stones, and minerals) coming from any corner of the ancient world (Nutton 1988; Guardasole 2006: 35–9). The short treatise *Peryplus of the Erythraean Sea* (first century CE) records the sailing itineraries from Egyptian ports on the coast of the Red Sea (including the Horn of Africa) to Pakistan and southern India. It lists a great variety of imported and exported items. Costly imports from overseas ports to Roman Egypt included such commodities as turquoise, lapis lazuli, sapphires, ivory, aromatics, myrrh, various drugs (cinnabar, indigo, nard), and lac dye. Roman Egypt, on the other hand, exported, among other commodities, purple cloth, saffron, realgar, orpiment, and antimony (Huntingford 1980: 122–42; Casson 1989: 39–43).

This roaring trade also emerges from the earliest alchemical recipe books written in Egypt between the first and fourth centuries CE. The four books of Pseudo-Democritus and the Leiden and Stockholm papyri often specify the regions of origin of the substances used (see Table 6.1).

Many varieties of earth used in alchemical processes (e.g. Samian, Kimolian, and Chian earths) were well known for their medical properties, already mentioned in the Hippocratic writings and often described in pharmacological treatises on simple drugs (Totelin 2016a). Italy appears among the sources for antimony (or stibnite), widely used as makeup and as a medicine for eye diseases. Alum from Melos (often called *melinon*) was highly valued in antiquity (Photos-Jones 2018: 7–10), while different local varieties of cadmia were considered, coming from Cyprus, Thrace, and Galatia. Cyprian cadmia was praised as the best kind by Pliny the Elder (*NH* XXXIV 103) and Dioscorides (V 74). A century later, Galen provides us with a detailed account of his journey to Cyprus, which he decided to visit in person to collect top-quality minerals (*On the Capacities of Simple Drugs*, IX 3 = XII 208–44 Kühn). He could gather from the local copper mines the so-called *diphryges* (lit. "twice burned"), cadmia, *misy* (copper–iron ore), and other medicines, thus avoiding the risk of buying adulterated materials. He collected *molybdainē* (lead sulfide) in a village between Pergamon and Cyzicus, meaningfully called *Ergastēria* (lit. "workshops"), where there was a lead mine (XII 230,1–5 Kühn).

The many recipes on the making of precious stones point to the existence of a thriving market of cheap objects – a kind of *bijouterie* that probably met the demand of customers who could not afford the purchase of precious gems (on jewels and gems in Greco-Roman Egypt, see Russo 1999). The Leiden and Stockholm papyri, on the other hand, never mention glassblowing, whose

TABLE 6.1 Local specialties listed in Pseudo-Democritus (*PM* = *Physika kai mystika*; *AP* = *On the Making of Silver*; *Cat.* = *Catalogues*) and in the Leiden and Stockholm Papyri (*P.Leid* and *P.Holm*).

Achaea	Dyeing flower (*PM* § 2)
Attica	Ocher (*PM* § 7)
Cappadocia (modern Turkey)	Salt (*PM* § 13; *Cat.* 3; *P.Leid.* 2, 83; *P.Holm.* 1)
Chalcedon (modern Turkey)	Antimony/stibnite (*PM* § 11)
Chios	Earth (*PM* § 19; *AP* § 2; *Cat.* 3; *P.Leid.* 5, 12, 16, 79, 83, 84; *P.Holm.* 1)
Cilicia (modern Turkey)	Saffron (*PM* § 18, *AP* § 6; *Cat.* 2)
Cyprus	Cadmia (*PM* § 10)
Egypt	Tabasheer (*P.Holm.* 38); dyeing shell (*PM* § 2); acacia gum (*Cat.* 2)
Galatia (modern Turkey)	Cadmia (*P.Leid.* 15); copper (*P.Leid.* 13, 29; *P.Holm.* 3, 4, 7); kermes (*PM* § 2: "worm" from Galatia; *P.Holm.* 156).
India	Indigo (*P.Holm.* 47, 63, 84); quartz (*P.Holm.* 35)
Italy	Antimony/stibnite (*PM* § 5, 12; *Cat.* 3); madder (*PM* § 2); pomegranate (*PM* § 2)
Kimolos	Earth (*AP* § 6; *Cat.* 3; *P.Leid.* 65, 78; *P.Holm.* 12, 124)
Koptos (Egypt)	Antimony/stibnite (*PM* § 6)
Laodicea (in Phrygia)	Alkanet (*PM* § 2; *P.Leid.* 97)
Lybia	Purple (*PM* § 2)
Macedonia	Malachite (*PM* § 14; *P.Holm.* 20, 35)
Melos	Alum (*PM* § 5)
Oasis of Amon-Rē (Egypt)	Salt (*P.Leid.* 10, 83, 84)
Palestine	Resin (*P.Holm.* 62)
Paros	Earth (*PM* § 19)
Phrygia (modern Turkey)	Stone (*P.Leid.* 93; *P.Holm.* 101, 104, 134, 154)
Pontos	Alkanet (*P.Holm.* 61); rhubarb (*PM* § 17; *Cat.* 2); red ocher (*Cat.* 1)
Samos	Earth (*P.Leid.* 12, 18)
Sardinia	Purple (*P.Holm.* 98)
Scythia (Central Asia)	Black dye (*P.Holm.* 76)
Sicily	Purple (*P.Holm.* 99)
Sinope (modern Turkey)	Red ocher (*P.Leid.* 16, 86; *P.Holm.* 14, 117, 119)
Syria	Dyeing flower (*PM* § 2); kermes (*P.Holm.* 125); lac dye (*PM* § 2)
Thrace (between modern Bulgaria, Greece, and Turkey)	Cadmia (*P.Leid.* 15)
Tyros	Purple (*P.Holm.* 154, 157)

introduction radically transformed the traditional craft of glassmaking and its market. Glass, in fact, became an extremely important and widespread material, used for the production of a variety of objects, from windows to mirrors, from architectural decorations to mosaics, from cups to vessels, which included small containers for medicines and cosmetics as well as pieces of alchemical devices (Beretta 2009: 64–8, 109–23). The demand for these objects dramatically

FIGURE 6.3 Stele known as Del Purpurarius. Roman civilization. Parma, Museo Archeologico Nazionale (Archaeological Museum) Palazzo Della Pilotta. Photograph by De Agostini/Getty Images.

increased in the second century CE, when, according to Fleming (1999: 60), "glassworkers had to turn out close to 100 million items annually just to keep pace with current demand." This is an incredible figure, especially if one considers that, for technical reasons, ancient glass workshops could not accommodate a large number of craftsmen (Stern 1999: 454–6).

Along with metals and stones, the Greco-Egyptian alchemical recipe books mention and assess a variety of dyes coming from different places (from India to Sicily, from Scythia to Italy), which were used as substitutes for the much more expensive Tyrian purple (see Table 6.1). This dye was produced from the glands extracted from certain shellfish, most notably the *Murex* snail. The glands, diluted in honey, were probably traded long distance in little glass flasks, such as the bottles depicted in the stele of a Roman *purpurarius*, C. Pupius Amicus (Marzano 2013: 149) (Figure 6.3).

Tyrian purple, also referred to as royal purple, had an exalted status, and clothes dyed with this expensive ingredient became symbols of aristocratic rank or imperial power. Under the Roman Empire, their manufacture, as well as the sale of the precious dyestuff, was partially controlled by the central authority; their use became restricted solely to the emperor only in the fourth century CE (Marzano 2013: 149–51).

The many substitutes listed by the alchemical sources point to the use of less expensive dyes in this lucrative industry in order to meet the high demand for purple textiles that prosperous customers wanted to ostentatiously display as signs of their affluence. Papyrological and archaeological evidence confirm this rich market in Greco-Roman Egypt (Martelli 2014c, with further bibliography). Here, dyers worked in specific workshops, which could be either independent buildings (equipped with all the necessary facilities) or a part of the house, as it is possible to infer from some contracts for leasing or selling *ergastēria* (e.g. *P.Oxy.* XIV 1648, second century CE). Some instructive examples are offered by the archive of Apollonios, a *strategos* (ca 113–120 CE) of the Apollonites Heptakomia (Upper Egypt). His family belonged to the upper stratum of Egyptian Greeks and administered a large weaving enterprise that involved many spinners and weavers, who worked under the supervision of Apollonios' sister, Aline, and his mother, Eudaimonis. They also supervised the purchase of expensive dyes, even though it is not clear whether the dyeing of fabrics was carried out within the weaving enterprise: perhaps dyestuffs were just bought by Apollonios' family and then delivered to the workshops of specialized craftsmen. Otherwise, we cannot exclude the possibility that professional dyers were temporarily hired by Apollonios to work in the workshops of his enterprise (Martelli 2014c: 118–20).

On the other hand, a tighter relationship between workshops specialized in the different phases of the textile industry has been detected in Roman workshops excavated in Pompeii. In particular, as recently pointed out by Flohr (2013b), dyeing workshops have been identified in houses that also accommodated either felt-making workshops or *fullonicae*. These houses probably belonged to investors who concentrated their efforts on the textile industry and "were expecting a relatively constant and substantial flow of orders. In other words, these workshops were not built to deal with private consumer demand .... Their scale, as well as their economic orientation, suggests some sort of involvement in supra-local networks of trade and exchange" (Flohr 2013b: 73).

CONCLUSIONS
Marco Beretta

In ancient civilization, trade was a necessity that gradually became a source of cultural exchange. In Egypt, civilization developed around the Nile valley, a narrow line of fertile land that stretched from the Mediterranean to the heart of the African continent, which brought its inhabitants into contact with several different traders, from the Phoenicians to the Nubians. Following military campaigns in the Near East and in Asia during the Twelfth Dynasty, specialized artisans migrated to Egypt and contributed to creating new manufactures of luxury commodities (such as glass imitations of precious stones) that were previously imported. Conversely, the celebrity of Egyptian medical remedies, highlighted by Homer and by several Greek medical sources, favored the trade of the most popular products, such as kyphi, and also, at a later stage, the emigration of physicians around the Mediterranean.

In Mesopotamia, trade was triggered by other factors. Thanks to the remarkable progress made in the textile arts, Mesopotamians established a flourishing trade with neighboring countries as early as the third millennium BCE. Metals, obsidians, lapis lazuli, oils, unguents, and textiles became the favorite commodities of an increasingly rich exchange with the regions of Anatolia, Syria, and Egypt. During the Greco-Roman period, trade remained an essential feature of exchange, although the rapid development of urban centers created rich markets, increasing demand for goods and leading to the rapid diversification of products. Specialized traders of iron, bronze, emetics, perfumes, drugs, roots, pigments, and stones suddenly appear in Greek literary sources, conquering a scene that highlighted their growing socioeconomic importance. The unprecedented expansion of the Roman territories during the Empire enhanced through the introduction of laws and edicts the roles of both trade and craftsmen.

CHAPTER SEVEN

Learning and Institutions: *The Invention of Chemical Recipes*

SYDNEY H. AUFRÈRE, CALE JOHNSON,
MATTEO MARTELLI, AND MARCO BERETTA

EGYPT

Sydney H. Aufrère

It appears that Egyptian craftsmanship had mastered specific procedures for the preparation of materials related to the sacred in order to always produce the same consistent product. However, unlike for Mesopotamia, Egypt left few traces of traditional recipes of liturgical perfumes requiring a long time to prepare (Bottero 2002; Plouvier 2010). Some recipes preserved on the walls of temples (Edfu, Athribis, Kom Ombo, Philae, and Dendara) give some details on the manufacturing processes used by the perfumers, the ingredients, and the order in which they were added. Inscriptions on the walls of Edfu's so-called laboratory (cf. Chapter 1) list the recipes for what was described as "finest ointment of styrax," "the *hekenu* ointment," "the *medjes* ointment," "the nine ointments of the Ennead," "the precious ointment," "the ointment of divine mineral," "the *kyphi*," and the "pure natron pellets of the Ennead"; they also describe the different varieties of frankincense and styrax (Aufrère 1998c). Besides these recipes, other recently published medico-magical texts give details of some operations, thereby making it possible to reconstruct how the embalmer prepared mummies in the workshop.

The principles used in the preparation of Egyptian recipes were highly technical. The study of recipes engraved on the walls of Edfu's laboratory shows that Egyptians used three types of preparations: fumigations, fat-based cooked ointments, and resinous ointments for the divine wood or stone statues to make them look like the representations that the Egyptians had of the gods. Indeed, as written in the "Memphite theology" (Twenty-Fifth Dynasty; Traunecker 2004), "This is how the gods entered their bodies, a mixture of all kinds of wood, of stone, of all types of faience, of all things which emerge from him and through which they manifest themselves." What follows here describes how perfumes were generally manufactured, even if the preparations do not correspond exactly to what we would consider to be perfumes.

Some recipes such as that for the preparation of the "precious mineral ointment" were probably for local use. It is said that this ointment had once been used in Coptos to anoint the statues of Min-Amon. It contained a mixture of precious metals, minerals, bitumen, resins, and oils, products imported from foreign countries like Punt or God's Land – in which the god Min, acting like a Bedouin, exploring the valleys of the Eastern Desert to the Red Sea in search of resources, was spreading his influence as "mineral prospector and collector of oleoresins." Over time, this ointment had come to be used throughout Egypt to embellish the divine status of the gods' statues (Aufrère 2005a: 242–6). However, the technical precision of these recipes, which described exactly how to follow them, made it possible to perpetuate the craftsmanship of perfumers. The *modus operandi* seemed to be always the same. The recipes first mentioned the name of the product and then specified the amount to be produced. This standard quantity was the *hin*, a unit of volume for liquids, namely 0.48 l. Why and for whom it was specifically prepared were then given. Then followed the chronology and the proportion of each ingredient to be progressively added. A recent study revealed the existence of seventy-one basic ingredients used in all these recipes (Aufrère 1998c: 60–2).

These operations were described with great precision. For example, the recipe for preparing a *hin* of *hekenu* ointment – principally made of premium frankincense oil mixed with styrax for divine anointing (Aufrère 2005a: 224–35) – required nine ingredients. Among these, there were two different qualities of frankincense and "the excellent wine of the Oasis." The latter contained one key substance that appeared several times, namely alcohol. It indicated a specific process, since oleoresins are soluble only in alcohol. The perfume was obtained after successive and long operations. Some ingredients required individual processing, such as extracting the flesh of the fruit of the *nedjem* tree using a twisted cloth and then cooking it in a cauldron, not to mention the nature of the fuel used (the latter is changed during the preparation as the ingredients were added) for the processing. Other ingredients are added during the following operations, at specific times of the lunar calendar (e.g. the new moon). The cooking time required was specified. Before cooking, the mixture was weighed, and then regularly reweighed during cooking. If too much weight was lost, some liquid (water or wine) was to be added very gradually. The number of operations required to prepare this ointment – probably one of the most precise recipes of Edfu's laboratory – amounted to around forty-two.

Time was therefore a fundamental parameter. The preparation of the *hekenu* ointment took between 487 and 492 days (Aufrère 2005a: 233), and that of the highest-grade extract of styrax required 303 days (Aufrère 2005a: 37). This extremely long preparation time gave the products greatly added value. As mentioned previously, the use had to be specified, because certain substances had to be used either fresh or dry. Chronologically, the material was first crushed then filtered through a sieve or a cloth to extract the vegetal debris, since perfumes were made with fresh or dry aromatics. Some preparations had to be kept warm during the procedure using a bain-marie. The vocabulary in perfume-making was always technical. It had nothing to do with the vocabulary used in religious or medical texts, which, from time to time, added a magico-religious formula when applying or absorbing a drug.

In general, the rule, even going back to the distant past, was to abide by the *modus operandi* of the preparation, presentation, and use that was declared in the Egyptian pharmacopoeia. Thus, the preparation of liturgical perfumes or recommended remedies always specified the ingredients, their quantities, how the medication was to be made, whether the ingredients had to be crushed or finely ground, how they were to be cooked, whether they were to be reduced, filtered, or not, and how they were to be administered. These products were sometimes used for fumigation.

Traditional recipes were more or less precise. The recipe for the divine mineral ointment given in Edfu's laboratory was purely technical, giving with precision the phases of preparation of the perfume. By contrast, the preparation of Papyrus Salt 825 evoked a fairly concise method of a similar perfume (see Chapter 1, p. 31), ascribing an etiological legend to it and assigning a symbolic value to each component. Thus, the use of brief legends (*historiolae*) became associated with technical know-how. However, depending on the status of the text, the information was sometimes delivered only little by little. Thus, the ingredients to prepare inks – amalgams of metals, oleoresins (myrrh, frankincense), plant or mineral products, and animal by-products of magical or symbolic value (Aufrère 2001b) – were listed in the magical texts, but without explaining in what order they had to be added, triturated or not, and whether heating was required or not.

Two aspects should be highlighted here: the need to keep the tradition "alive" by copying the texts and the need not to disclose the secrets they contain to the uninitiated. The continual copying of the ancient texts ensured their preservation. The Ethiopian king Shabaka (716–702 BCE) of the Twenty-Fifth

Dynasty ordered that a document written by elders and known as "Memphite Theology" be copied, as it was partially eaten by worms. The entire text was gradually reconstituted and engraved on a hard-stone support. This was considered a work of piety that allowed the sovereign to keep his name alive for eternity. It is important to note that there are several examples of texts copied despite gaps in the original. When a gap was observed, the scribe then wrote on the papyrus "gap found" (*gem ush*; i.e. text was missing) and continued copying. Elders, like the priest dresser (stolist) of Min at Panopolis (Akhmîm), Petarbeskhenis (Derchain 1987), a contemporary of Emperor Hadrian, encouraged young people to copy the ancient texts and to draw inspiration from them.

It was fundamental not to divulge the contents of sacred texts, because keeping them "secret" (Traunecker 1989: 108–10) gave added prestige to the liturgical or technical knowledge that was considered to be esoteric. The notion of "secret work" was frequently highlighted. According to the texts, it was important that knowledge be accessible only to authorized personnel, otherwise the processes involved would lose their mystery and hence their effectiveness. The engraving on the funeral stele of the artist Irtysen (contemporary with the Eleventh Dynasty, ca. 2160 BCE) reads that the mastery of his art be entrusted to his eldest son only, whom he considered as an initiate and therefore competent (Delamare 2007: 21; Stauder 2018a: 249). Secrecy veiled all complex preparations requiring a particular process. In the medical papyrus of the Louvre inv. no E 32847 is written the following recommendation: "No eye sees, no ear hears about (the composition of) your ointment, except that it protects the patient and the exorcist. The magic formula used by the embalmer priest who masters his art remains a well-kept secret" (Bardinet 2017: 223–4).

In yet other documents, instructions suggested that a magical compilation (Cairo, no. CGC 58027) dating from the Ptolemaic period meant to protect the royal chamber and relating to the composition of perfumes could not be seen by "any individual except by the king himself, the magician ritualist and perfumer in the House-of-Life" (Colin 2003: 95–6). One of the recipes kept in Edfu's laboratory explicitly says the following about the divine recipe mineral ointment: "It is a secret that no one should see or hear, but which the old man passes on to his child" (Aufrère 2005a: 243). Moreover, the recipes of sacred perfumes reproduced in Edfu's laboratory were supposed inaccessible to those who could not enter the sanctuary. As mentioned, the sacred library of Edfu cited a book entitled "List of all secret recipes of the laboratory" (*Edfou* III 348, 1; Aufrère 2005a: 218–19). The death penalty could be imposed on anyone who did not honor this secrecy. Thus, the Papyrus Salt 825 (XVIII 1) said: "Whoever would disclose this, he would die by murder, for it is a great mystery." Nevertheless, some interdictions simply meant that some operations had to be done in private. "That this be neither seen nor watched," said the author of Papyrus Louvre, inv. no. 32847 (VI, 19, 2–4), who added, "It is only if the coffin has neither been seen nor watched that it will be entirely decorated with a coating of gold and that the august deceased will be brought into it" (Bardinet 2017: 225). Indeed, the gold coating – the most sacred of all precious materials (see Chapter 1) – on the coffin was an operation that had to be preserved from the impurity of having been seen by others. Hieroglyphic texts were not the only ones used to impose secrecy. The author of Ifao's Coptic Medical Papyrus used an encrypted alphabet to write the names of products that the physician wanted to keep secret (Chassinat 1921). Although one may think there to be a distant connection, this encrypted alphabet had nothing to do with the sort of secrecy that permeated religious practices.

It is reasonable to think that the procedures to prepare these recipes, or at least some specific operations, were recited by a specialist. This was especially the case during the *Ritual of Embalming* (Goyon 1972: 26–7), a rite supposedly able to restore the vital functions of the dead. Whatever the nature of those products, including those intended for mummification, the texts were recited by a ritualist–magician. The texts of the Greek period referred to these specialists as "hierogrammats" or "pterophors." Reading the texts aloud, the ritualist–magician provided a prompt for the practitioners, without participating in the technical operations reserved for specific personnel with pastophor status – that is to say, not constrained to the same degree of purity as the high-ranking priests. Reading the texts aloud specified the operating mode, added solemnity to the rituals, and supposedly changed the intrinsic quality of the results for which the products were made and the operations for which they were used. The written word prevailed on the experimental approach, as seen in the practice of medicine (Aufrère 2001c: 104–6).

However, reading the recipes of Edfu's laboratory implies that their technical details had to be recalled by the perfumers preparing them, and the recipes acted like a sort of prompt or memory aid. It was undoubtedly the complexity of these preparations, their highly ritualized status, and the necessity of following with precision the conduct of operations in order to guarantee their efficacy that explains why the clergy insisted on keeping them in writing. The technical information discovered in these written texts prefigured the recipes of the alchemical papyri, in particular those of Leiden and Stockholm, which contained recipes for silver and gold, inks, stones, and dyes (Halleux 1981: 35–52). Concerning dyes and metallurgy, about which little is said in the texts in Edfu and Dendara, it is possible that texts of a similar nature may have been written and kept in the Houses of Life so named by the Egyptians, instigating the search for certified chemical processes – alchemy – to artificially create precious metals (gold or silver;

see Chapter 1), whereas the Egyptians wanted to use only genuine precious metals intended to connote the divine.

It would seem that the memorization of sacred texts was greatly favored. In his *Stromata* (VI, 4, 35–8), Clement of Alexandria gave information showing that the high-ranking priests and pastophors (priestly personnel in charge of tasks that require a relationship with the public) of a temple not only were responsible for the maintenance of a number of books, but were also expected to "know them thoroughly," to learn them by heart and memorize their contents. Thus, concerning the horoscope priest: "It is necessary that he has constantly present in memory the astrological treatises of the books of Hermes, four in number" (*Stromata* VI, 4, 35; Aufrère 2001c: 102–3).

The Egyptian perfumers also had access to information describing the products in terms of the qualities of oleoresins according to their origin, the season, and the way they were harvested. Although summarized on a product-by-product basis, such information, although not scientific, was significant because it avoided using products considered unsuitable for the gods. In the laboratories of the temples of Edfu and Athribis (Baum 1994a; Chermette and Goyon 1996; Baum 1999; Colin 2003; Aufrère 2005a: 253–9), there were two parallel lists describing the varieties of frankincense (𓍹𓃀𓈖𓏏𓏭𓆰 *ânty*) and styrax (𓍹𓈖𓏭𓆰 *nenib*). Recently a list for myrrh (𓍹𓆑𓏭𓆰 *khel*) was identified in the temple of Athribis (Incordino 2016: 152–4). This type of presentation of information became a general model, which, by modifying what needed to be modified, described anything related to natural history in ancient Egypt. The diagram of these descriptions first specified the name, then the color, the aspect (viscous, soft, hard), the place where it came from, how it was presented (like tears, as a foam), the smell (pleasant or not), and other intrinsic qualities.

It should also be noted that the same model applied to products not used for liturgical rituals, pinpointing products of poor quality considered to represent the "eye of Seth." Indeed, any reference to the "eye of Seth" indicated a secondary product and whether it was for medical use. It was quite possible that such a model had been applied to many other product lines to inform authorized personnel on how to use them, notably in the medical field. Although several of these lists, much more precise than the *onomastica* vocabulary (see Chapter 2, pp. 55–56), have disappeared, it is easy to postulate that they were necessary to know not only the pharmacopoeia but also the chemical substances they contained and their qualities.

This is what emerges, for example, from the examination of the fragments of the *onomastica* of Tebtunis, which provided information on the origins of exotic products that are difficult to recognize, such as "(country of) Heh: mountain of gold" and "(country of) Teferer: lapis lazuli mountain," etc. (Osing 1998: 107). These examples are sufficient to demonstrate that specialists had at their disposal an extensive encyclopedic paratextual literature allowing them to stay close to the realities of the physical world and to be aware of the intrinsic qualities of the products they used in order to excel in their areas of expertise. Some of these products were potentially dangerous. Likewise, the venoms of snakebites were described and the chances of avoiding the toxic effects of some of them assessed (Sauneron 1989).

Specifically written for the use of specialist personnel at a given time, the sacred texts showed how important it was to ensure that they did not fall into oblivion. These texts could gradually become obsolete unless the changes made over time were explained. In summary, the technical vocabulary of a product or its know-how evolved over time, and what was understandable at one time was no longer so 100 years later. It was therefore necessary to explain these changes by making commentaries and glosses – sometimes comprising a sophisticated metatextual system – in order to update traditional knowledge and thus avoid misinterpretation by specialized practitioners.

One of the best-written texts, the Edwin Smith Surgical Papyrus, included technical explanations of very high quality (Bardinet 1995: 493–517), implying the existence of similar metatextual literature in technical fields. Such glosses are also found in the recipes of liturgical perfumes. This is the case for *kyphi*, for which the author gives the names of equivalents having ancient names, the use of some of which had become obsolescent, replacing them with names that were easier to understand. Thus, the name of the plant "Ethiopian reed" was replaced by "feather of Nemty," and that of the aromatic product "*agaiu*" (Aufrère 2005a: 250) by "mint." We have already seen that the texts in the "Mansion-of-Gold" (Dendara) contained glosses giving equivalents to the traditional names of materials that were not immediately comprehensible to contemporaries (see Chapter 1, p. 27).

In Roman times, when the meanings of the words attested to in the *onomastica* began to decay and the explanations given were no longer clear, a system of transliteration was developed using the Greek alphabet and letters formed by demotic signs (the so-called Old Coptic writing) to write in full the meaning that had been lost. These terms could also be transcribed into demotic (cursive writing, much used until the fourth century CE). These additions were presented as supralinear glosses, which allowed Egyptian scholars highly experienced in the elaboration of commentary – probably hierogrammats – to understand the textual follow-up word by word. The reading of technical texts was, theoretically speaking, facilitated by these glosses, which are interesting in many ways. They show, for example, on what model a large amount of technical knowledge, not to mention hermeticism, was gradually transmitted in Egyptian–Greek collaborative circles, bringing together those who boasted of being "philosophers," literally "friends of knowledge" (Jasnow-Zauzich 2005: 13–5; Aufrère 2016a: 255–8).

FIGURE 7.1 Text of the manufacture of the statuette of Osiris-Khentymentiu during the Khoiak Festival. Western Osirian chapels, inner courtyard, eastern wall. Temple of Dendara. © Sydney H. Aufrère.

FIGURE 7.2 Offering of oliban to Hathor of Dendara. Laboratory, northern wall, first register. Temple of Dendara. © Sydney H. Aufrère.

MESOPOTAMIA

Cale Johnson

The transformation, transmission, and revaluing of procedures for manipulating objects and ingredients drawn from the natural world – often spoken of as recipes – is the most important domain for identifying and tracking scientific thought in the ancient world. In considering the role of recipes in Mesopotamia, however, we need to carefully distinguish between a sometimes unconsciously memorized sequence of actions and the textualization of a sequence of actions in written form. Many technical and medical/pharmaceutical procedures existed for centuries, if not millennia, before they were encoded in writing. Moreover, once standardized compendia of recipes emerge, in the Ur III and Old Babylonian periods (ca. 2112–1595 BCE), they often present us with only a few crucial elements of the procedure and little or no specific information on quantities of ingredients or details of procedure: the full recreation of any procedure often requires much additional information that is not present in the text. The medical/pharmaceutical recipes from Assurbanipal's library at Nineveh (subsuming several physical locations on the citadel in the seventh century BCE; see Robson 2013) often use a placeholder such as "one shekel" for each ingredient, even when we know from other, non-library texts that much more specific quantities of each ingredient were known to specialists. We must assume that these schematic recipes, particularly when embedded in large, multitablet compendia, were not meant to be used in isolation. Students would, no doubt, have memorized large cross-sections of these materials as part of their textual and disciplinary training, but in the relatively few examples of recipe collections that were involved in practical contexts, we find much more specific quantities, means of administration, as well as comments on the effectiveness and seasonal appropriateness of specific remedies, global annotations, and qualifications that are almost never found in the library copies (the premier examples of this are in the medical school texts that Finkel published in 2000; see also Stadhouders and Johnson 2018).

The Marduk-Ea formula is the best-known way of framing technical (including magical) knowledge in the cuneiform tradition (Falkenstein 1931; Cunningham 1997: 24–5, 115–21; Ceccarelli 2015). This literary pattern, in which two deities, father and son, prototypically Enki/Ea and Asalluhi/Marduk, discuss the illness of a human being, both authenticates the incantation-plus-ritual that Ea eventually hands over Marduk as "of the gods" and also sets out the role of the incantation-priest who will recite the incantation and lead the performance of the ritual. Just as Marduk witnesses the human being suffering from the disease and then travels to his father Ea, seeking a treatment, the human incantation-priest claims to act in the stead of the junior partner in the formula, namely Marduk. Later on, in the first millennium BCE, with the emergence of groups of physicians who were less interested in ghosts and demons as the cause of illness, we occasionally find medical compendia in which parodies of the Marduk-Ea formula appear, presumably an effort on the part of pharmaceutically oriented physicians to set themselves apart from the ghost- and demon-driven models of disease etiology favored by the incantation-priests (Johnson 2018). Even when groups of specialists disagreed about the proximate cause of illness or the most effective treatment, the formal properties of compendia in the legal tradition and the materials associated with the incantation-priest provided long-term models for the organization of sets of depersonalized case histories into technical handbooks and compendia (Johnson 2015).

The earliest recipes, often hidden away as unlabeled "ritual instructions" in collections of incantations in southern Mesopotamian traditions, are found in the Early Dynastic III period (ca. 2600–2400 BCE); the most recently published examples are several compendia tablets from the Schøyen Collection, published by George in 2016. The first of these tablets, MS 4549/1, includes both the earliest reference to Asalluhi (Sum. ^{d}asar) in the context of a Marduk-Ea formula, and also a small subset of incantations (vii 5–x 4) against stomach illness (Sum. ša$_3$.gig), while the third of George's tablets (MS 4550 obv ii′ 3′ and following) likewise includes a section on stomach-related illnesses (ii), but no mention of Asalluhi. Several of these texts end with the patient drinking a specially prepared liquid ("water of Tigris and water of Euphrates" in MS 4549/1), presumably the final step in the production of these "drugs." This family of incantations has been found in Abu Salabikh (IAS 549) and among the Sumerian incantations that were copied in Ebla in northwest Syria (VAT 12524 = Krebernik 1984, no. 11). More recently, a couple of Semitic-language texts written in Ebla, still from the end of the Early Dynastic III period (ca. 2400 BCE), present us with the earliest recipe collections, primarily TM.75.G.1623, in which individual drugs are paired with the illnesses they are meant to treat: "the name of the plant is 'snake herb': it is a treatment for swelling, bile, incontinence (?) and 'hand of a god'" (Fronzaroli 1998: 227). Even here, where the focus is clearly on pharmaceutical remedies, only one or two steps are mentioned, and these steps are invariably part of the administration of the drug ("wrap it up and give it to the sick person to eat …"), rather than its formulation (Fronzaroli 1998: 228).

As recipes focusing specifically on medicine and other technical practices become more complex, and especially when scribes begin to gather them into multi-recipe collections in the Ur III and Old Babylonian periods, the obvious question was one of organization. In the Ur III period, when our earliest casuistically formulated legal compendia come into existence, such as the Codex Urnamma (Civil 2011), we also find our first evidence for medical prescriptions assembled into collections in a nearly unique Ur III tablet that Miguel Civil published in 1960: CBS 14221.

FIGURE 7.3 The Ur III medical tablet. Courtesy of Science History Images/Alamy Stock Photo.

This collection includes no descriptions of symptoms or illnesses and consists solely of medical recipes, organized in terms of the means of introducing the drugs into the body of the patient: poultices and bandages on the obverse of the tablet, medicines to be drunk in the first half-column on the reverse, and in all likelihood suppositories for the last two columns. It is really here in CBS 14221 that we first see the type of step-by-step, second-person-oriented recipe that dominates the technical literature for the rest of cuneiform history. In lines 46–54, we get a typical example: "you should crush (Sum. u_3.gaz) a.na.aš.du-plant, branches of camelthorn, seed from ..., you should pour (Sum. u_3.de_2) it in beer, you anoint (Sum. i_3.[bi_2.še]$š_4$) him (= the patient) with pure (or sesame) oil (and) you hang it (Sum. ab.la₂.e) on him (as a poultice)" (Civil 1960: 60, prescription 1.4). Other than this large compendium (and one much smaller tablet), the evidence for procedural knowledge like this in the Ur III period is limited to the numerous tablets that list precise amounts of raw ingredients and finished products, as for example in the soup-, beer-, and perfume-making records, but do not directly list the steps in the production process (Stol 1990; Brunke and Sallaberger 2010; Brunke 2011). More detailed procedures are also implicit in a few Ur III texts, such as UIOM 892, the metal-alloying problem text that we looked at in Chapter 2, but in that case it is calculations of the different ingredients and the amount of waste products that drive the text, not step-by-step procedure.

The most important examples of step-by-step procedure emerge in the Old Babylonian period: these are well known in the realms of mathematics and cooking recipes, but there are also large compendia of medical prescriptions from this period that remain largely unpublished (Irving Finkel will publish, in future, a number of British Museum tablets that will greatly expand this corpus). In all three of these domains, second-person directives ("you do ..." or "you should do ..."), rather than imperatives, dominate the literary form of these materials, but first-person elements do occasionally appear as well. This contrast is most clearly seen in mathematical problem texts from the Old Babylonian period, where the problem is framed by someone in the first person, who can easily be assimilated to the role of the teacher, but the steps in the actual solution procedure are all in the second person, corresponding to the student (see, for example, the excavated pit problem in Høyrup 2002: 145, no. 23; Friberg 2007: 58–61). The Old Babylonian cooking recipes (YOS 11, 25–7) edited by Jean Bottéro in 1995, are almost entirely in the second person except for a short section in which first and second person alternate (YOS 11, 26 i 50–ii 20, already highlighted in Bottéro's earlier 1987 paper):

> I extract gizzard and giblets and after I have cleaned well (the pigeon), YOU place it to soak in water. I then cleave and I skin the gizzard, and after I bring water to boil, YOU place gizzard and giblets in a cauldron ...
> (Bottéro 1987: 18, corresponding to Bottéro 1995: 74, 151–2; the English translation of these texts added at beginning of Bottéro 1995 renders all instructions, rather misleadingly, in the [second-person] imperative)

This is not the carefully structured alternation between a teacher posing the problem in the first person and the student described as solving it in the second person that we see in mathematical texts, but in all likelihood, as Bottéro suggests, it is evidence for different types of textual traditions that were assembled into a single text by the editor. This type of variation between second- and first-person procedural instructions does not seem to occur in the medical texts. Crucially, in these Old Babylonian step-by-step procedures, the epistemological framing provided by the performatively and theologically complex picture of Marduk/Asalluhi and Ea/Enki is replaced with a bare-bones social scene in which an unnamed "teacher" addresses a similarly unnamed "student" in the second person.

The absence of any kind of authorization in these step-by-step procedures is something of a conundrum. None of these texts include historiolae or other incantations that might provide a link to a deity or other source of authority, nor does a Marduk-Ea-style frame appear in any of these texts. One might guess that the use of a properly casuistic formulation – "if ... then ..." – points to a broadly scientific milieu for these materials, but this type of formulation is regularly found only in the medical texts. The mathematic and culinary texts are more likely to start with a simple descriptive label like "an excavated room" or "gazelle broth," but we should probably take the phraseological variety in YOS 11, 26, and 27 as the norm: alongside simple labels like "gazelle broth" and occasional uses of *šumma* "if" (preceding a hypothetical situation), we also find temporal phases using *inūma* "when" and purpose clauses consisting of the preposition *ana* plus an infinitive. These last two alternatives ("when ..." and "[in order] to ...") occasionally appear in later medical compendia (and the purpose clause dominates in even later Aramaic materials), but temporal and purpose clauses are predominant in YOS 11, 26, and 27.

We have no direct social contexts for any of the mathematical and culinary step-by-step procedures in the Old Babylonian period (although Robson 2008 offers a broader social history of mathematics), but we do have, already in the Old Babylonian period, a parody known as *At the Cleaners*, in which a step-by-step procedure for laundering a garment is embedded in a dialogue between a fuller and his client (see Wasserman 2013 for a new edition and detailed study). The humor, in as far as we moderns can comprehend it, lies in the fact that the client adopts the first-person role of the teacher and addresses the actual expert (viz. the fuller) as if he were a student in the second person: "What I instruct you, do not lay aside, / Your own (ideas), you should not do! / As for the hem of the garment, you will lay down the selvage ..." (Wasserman 2013: 274). But of course the deeper incongruity, which is only indirectly hinted at in a text like *At the Cleaners*, is that a written step-by-step procedure like this might lead a scribe or littérateur to the absurd conclusion that they could instruct someone about how to launder a garment by *reading through* a set of instructions. Indeed, the real goal – practical or literary – of step-by-step procedures like these is often unclear: any full-time specialist in cleaning clothes, cooking soup, or even formulating medical prescriptions probably had little use for these texts as an actual set of instructions. Like the mathematical texts, however, these procedural statements were almost certainly first produced as a kind of intellectual experiment: can we codify cooking in the same way the incantation-priests codify their ritual practices and incantations? Only once they had been textualized did other uses come to mind, whether defining the content of a discipline or, as in *At the Cleaners*, providing us with a parody of overweening scribal conceit.

It has long been recognized that the growth and elaboration of large-scale compendia largely took place at the very end of the Old Babylonian period or, more often, in the subsequent Middle Babylon period. As Rochberg puts it in her classic paper on "Canonicity in Cuneiform Texts":

> The process by which the celestial omen series Enūma Anu Enlil or any other omen series reached its final form is nowhere explained or even mentioned in our sources, but is likely to be the work of Kassite period transcribers and editors, since many representative texts of the scholarly tradition, omens, or lexical lists, emerged from the library of Tiglath-Pileser I (1115–1107 BC) in the form in which they are later attested in Neo-Assyrian and Neo-Babylonian copies.
>
> (Rochberg-Halton 1984: 127)

In contrast to biblical notions of canonization, Rochberg argues for a much more limited and gradually evolving pattern of "text stability and fixed sequence of tablets within a series" (1984: 129). This process advanced at very different rates in different bodies of technical or scientific literature; major compendia that stood at the center of the work of the incantation-priest such as *The Diagnostic Handbook* and the incantation compendium known as *Evil Demons* (*Utukkū Lemnūtu*) were already being serialized at the end of the Old Babylonian period but only reached their received form in the early first millennium BCE (Geller 1985; Geller 2016: 3–26; as well as the catalogs and papers collected in Steinert 2018).

The latter phases of this type of editorial work, during the last centuries of the second millennium BCE, are epitomized by Esagil-kin-apli's re-edition of *The Diagnostic Handbook* and *The Physiognomic Corpus*, a re-edition that is probably as close as Mesopotamia ever comes to an explicit model of editorial refashioning. Esagil-kin-apli was the chief royal scribe under Adad-apla-iddina (reigned 1068–1047 BCE), and from this august vantage (the most important institutional role for a scribe, after half a millennium of intense editorial work), he provides us with a unique statement of editorial method sandwiched in between the catalogs of *Sakikkû* (*The Diagnostic Handbook*) and *Alamdimmû* (*The Physiognomic Corpus*; see Heeßel 2000: 105; Schmidtchen 2018a: 147); he describes his source materials as "tangled threads … lacking an edition," and goes about, in as far as possible, reordering the materials anatomically from head to foot (lines 51–2 in Schmidtchen 2018b: 316–17). These two lines are really a codicil attached to the total of forty tablets and 3,000+ entries rather than the formal statement of the editor, which starts in line 53 (the first line on the reverse of Ms B). But Esagil-kin-apli was not really speaking truthfully here; earlier editions of major components of his bipartite series certainly did exist. This is already evident from the subseries in *The Diagnostic Handbook*: the forty tablets of *The Diagnostic Handbook* are subdivided into six subsections, and subsections 4 and 5 largely consist of existing compendia that have been incorporated en masse into Esagil-kin-apli's edition (Heeßel 2000: 107). More importantly, Heeßel published a tablet of materials from *The Physiognomic Corpus* (*alamdimmû*) in 2010 (viz. VAT 10493+) that describes one section as part of "the older (series) *alamdimmû*, which Esagil-kin-apli did not supplant," suggesting that an earlier series of the same name existed in Assur and that the scholars in Assur refused to accept Esagil-kin-apli's new edition (Heeßel 2010: 154).

It was sometime in between the incipient serialization at the end of the Old Babylonian period and the competing compendia in the eleventh century BCE that the next wave of Akkadian step-by-step procedures were transmuted into textual form. The glass- and perfume-making texts, which have long been recognized as quintessential precursors for the history of chemistry, were first inscribed in clay at some point in this period, although the bulk of the glassmaking texts are only known from the first-millennium BCE copies recovered from the Library of Assurbanipal.

FIGURE 7.4 Glass and alabaster jar belonging to the ancient Assyrian King Sargon II. Dated eighth century BCE. Photograph by Universal History Archive/Universal Images Group via Getty Images.

Thanks to a colophon in one of the perfume-making texts (KAR 220), we know that it was inscribed during the reign of Tukulti-Ninurta (1233–1196 BCE), while the earliest glassmaking text is a seemingly Middle Babylonian tablet (BM 120960) published by Gadd and Campbell Thompson in 1936 (see Figure 1.5 in Chapter 1). Like most scholarly texts from the second half of the second millennium BCE, there are few if any contemporary duplicates for these materials, and their interpretation is largely framed in terms of how they relate to later, first-millennium BCE materials from Assurbanipal's Library or the Late Babylonian materials from Uruk.

All the more reason, therefore, to emphasize the cryptic orthography of the Middle Babylonian glassmaking text published by Gadd and Campbell Thompson. Martelli and Rumor, for example, have pointed out that linguistic-cum-orthographic puns in cryptic texts like this provide a crucial link between Mesopotamian sources and the recipes for dyeing metals associated with Pseudo-Democritus, especially a dyeing bath that includes "copper rust and bile of mongoose and vulture" (2014: 46). These are clearly *Decknamen*, and if we look carefully at the Akkadian term *erû* (meaning "eagle" but probably corresponding to "vulture" in Greek) in the Middle Babylonian glassmaking text, we can see the basis for its cryptic character. This raises the possibility that two different orthographies for *erû* "copper" were being used to differentiate the raw material (written with the usual logogram URUDA) and the label for the colored glass ($TI_8^{mušen}$, which normally means "eagle" but puns here on "copper," both of which are *erû* in Akkadian). These orthographically distinguished homophones seem to underlie "(the bile of) the vulture" discussed by Martelli and Rumor. Be that as it may, Oppenheim was certainly wrong to claim that this, the earliest glassmaking text known, was not deeply encoded (Oppenheim 1966; Oppenheim 1970: 59–65; Martelli and Rumor 2014: 48 and n. 42).

The vast majority of the cuneiform glassmaking texts, just like any other group of texts in the first millennium BCE, are from the Library of Assurbanipal (reigned 668–627 BCE). Robson has emphasized that the "Library of Assurbanipal" is largely defined by its excavation in the nineteenth century (and role in present-day Assyriology):

> It was dug by Austen Henry Layard and others in 1847–55, well before the advent of recorded, stratigraphic archaeology, yielding around 30,000 tablets and fragments. This first large discovery of cuneiform tablets made its way to the British Museum, where it is still housed today.
>
> (2013: 42)

Although conceptualized by generations of Assyriologists as a single entity, Reade (1998–2001) shows that the tablets likely derive from a number of institutions and archives in eighth- and seventh-century BCE Nineveh, ranging from

Sennacherib's Southwest Palace or Assurbanipal's North Palace to the temples of Nabu and Ishtar. It is, therefore, small wonder that nearly all glassmaking manuscripts stem from this massive yet somewhat artificially unified library.

The real surprise, however, is that the glassmaking recipes from Assurbanipal's Library were already heirlooms in the seventh century BCE, presenting us with a battery of techniques more at home in the thirteenth or twelfth century BCE than the seventh century BCE. Different levels of glassmaking technology in these texts were already noted in Chapter 2, but as Robson perceptively notes,

> ... glass-making technology had long-since moved on, with the introduction of casting and cold-cutting in about 900 B.C. ... Around the same time, it became fashionable to add antimony to the glass mixture, making it an almost transparent shade of yellow or green in imitation of rock crystal ... Why, then, were Neo-Assyrian royal scribes continuing to copy detailed instructions for an out-of-date technology?
>
> (2001: 51–2)

Out-of-date and encapsulating at least two levels of glassmaking technology, the glassmaking texts are only known from library copies: they are not attested in personal archives or school contexts, and unlike the medical prescriptions, which were extracted, adapted, and specified in certain Late Babylonian contexts (see Finkel 2000: 140–1), the glassmaking recipes never appear in an applied or practical situation. There is slim but real evidence for a glassmaking compendium entitled "Door of the Kiln" (Oppenheim 1970: 25; Robson 2001: 53), but when we turn to later traditions, namely literary royal letters from the Late Babylonian and Seleucid periods, which seek to memorialize Assurbanipal's tablet acquisition scheme (Beaulieu 2010), there is no specific mention of technical procedures like these (Frame and George 2005). The fact that glassmaking recipes do not appear elsewhere in the first-millennium BCE scholastic enterprise that compiled Middle Babylonian sources into Neo-Assyrian compilations – compilations that then often made their way to Seleucid-period Uruk in one form or another – suggests that they were never central to native intellectual histories in the first millennium BCE.

GRECO-ROMAN WORLD

Matteo Martelli

Small workshops in classical Athens were run by lone craftsmen whereas larger businesses employed slaves, if the owners could afford to buy them as fellow-workers (Acton 2014: 272–74). In both cases, in the archaic and classical periods, both textual and archaeological sources show that crafts such as weaving, pottery, and simple metalworking were taught in family workshops, being transmitted from father to son and, in some cases, to extended kin (Hasaki 2013). The apprenticeship could last for several years, depending on the complexity of the craft being learned: almost ten years, for instance, were necessary to properly master pottery production, from the first steps (e.g. collecting and preparing the clay) to the most advanced skills, such as learning how to use wheels and kilns as well as how to decorate the produced vases (Hasaki 2013: 193). In the case of sizable enterprises employing gangs of slaves – such as the knife factory that Demosthenes inherited from his father (*Or.* XXVII 9) – skilled slaves, even without being freed, could sometimes manage the activities of the workshops on behalf of their wealthy owners. Freed slaves – whose names have been recorded, for instance, in the fourth-century BCE Attic "Manumission bowls (*phialai*)," which mention wool-workers, goldsmiths, gem-cutters, and blacksmiths (Jordan 2000: 99) – could also set up their own businesses and continue practicing the professions they learned in the workshops of their former masters (Acton 2014: 281–8). From a speech of the classical orator Hyperides (fourth century BCE), we know that Athenogenes, an Athenian citizen of Egyptian origin with two generations of perfume-sellers behind him (Hyp. III 19), owned three perfumery workshops in the city, one of which was run by the slave Midas and his two sons. The extraordinary price of thirty *minae* for which they were sold (the average cost for a slave was of three to four *minae*) might be justified by their exceptional skills, which it is likely they developed after a long apprenticeship (Acton 2014: 243).

Learning a craft would not only take a long time, but it could also be exhausting and painful. In a letter inscribed on a lead tablet found in the Athenian Agora (inv. no. IL 1702, fourth century BCE), Lesis, an apprentice in a workshop for metalworking (*chalkeion*), describes to his family the severe punishments he suffered there: he was afraid he would die, being always whipped, tied up, and treated like dirt by his wicked masters (Jordan 2000; Hasaki 2013: 185–6). A similar scene seems to have been represented in a fifth-century BCE *skyphos* (wine cup) found at Abai in Lokris (National Museum, Athens, inv. 442), where a pottery workshop is depicted: above a painter sitting at the wheel we see an apprentice who

> … has been suspended face down from the ceiling. His left foot is tied against the ceiling itself; his right foot hangs lower, from a cord … Another cord from the ceiling is around his neck, strangling him so badly that his tongue hangs out.
>
> (Jordan 2000: 100)

Lesis asked his family to come to his masters and find something better for him.

The apprentice–master relationships were regulated by agreements, which, according to later papyri, could imply two possible conditions: apprentices either worked in the workshops by day and spent the night in their homes or they lived with their master until the end of the training period (Jordan 2000: 99). Almost forty apprenticeship contracts (*didaskalikai*) have been discovered in papyri from Greco-Roman Egypt; even though they mainly deal with the textile industry (especially weaving), other professions are also recorded, such as nail-making, coppersmithing, or building (Bergamasco 1995). Petosirirs, for instance, sent his son to the master Herakleides, a coppersmith (*chalkotypos*), submitting, at the same time, an application to register his son in the lists of the quarter where Herakleides' workshop was located (*PSI* VIII 871, first century CE). In Greco-Roman Egypt, in fact, apprentices were registered in groups officially recognized by local authorities, who could tax them. Craftsmen had to pay regular taxes (*cheironaxion*) on the trade produced by their workshops, which could benefit from the work that young apprentices were doing while learning the art. In some cases, however, craftsmen committed to teaching their art could apply for tax reductions in return for the service they provided to the city (Freu 2011: 28).

As regulated by apprenticeship contracts, masters were expected to provide their apprentices with clothing and food as well as to pay them wages, which could be adjusted to the different phases of the training; more advanced students contributed more effectively to the output of a workshop, and they were to be paid accordingly. In return, students were obliged to obey their masters and work full time in their workshop (in some cases, holidays were stipulated in the contracts), while fines were imposed if they decided to quit abruptly. The length of the training period could vary significantly, from six months to several years, depending both on the complexity of the art being learned and on the range of skills the students wished to acquire. Among the students, papyri mention both sons of craftsmen and slaves. In the latter case, patrons invested in their slaves, who were sent to a master and, after learning an art, could set up a lucrative business for their owners. On the other hand, craftsmen could also decide to send their own sons to the workshops of other artisans in the same trade, so that the sons could extend their set of mastered skills beyond a family tradition. Moreover, similar apprenticeships could also strengthen the links between different families, professionals, and masters of various workshops. As recently pointed out by Liu (2017: 219), "both apprenticing one's son to a fellow practitioner, and the acceptance of that child from a fellow practitioner as an apprentice, can be interpreted as tokens of trust, especially since they involved sharing trade secrets as well as information about clients, credit, and supplier networks" (see also Freu 2016).

Artisans often associated themselves in professional corporations or guilds – the *collegia* in Latin. Papyri from Greco-Roman Egypt often mention groups of various craftsmen working in the same field, which could include only a few members (Van Minnen 1987: 48–72). Latin inscriptions, on the other hand, testify to the status of Roman corporations that, in some cases, could remain active for various decades, such as the *collegium fabrum tignariorum* (builders and carpenters) and the *collegium aromatariorum* (pharmacists) in Rome (Liu 2017: 206–12, with further bibliography). Guilds both served the economic interests of their members (collective payment of taxes, mutual support in case of financial difficulties) and promoted civic, social, and religious activities. A papyrus from Tebnytis (*P.Tebt.* II 287, second century CE), for instance, preserves the official record of an appeal that fullers and dyers of the Arsinoite nome (district) made against what they held to be an undue amount for the tax upon their respective trades. The guild of silversmiths in Ephesus gathered in the theater of the city when their interests, mainly linked to the making of small silver shrines of Artemis, were threatened by some Christian followers of Paul, who denied the sacred nature of these objects. To their choral chant in the theater of "Great is Artemis of the Ephesians," the city's *grammateus* responded by reassuring them that the city, as a guardian of the temple of Artemis and her images, would protect their trade (*Acts* 19: 24–40; see Venticinque 2016: 167–9). In the mid-fourth century CE, four inscriptions from the Egyptian temple of Deir el-Bahari testify to the regular sacrifices of donkeys that a guild of ironworkers from Hermonthis offered to the Egyptian demigods Amenhotep and Imhotep (Łajtar 1991).

Less clear is the role played by corporations in apprenticeships; some scholars have supposed that they could have acted as intermediaries between students and masters (Van Minnen 1987: 69–71; Freu 2011: 28–9). Fullers and dyers are mentioned together in a list (probably a tax list) given in a papyrus from Tebnytis (*P. Carlsberg* 53; second century CE). The bottom of the papyrus deserves particular attention: after mentioning a dyer called Tyrannos Kroniōnos, the list goes on (ll. 12–3) to give the name of "Orsenouphis, son of Heron, supervisor (*hēgoumenos*, "leader") of the weavers in Tebtynis, for the screening of apprentices." As already pointed out by Van Minnen (1987: 70–1), Orsenouphis was probably leading a guild of weavers that was somehow involved with the training of new apprentices (Freu 2011: 35).

The apprenticeships of dyers probably followed similar rules and took place either in the workshops of external masters or in family workshops, where the art of dyeing was passed on from father to son (Wipszycka 1965: 146). Justinian's *Digest* XXXII 91,2 (from the jurist Papinian) mentions a workshop (*taberna purpuraria*) willed from father to son, while in a sixth-century CE papyrus (*CPR* X 124, ll. 4–6) two dyers (*bapheis*), father and son, are mentioned. On the contrary, no ancient source seems to mention supervisors for the screening of apprentices with reference to a guild of dyers.

The inscription *Alt. v. Hierapolis* 227 (third century CE) must be mentioned at this point, since the text records a certain M. Aurelius Diodorus Corescus who left an amount of 3,000 *denarii* to the guild of purple dyers (perhaps his colleagues), whose leader was required to burn poppies on his tomb every year. But if they neglected to do so, the money had instead to go to the *ergasia thremmatikē*, a term that has been interpreted by some scholars as a *collegium alumnorum* or an association of slaves' children. This hypothesis, however, was severely criticized by Van Minnen (1987: 71, n. 142).

In the alchemical literature from Greco-Roman Egypt, the motif of the transmission from father to son was inherited and reshaped in various narratives about the original revelation of the alchemical art. In the first century CE, (Pseudo-)Democritus is said to have received his initiation into the dyeing arts – which included, as already pointed out (see, for instance, Chapter 2), the making of gold, silver, and precious stones, along with purple dyeing – from the Persian master Ostanes in the Egyptian temple of Memphis: even though Egyptian priests and other apprentices were taught by the master, Ostanes kept his books hidden only for his son (Martelli 2013: 82–5). Likewise, the angel who revealed the secrets of gold- and silver-making to Isis made the goddess swear to pass these instructions only to her son, Horus (*CAAG* II 33–4). In a more historical context, Zosimos of Panopolis addressed most of his alchemical teaching to the wealthy woman Theosebeia, who is sometimes introduced as the "sister" (*adelphē*) of the alchemist (*Souda* Z 168; Hallum 2008). The term probably refers to an alchemical kinship, stressing the tight relationship between teacher and pupil, as one can infer from Zosimos' complaints when Theosebeia temporarily joined another group of alchemists led by his rival, the Egyptian priest Neilos (*CAAG* II 190; see Chapter 5, p. 159).

All these examples introduce a second crucial element in the alchemical apprenticeship: alchemy was, first of all, a craft that was to be learned from books. As we have already seen (see Chapter 4, pp. 131–2), according to Zosimos' account of the Enochian myth, a book entitled *Chēmeu* – from which the term *chēmeia* (i.e. "alchemy") derived – was the main secret revealed by the fallen angels (Martelli 2014a). Zosimos himself often urges his pupil Theosebeia to study the books of the ancients, which represented the basis for any alchemical training; their reading always supported the personal teaching of the master, who was expected to guide his pupils through the technical instructions left by the founders of the discipline.

The books of the ancients – such as Pseudo-Democritus' alchemical writings, to which Zosimos always refers – certainly contained recipes, which may be identified with the smallest units of any alchemical text (Halleux 1979: 74). Collections of recipes, on the other hand, started circulating in Greece centuries before the first appearance of alchemical books. Medical recipes, for instance, are included in various writings of the Hippocratic corpus, either in sections devoted to specific diseases (e.g. *Diseases* I and II; *Internal Affections*) or in catalogs, such as in *Diseases of Women* I and II. Probably based on earlier small collections selected and reorganized by the Hippocratic authors, these recipes omit crucial information (e.g. instruments to be used in preparing the drugs and quantities of single ingredients). They have therefore been interpreted as "short *aide-mémoires*, which needed to be supplemented through oral explanations; these recipes could not, and did not intend to, replace the oral word" (Totelin 2009: 258).

A few decades later, the philosopher Theophrastus included various recipes for the making of perfumes in the second part of his treatise *On Odors*. These recipes seem to derive from direct contact with the experts working in Athenian workshops. Theophrastus himself was the son of a fuller, and firsthand experience in his father's workshop may explain the analogies he drew between wool-working and perfume-making (*On Odors,* 17). Indeed, Theophrastus' account is based on direct observation of workshop procedures and an ongoing exchange with perfumers, whose different opinions on the preparations of perfumes are reported by the philosopher. However, the author did not mean to write a handbook for perfume-makers (Reger 2005: 257–9).

More technical, on the other hand, are the recipes for perfume-making collected in the treatise *De materia medica* by the first-century CE physician Dioscorides, one of the heirs of the rich and sophisticated pharmacological tradition developed by Hellenistic physicians. This tradition represented the basis for Galen's writings on the composition of drugs. Galen claims to possess the best selection of recipes in the Roman Empire, which contained expensive formulas assembled from all over the world (*On Avoiding Distress*, 31–7). In his *De materia medica*, Dioscorides included the recipe of the precious Egyptian incense called *kyphi* (Diosc. I 25), which is also recorded both in hieroglyphic inscriptions (e.g. in the temples of Edfu and Philae) and by Plutarch, at the end of his treatise *On Isis and Osiris* (383e). In this work, after listing all the ingredients of the perfume, Plutarch specifies that they were not mixed at random, but according to the sacred writings that were read to the perfumers as they were mixing the ingredients. Therefore, "sacred writings" (*hiera grammata*), to be identified with written formulas for *kyphi*, seem to have been used in ancient workshops, perhaps read aloud by a master to his fellow workers. The use of books in a workshop is apparently confirmed by a famous fresco in the *House of the Vettii* in Pompeii, which depicts various cupids performing the different steps necessary for the preparation of a perfume. In the middle of the fresco, a scene displays a papyrus scroll (perhaps a recipe book) next to a little flask into which a cupid pours the perfume (see Chapter 3, Figure 3.6).

A similar use in workshops has been proposed for the two alchemical recipe books preserved in the Leiden and Stockholm papyri. As already seen (see Chapter 2), the two papyri collect more than 200 recipes on a variety of dyeing and metallurgical techniques, along with instructions on how to produce artificial gemstones and make golden and silver inks. Written in a codex format, they might have allowed potential users who looked for specific recipes to browse through the collection more easily than in a papyrus scroll. Although the lack of stains or signs of damage (which, presumably, would have been left by their use in a workshop) has led scholars to consider the papyri mainly as copies for libraries (Halleux 1981: 27), one cannot rule out that they were simply read aloud by a master in his workshop (like in the case of the *kyphi* formulas) or were based on earlier workshop copies (Clarke 2013: 13–8).

The wide range of techniques considered by the compiler(s) of the two papyri, however, implies a broad umbrella of competencies, which were scattered among different categories of experts, such as metallurgists, makers of fake precious stones, glassmakers, or dyers (Martelli 2011: 280–1). As discussed above, each of these craftsmen developed a high degree of specialization through long and demanding apprenticeships, so that specialized craftsmen in antiquity probably never committed themselves to learning and practicing such a variety of *technai* ("arts"). Obviously, we cannot exclude the possibility that more specific recipe books devoted to single arts were circulating among specialized craftsmen. For instance, *P.Iand.* 85 (Halleux 1981: 158–60; first to second centuries CE) preserves recipes focused on purple dyeing only, while Pliny the Elder (*NH* XXXVII 197) informs us of *commentarii* devoted to the making of precious stones. In recipe 53 of the Stockholm papyrus (Halleux 1981: 125), we find an interesting reference to unspecified writings (presumably on the making of precious stones):

> Corroding and opening up of stones. Grind alum and melt it carefully in vinegar. Put the stones therein, boil it up, and leave them over night. Rinse them off, however, on the following day and color them as you wish by use of the recipes (*graphai*, lit. "writings," perhaps recipe-books) for coloring.
> (transl. by Caley 1927: 987)

The Leiden and Stockholm papyri, in fact, preserve a few references to their sources. In the Leiden papyrus, recipe 89 reads (Halleux 1981: 105): "How to fix Alkanet. Urine of sheep, or *komaris*, or henbane as in the recipe 55." Perhaps we find here a reference to recipe 147 of the Stockholm papyrus, which gives the same instructions for fixing alkanet. If we only consider the section on purple dyeing of the Stockholm papyrus (recipes 89–159), recipe 147 is no. 58 of this section, a position that is very close to the numbering given in the reference that we find in the Leiden papyrus. Perhaps a smaller recipe book on purple dyeing was among the sources used by the compiler of the two papyri (Halleux 1981: 105, n. 3). Indeed, recipes were organized around specific areas of expertise in the four books of Pseudo-Democritus, and these short procedural texts continued to be the major vehicles for the transmission of alchemy in the following centuries. In late antique and early

Byzantine recipe books we find various references to the tools of craftsmen, such as the furnaces of glassmakers (*CAAG* II 307–8) and goldsmiths (*CAAG* II 305), along with their crucibles (the Leiden papyrus, § 68 in Halleux 1981: 100). If the identification between these artisans and ancient alchemists is problematic (Halleux 1981: 28; Martelli 2011: 280–5), a certain overlapping between their areas of expertise – maybe implying direct contacts and mutual exchanges – seems to emerge from the available sources.

CONCLUSIONS
Marco Beretta

The art of making recipe books containing instructions on how to make perfumes, glass, and other chemical products seemed to have originated in Egypt and Mesopotamia and it was successfully embodied in the Greco-Roman technical literature and was equally successfully transmitted to early modern times. If we consider the completely different material means by which literary texts were composed in these three civilizations, the continuity of this tradition is all the more remarkable. Although Egyptian civilization left few traces of chemical recipes, the surviving ones preserved on the walls of temples contained extremely detailed technical operations and, unlike medical texts, were mostly devoid of any religious references. The record of chemical recipes was important to ensure the transmission of knowledge to future generations but, at the same time, it was crucial to keep this knowledge under the control of a restricted circle of people (priests and scribes). The tension between public record on the one hand and secrecy on the other was destined to accompany chemical recipes until the end of the seventeenth century.

In Mesopotamia, compendia of recipes did not contain all the technical know-how displayed in the artifacts produced in the laboratory, thus showing a gap between the practical knowledge preserved in the workshop and its literary record. A feature that was destined to play a crucial role in the composition of recipes was the use of the second person in procedural instructions.

Greco-Roman artisans specialized in guilds and transmitted their knowledge within a more standardized set of rules, both legislative and literary. In the alchemical literature of the Greco-Roman period, the Egyptian and Mesopotamian traditions were enhanced within a theoretical framework that tended to give a higher status to chemical arts. Many recipes recorded in hieroglyphic inscriptions ended up in recipe books such as the Leiden and Stockholm papyri and thus provide an extremely interesting testimony of the circulation of chemical knowledge.

CHAPTER EIGHT

Art and Representation: *The Iconographic Imprinting of Ancient Chemical Arts*

SYDNEY H. AUFRÈRE, CALE JOHNSON,
MATTEO MARTELLI, AND MARCO BERETTA

EGYPT

Sydney H. Aufrère

Egyptian iconography included painted scenes, low-relief carvings, and stone statues or groups of wooden statues, showing how Egyptian craftsmen had mastered the chemical processes involved in the transformation of a large range of materials. Although here we treat only the most significant parts of the techniques used from the Old Kingdom to the Greco-Roman period, our discussion nonetheless provides an overview of the processes involved in such activities as bread-making, beer-brewing, drying, and salting, as well as metallurgical procedures, including goldsmithing, along with the manufacture of papyrus, perfumes, and aromatics. None of the scenes of processes used in ancient times was meant to fulfill a didactic role for craftsmen, since they were traditional parts of everyday life. The aim of this iconography was simply to keep the memory of these important activities alive forever, since they had provided the owner of the grave and his entourage with a high quality of life.

It seems that bread-making and beer-brewing (Samuel 2000) were carried out on a daily basis in the same workplace, be it for domestic use or on an industrial scale. To begin, both activities required mastery of the process of grinding cereals to obtain flour (Murray 2000). Women were in charge of the millstones used to grind cereals. Bakers then worked the dough on tabletops. Different processes to make bread and cakes were known and applied, depending on the season. Traditionally, bowls of leavened dough were placed in baked earthenware molds heated over a fireplace. The iconography shows the diversity of these molds along with slices of bread burned on the surface due to overheating. In the New Kingdom, round flat leavened dough was prepared and then slapped onto the walls of the oven, very much like in Egypt even today. The quality of cereals employed and the large number of recipes and cooking processes available permitted the making of a great variety of breads, examples of which have been found by archaeologists in the tombs of workers excavated in the village of Deir el-Medina.

According to pictures found in the tomb of Ti, the most common beer, of which there was a great variety, was brewed from ground barley (*besha*) and malt. Water was added to the barley, followed by wheat flour to prepare the dough, then this was baked. Loaves of baked dough were then triturated in water (the strength of alcohol depended on the amount of water added) in a large colander placed over a wide-bottomed jar used to gradually trap the liquid. The liquid was poured into vessels using a spout; these containers were then hermetically sealed and the liquid left to ferment. This recipe was so easy to follow that it is still used today to produce a beer called *buza* in Nubia or *merissa* in the Sudan (Helck 1971). The proportions of the ingredients were indicated (Helck 1971: 33–5), and a recipe for this same beer (called *zythos/zythum*) was also included in the work of Zosimos of Panopolis (third century CE; Helck 1971: 40; Mertens 1995: lix, n. 16; Nelson 2001: 124, 396) as well as in rabbinic literature (Bondi 1895). It seems that, depending on the amount of water added, state-owned breweries produced beers with different strengths of alcohol, which were measured in *pesu*; beers could be of small or high *pesu*, based on mathematical calculations (*Papyrus Rhind*, *Papyrus of Moscow*) that were also applied to mixtures of beers of different *pesu* (Couchoud 1988). But for everyone else, this recipe was simply for making bread and beer (Michel 2010).

Wine-making comprised three stages: grape harvesting, crushing, and juice fermentation in jars (tomb of Nakht and Khaemuaset; Murray et al. 2000). Scenes of grape harvesting and of wine-making appeared early in the Old Kingdom, as shown in the tombs of Nyankhpepi or Mereruka. These tombs feature bunches of grapes that were cut from the climbing vines and collected in baskets. Full baskets were then emptied into tanks where barefoot men are depicted holding on to a central wooden bar while treading the grapes. The juice flowed into an underlying basin. In order not to lose any remaining must and to conclude the destemming, the residues of the treading were placed in a large sack, the ends of which were twisted by four men with the help of two wooden sticks, with a fifth man maintaining the spacing between the two sticks. The must thus squeezed also went into the jar, and all the must collected was then poured into fermentation jars and sealed with clay plugs (Varille 1938: 23–4, pl. XV).

At the end of the Thirty-First Dynasty, the tomb of Petosiris at Tuna el-Gebel, in Middle Egypt, added further details to this iconography by depicting other scenes of grape harvesting; the bunches are shown being harvested from climbing vines. According to the captions that accompanied the iconography, the operation proceeded nonstop from early morning when the dew was abundant, throughout the day, and only stopped at nightfall. The bunches of grapes were then deposited into a tank in which the grape-pickers, holding on to a fixed transverse wooden bar, treaded the grapes. The must flowed into a basin located below the treading tank, as shown by archaeological remains associated with the Mareotide vineyards (Redon et al. 2016). Must was then transferred into wide jars, which were counted by scribes and then brought to the master (Lefebvre 1923: 16–18, inscriptions nos. 43–4, pl. XII; Lefebvre 1924: 60–3; Cherpion et al. 2007: 58–63, scenes 56a–c).

Regarding the preservation of fish, the packing operations that took place after fishing are pictured in the mastabas of the Old Kingdom. Fish, degutted and visibly sliced at the dorsal ridge, were suspended for drying in the open air to ensure preservation (Vandier 1969: 635–58). It is very likely that sea salt and natron were used. It appears that ovaries of mullet fish, easily identified by their silhouette, were washed in salt water, the eggs collected and cured, and then finally packaged between two small wooden plates (Vandier 1969: 643–8).

From the Old Kingdom onwards, in the field of crafts, the work of metallurgists has been best explained by didactic scenes, while the knowledge of the technology of glassware and faience processing is only archaeological. From an iconographic point of view, only a few operations of the technology employed were depicted in pictures that had no didactic or practical value, because this knowledge had to be kept within specialized circles. In the New Kingdom, pictures in the tombs of Nakht and Rekhmirē' (TT 100) summarize the operations involved in metallurgy as follows: basin-shaped crucibles were placed on hearths into which the imported bronze oxhide-shaped ingots or loose ore were put to be smelted. The hearths were fanned to the required smelting temperature by means of air blasts using two cylindrical leather bellows strapped to the feet of a person (several could work at the same time) who operated them by alternately lifting one foot after the other. The crucible containing the molten metal was lifted by the metallurgists with the help of two tree branches, and the molten metal cast in a lost-wax mold was fed by several funnels, which were used to cast a temple gate several meters high for the sanctuary of Amon-Rē'. In view of its purpose, this operation required mastery of the technique by the craftsmen (Garenne-Marot 1985).

Concerning goldsmithing, scenes in the tombs of Mereruka as well as of Nyankhkhnum and Khnumhotep at Saqqara (Fifth Dynasty) show that gold was melted after kindling the fire with blowpipes, through which workers blew air into an opening on one side of the hearth. The workers – dwarfs renowned for their skills – then presented the result of their work to the owner (Montet 1952a; Montet 1952b; Pons Mellado 2005; Dasen 2013: 89).

The tomb of Petosiris contained several scenes depicting the work of metallurgists and goldsmiths (Lefebvre 1923: 8–12, pl. VII–IX; Lefebvre 1924: 52–5; Cherpion et al. 2007: 34–5). However, the iconography is not sufficiently detailed to allow us to grasp the full meaning of the operations. In one scene, a coppersmith is shown using a pair of pliers to hold a piece of copper on a stone anvil while another is seen hammering it with a small stone. The caption states: "Men working copper to make the home of their master shine through their work." Other scenes included goldsmiths hammering containers into shape, or smoothing them, on a side arm fixed to an anvil using a rudimentary mallet. Another operation shows silver and gold objects being polished. In order to avoid theft of precious metal, the objects were weighed on a balance with a mobile weight indicator, their weight recorded by scribes, and then the objects were locked in a wooden chest stored in the treasury.

Scenes depicting the preparation of perfumes are only meaningful when explanations are given in the texts. Thus the verb *nud*, "to cook" in Egyptian (*Wb* II: 226,8–9), referred to aromatists, called *nudu*. The Egyptian hieroglyph (*Wb* II: 226,10–11) depicted both the determinatives of the flame rising from a brazier and that of the aromatist preparing the perfume. Indeed, the preparation and quality of perfumes lay in the cooking of oils and fats on stoves for long periods of time. Etymologically speaking, the aromatists were therefore mainly "cookers" who had to strictly adhere to the cooking stages indicated.

The appellation of perfumers was given according to the different contexts in which they worked. The word '*ānt* (cf. "frankincense," '*ānty*) showed that they were not preparers of incenses as such, but simply perfumers (Colin 2003). Moreover, according to its etymology, the texts mentioned the presence of a "perfume attendant" (*iry-'ānty*; Colin 2003), a word suggesting that there was a hierarchy among aromatists. The second part of the composed word ('*ānty*) stands for "frankincense," generically referring to varieties of gum resins. The attendant was probably a person who had knowledge of both the essential components – a wide range of incenses, myrrh, and styrax – listed in the recipes of perfumes reproduced on the walls of temples' so-called "laboratories" and also of the various aromatics used in their preparation. The word "frankincense" ('*ānty*) in the title of "perfume attendant" embraced the entire spectrum of aromatic gum resins, namely sixteen categories, fourteen of which were of high quality and two of lower quality. In other words, the attendant supervised the production of perfumes, ensuring that books giving the recipes of perfumes were kept safe. His knowledge allowed him to distinguish between the different qualities of aromatic resins available on the market and thus prevent fraud, which was extremely frequent at the time. It is likely that the aromatists or cookers worked under his direction.

The elaborate techniques necessary to prepare perfumes or dyes, using both vegetable and mineral substances, implied the use of important and unknown chemical reactions later perfected by alchemy. Nonetheless, this in no way justifies the claim that ancient Egyptians mastered the indispensable distillation process, which was first documented only in the second century BCE (Forbes 1950; Kochmann 2014).

Scenes of flower-picking and perfume-manufacturing operations are scarce in Egyptian iconography. What depictions exist consist of succinctly presenting the process between picking and the final presentation of the finished product to the owner. There are a few scenes of lotus flower harvesting for the preparation of lotus ointment (Vandier 1969: 453–5). The plant grew in shallow marshy areas. After picking, the flowers were bagged for transport to the place of manufacture. In several tombs, scenes of perfume-making are pictured. In the tomb of the Vizier Kagemni (Fifth to Sixth Dynasties, 2510–2200 BCE) in Saqqara, four perfumers are pictured mixing oil and perfume in cylindrical travertine pots, the process being explained by captions. Several New Kingdom tombs of Thebes provide more information. In that of Amenmes (TT 89) at Sheikh Abd el-Qurna, two men are seen filtering a substance with a strainer. The liquid is collected in a basin and then transported to two other perfumers responsible for overseeing the cooking of the oil in a container on a hotbed of embers (Shimy 1998: 336–7; El-Shahawy 2010: 131). In an anonymous grave (TT 175), scenes depict the pounding of the aromatics and the cooking of fat, a rare scene shows cold or hot enfleurage, and another one shows the filtering of the liquid by twisting a sack and the maceration of scented wood chips (styrax?) in oil (El-Shahawy 2010: 138). In the tomb of Qenamon (TT 93) in Sheikh Abd el-Qurna, scenes show the mode of preparation of a mixture of various ingredients in a cooking pot placed on a stove and a man stirring the mixture with a stick (El-Shahawy 2010: 131).

The so-called *"relief du lirinon"* conserved in the Louvre Museum (inv. no. E 11377; Twenty-Sixth Dynasty, 672–525 BCE) evokes the sequence of operations to realize the essence of lily. Women are seen picking lilies in a field and putting them in baskets to keep them fresh. The flowers are then macerated in oil. The macerated flowers are then placed in a linen sack, and women are seen twisting the sack with the help of sticks to release the oil. According to a description by Pliny (23–79 CE; Plin. *NH* XXI 5), the "lily oil" or *lirinon* oozed into a jar placed on a support. The final product was then presented to the owner, Pairkap, also known as Psametikmerneith (Bénédite 1921; Desroches-Noblecourt 1975).

The tomb of Petosiris in Tuna el-Gebel shows a rare set of scenes of perfumers at work with specific but traditional instruments. A succession of operations are depicted: sorting, pounding, cooking, and checking the quality of the products, most with faulty captions or explanations that we will briefly try to interpret. Roughly speaking, the sequence was as follows: the contents of a sack of aromatics were spread on the ground where men are pictured sorting them out under the watchful eye of a scribe. These aromatics were then pounded in high-edged mortars or crushed on low tables. According to the recipes given in Edfu's laboratory, the frankincense was first ground with a mortar. The aromatics and products of Punt were then added and the mixture was cooked in a cauldron resting on a cylindrical stove; a man is seen beating the firebox with a stick to keep the temperature constant while another man stirs the preparation with a stick. A caption clearly identifies this activity: "The perfumers who make frankincense with a sweet perfume." The end product is then seen being poured into jars, then packaged and presented to the owner of the tomb for his use (Lefebvre 1923: 14–15, inscriptions nos. 37, 41, pl. X–XI; Lefebvre 1924: 58–9; Cherpion et al. 2007: 41–2, 47–9, inscriptions nos. 42–3). However, neither these scenes nor their captions fully cover the extreme complexity of these operations over time, which are clarified in the recipes kept at Edfu, Dendara, Kom Ombo, and Philae. (cf. Chapter 7).

MESOPOTAMIA

Cale Johnson

Just as the signs or characters in one writing system can be coincidentally identical to the graphemes in another (meaning that the significance of any grapheme must be rigorously demonstrated in its originating context), the use and significance of tools, installations, and other artifacts must likewise be demonstrated. The simple fact that a sieve can be used in a specifically chemical process, for instance, does not in itself demonstrate that this was its primary use: it could just as well (and more parsimoniously) have functioned as a kitchen colander or steamer. If we peruse the secondary literature on the history of chemistry in Mesopotamia, many of Martin Levey's numerous shorter papers amount to arguments from the shape or possible function of vessels alone. Unusual forms of pottery, whether a vessel with a very low spout (Levey's "drip bottle") or the double-rimmed vessel from Tepe Gawra that appears in a number of his articles, could well have been developed for a specifically chemical process, but Levey provides no evidence for this from archaeological context (including residues) or iconography.

It is useful to keep Jean Bottéro's work on cooking recipes firmly in mind whenever we need to review Levey's proposals: is there any possible use of an unusual vessel for cooking a particular type of food? The most promising example of specifically technical artifacts, which we looked at in Chapter 3, are the copper-working tools from Tell edh-Dhiba'i, in the suburbs of present-day Baghdad. Christopher Davey's work on these materials presents us with an ideal demonstration. They were excavated from a site that included waste products from the production and manipulation of metals, and some of the tools themselves, such as crucible 4, were found to contain residues of "copper, tin, arsenic and traces of iron and nickel. The crucibles were therefore used for the melting of bronze" (Davey 1983: 174). Davey goes on to describe technological parallels from Egypt and surveys the iconographic evidence for the particular form of crucible found in Tell edh-Dhiba'i, namely the scene from the Tomb of Mereruka at Saqqara (Davey and Edwards 2007: 147).

As was the case with Davey's work on the crucibles from Tell edh-Dhiba'i, work on technological procedures in Mesopotamia and Syria must often look to depictions of these processes from outside of the region, typically to Egypt or even the Greco-Roman world. There are only a few depictions of "production processes" in reliefs and on cylinder seals from Mesopotamia, and these are, for the most part, early and focus primarily on agricultural or food production activities rather than specifically chemical ones. The Early Dynastic IIIb (ca. 2500–2400 BCE) milking scene from the Temple of Ninhursag in Ubaid (IM 513) is the most famous example, including the milking of the cows and the processing of dairy fats in vessels that correspond to the proto-cuneiform signs $UKKIN_a$ and DUG_b (Gouin 1993: 136–7; but see Englund 1998: 159–60 for the proto-cuneiform parallels).

FIGURE 8.1 Temple frieze from Ubaid. Courtesy of www.BibleLandPictures.com/Alamy Stock Photo.

FIGURE 8.2 Late Uruk cylinder seal impression of women working. Courtesy of The Metropolitan Museum of Art.

There are, however, even earlier depictions of craft or processing activities in cylinder seals from Iran in the second half of the fourth millennium BCE, such as pottery-making or weaving, although it is often difficult to identify the specific activity or the tools involved.

Although there are a few other images of processing activities such as butchering or the processing of dairy fats, the only other major sources of iconographic representations of these kinds of activities are found in depictions of military camps from the Neo-Assyrian reliefs. The workshops of the encamped military were obviously a point of pride for Neo-Assyrian rulers, and they appear in a number of different reliefs from their palaces. The best known of these is the circular depiction from the throne room of Assurnasirpal II's Northwest Palace (865–860 BCE). As Bottéro described it:

> a bird's eye view revealing the ground plan of a fortress and a cross section showing the activity within four chambers of the building. The four scenes include (beginning in the upper left and continuing counter-clockwise) a person opening wine jars to let them breathe, two people in the butcher's shop where a sheep is being dressed, a baker tending his oven and two women preparing various foods.
> (Bottero 1985: 39; see the detail photograph in Brereton 2018: 91)

Other depictions from Neo-Assyrian reliefs, such as the scenes from camp life in IM VA 965, are less iconic in their framing of the context, but the representations are very similar (Marriott and Radner 2015: 136). A rare depiction of nonmilitary life comes from contemporary (ca. 800–600 BCE) Iran,

FIGURE 8.3 Workshops in an Assyrian military camp. Relief from Royal Palaces of Nineveh, ca. 645 BCE. Photography by DEA/G. DAGLI ORTI/De Agostini via Getty Images.

in the relief fragment known as "the Spinner," in which an elite woman spins thread in what looks like a ritual or performative context. Of course, for our purposes here the salient characteristic of these depictions is that none of them relate directly to specifically chemical procedures.

If only to provide a foundation for what follows, it is best to start with the deities associated with the kiln, the vat, and the womb. For the kiln, or more broadly the oven, the deities of torch and censer, namely Gibil (a.k.a. Girra) and Kusu, are probably the most relevant. The bulk of the incantations associated with Gibil and Kusu focus on cultic purification of one kind or another (often summarized under the rubric of *Kultmittelbeschwörungen* in the specialist literature), and in Michalowski's foundational "Torch and Censer" paper (1993), there are numerous sites for this act of purification:

> Gibil raised his judicious head towards the heavens.
> He takes hold of water from the holy teats of the heavens;
> That water purifies the heavens, it cleanses the earth!
> It cleanses the ox in the stall,

> It cleanses the sheep in its pen,
> It cleanses Utu at the base of the heavens,
> It cleanses Nanna at heaven's zenith!
>
> (Michalowski 1993: 156)

But the more relevant part of this same incantation tradition is the role of the oven. In an incantation dedicated to the god of the censer, Kusu (but beginning with the god of the torch, Gibil), we hear of a list of ingredients ("cedar, cypress, zabalum, and boxwood, / white wool, black wool, black kin-tree and white kin-tree, / a string of dried apples and dried figs tied to a long string, / ibex horn and ghee") that have been purified by the torch. After a break, the incantation pronounces that "the oven has been purified, the oven has been made holy." The censer god Kusu then puts "abundant sheep and oxen into the mighty oven," and the incantation concludes with the cleansing of pen, stall, the base of heaven, and its zenith that we looked at a moment ago.

And since the deity at the center of the glassmaking ritual from Oppenheim's Alpha group is almost certainly Kusu, it is little wonder that there are numerous parallels between the Kusu Hymn and ritual for the foundation of a kiln. Here is Oppenheim's translation, with Kusu substituted for Kubu:

> As soon as you have completely finished [putting the kiln in place], you place (there) the [Kusu]-images, no outsider or stranger should (thereafter) enter (the building), an unclean person must not (even) pass in front of them (the images). … On the day when you plan to place the "stone" in the kiln, you make a sheep sacrifice before the [Kusu]-images, you place juniper incense on the censer and (then only) you make a fire in the hearth of the kiln and place the "stone" in the kiln.
>
> (transl. after Oppenheim 1970: 32–3, all occurrences of $^d ku_3$-*bu* in this passage are probably to be read as $^d ku_3.su_{13}$)

Ovens and kilns often function metaphorically as contexts of transformation, much like the vat and womb elsewhere in Mesopotamian literature, as, for example, in the phrase "the oven of humankind" in *The Class Reunion*, line 72 (Johnson and Geller 2015: 195–6).

We will turn to the beer-making vat below, but here we should also look at the deities and iconography associated with the womb. The pregnancy- and birth-related goddesses have been admirably surveyed in Frymer-Kinsky's *In the Wake of the Goddesses* (1992) and Stol's *Birth in Babylonia and the Bible* (2000), and need not be rehearsed here. More recently, however, Ulrike Steinert and Erica Couto-Ferreira have devoted a number of studies to the metaphors associated with the female body, and in particular the womb.

Here we see the mother goddess symbolized with the heads of babies sprouting from her shoulders, and on either side the so-called "omega"

FIGURE 8.4 Terra-cotta plaque with omega and Kubu. Courtesy of www.BibleLandPictures.com/Alamy Stock Photo.

symbol above the crouched and emaciated body of an Akk. *kūbu* "fetus-demon." As Steinert summarizes the situation:

> The central symbol of the mother goddess ... is the symbol resembling the Greek letter Omega, depicted on numerous seals from Mesopotamia, Syria, Anatolia and the Levant ... The identification of the "omega" symbols as an emblem of the mother goddess was already established by William Hinke (1907: 95, 121) who noticed that on the kudurrus the symbol regularly appears in the fourth position, beside the senior divine male triad (Anu, Enlil and Enki), which corresponds with the position of Ninhursaga/Ninmah in some of the lists of deities in the curse section of the kudurru inscriptions.

> Although a number of interpretations of the symbol have been suggested, Henry Frankfort's reading, corroborated by Egyptian parallels, as the schematic representation of the bicornate uterus of a cow remains the most generally accepted.
>
> (Steinert 2017: 207)

The omega symbol, therefore, as a representation of the reproductive system of a cow, ties together the many different metaphors involved in childbirth incantations and rituals, such as the frequent use of bovine imagery.

The dominant metaphor for the womb in literary sources is agricultural, and this metaphor finds its prototypical expression in the myth known as *Enki and Ninhursag*; the womb (represented by Sum. ša$_3$ "belly, womb, insides") combines with semen (represented by Sum. a "water, semen") to form, in the minds of the literati, Sum. a.ša$_3$ /ašag/ "agricultural field." And it is, after all, in *Enki and Ninhursag* that we find the earliest etiology for difficult childbirth (and the role of the midwife) in Mesopotamian literature. The text plays with the orthographic ambiguity of the two cuneiform signs ŠA$_3$ and A in various ways, but most dramatically in the account of the matrilineal genealogy in which Enki, the god of fresh water, wisdom, and the technical crafts, fathers a line of female goddesses, each with a "womb" (Sum. ša$_3$) that is an order of magnitude smaller than her predecessor. This leads invariably to the figure of Uttu, the goddess of weaving, whose womb is too small to give birth, obligating the mother goddess to extract Enki's ever-fecund semen (see Chapter 4).

In the technical literature, however, primarily involving incantations found in first-millennium BCE medical compendia, a number of metaphors are used to describe the womb as a container filled with liquid: a waterway in a dammed-up meadow, the image of a waterskin, and that of a fermenting vat (Steinert 2013: 10). These metaphors were often extended to express the dangers of a leaky womb, such as in BAM 237 ii 1′–6′ and parallels, in terms of "a fermenting vat whose stopper is defective or a water-skin whose knot and drawstring are failing" (Steinert 2013: 10). Similar metaphors are found in compendia dedicated to the treatment of illnesses of the digestive tract, and there are numerous other similarities between these two groups of materials. Perhaps the most important feature of the metaphors used to describe these unseen domains within the body is that they are conceptualized as containers for liquids, often fermented alcoholic beverages, but they are not conceptualized primarily in terms of cooking, and are therefore quite different, in this respect, from Greco-Roman materials (see Stol 2006).

In both of these domains, literary and technical (although much more transparently in the technical materials), the dominant metaphor, in terms of the George Lakoff-Mark Johnson conceptual metaphor theory, is THE BODY IS A CONTAINER (by convention conceptual metaphor theory puts distinct metaphors in capital letters). This idea is at the center of a number of the contributions to *The Comparable Body*, a conference volume edited by John Wee that appeared in 2017. Indeed, Steinert's contribution to *The Comparable Body* works its way through THE BODY IS A CONTAINER schema, as applied to the Mesopotamian textual record, in a thorough and theoretically sophisticated way. One of the main ideas of conceptual metaphor theory is that nonbodily domains can be used as models to conceptualize unseen processes within the human body, and THE BODY IS A CONTAINER is a straightforward first step in such a conceptualization. At the same time, however, it also seems that processes of fermentation were part of the conceptualization of transformation in Mesopotamia. No doubt, there is, hidden in among the container metaphors, a native model that used fluids (ranging from semen and alcohol to medicines in a liquid carrier) in more or less well sealed containers in order to represent internal states (Johnson 2017). Whether semen added to the womb, an alcoholic beverage introduced into the stomach, or a medical "potion" given to a patient, all of these procedures probably draw on a shared conceptual model.

In both Classical Sumerian mythology and a wide array of later incantations, the womb (in later texts, Sum. ša$_3$.tur$_3$ = Akk. *šassūru*, but in the Sumerian mythological texts, still simply ša$_3$) always functions as the premier transformative environment. Yet as Steinert has emphasized, analogous transformational contexts from beer-making and metalworking were often used to describe processes of transformation – above all, the beer-making vat, the smelting oven, or the crucible of the metalworkers, and the human digestive tract or "stomach" (also denoted by ša$_3$ in Sumerian). Each of these domains deserves a place in the early history of Mesopotamian conceptualizations of chemical reactions, but other than the womb, to which we have already referred, the beer-making vat played an especially powerful role in how Mesopotamians conceptualized the transformation of a series of raw materials into a new kind of substance. The fermentation process and the consequent creation of alcohol represented one of the most important (bio)chemical reactions in early Mesopotamian thought. The beer goddess Ninkasi is first attested in the earliest god lists in the mid-third millennium BCE, and *The Ninkasi Hymn* (first edited in Civil 1964, now re-edited in Sallaberger 2012) is one of our most important descriptions of a production process in Sumerian literature.

The Ninkasi Hymn 13–36
13. Your sourdough starter (lit. sprouted dough), formed with an enormous shovel,
14. The dough, with spices and honey mixed into the "well,"
15. O Ninkasi! Your sourdough starter, formed with an enormous shovel,
16. The sourdough, with spices and honey mixed into the "well,"
17. Your sourdough, baked in the huge oven,

18. It is smooth like the feel of hulled grain,
19. O Ninkasi! Your sourdough, baked in the huge oven,
20. It is smooth like the feel of hulled grain,
21. Your malt, covered with meal and with water poured in,
22. It is undulating vermin,
23. Ninkasi, your malt, covered with meal and with water poured in,
24. It is undulating vermin,
25. Your mash, with water added to the vessel,
26. It is the rising and sinking of waves.
27. Ninkasi, your mash, with water added to the vessel,
28. It is the rising and sinking of waves.
29. Your finished mash, spread out on huge reed mats,
30. The contents cooled (and) the … appeased,
31. Ninkasi, your finished mash, spread out on huge reed mats,
32. The contents cooled (and) the … appeased,
33. Your big (mass of) dried beer mash, ready to be formed,
34. Moistened with honey and wine,
35. Ninkasi, your big (mass of) dried beer mash, ready to be formed,
36. Moistened with honey and wine.

(transl. after Sallaberger 2012)

When Miguel Civil, undoubtedly one of the most important Assyriologists of the twentieth century, edited this text in 1964, more than fifty years ago, he could only arrive at a somewhat impressionistic description of what the hymn depicts: "the various steps of the brewing process … described in a poetic, but clearly recognizable, way" (1964: 67). Civil referenced "beer-making recipes" (viz. quantifications of ingredients and products, but not step-by-step processes), and one could easily infer from his edition that *The Ninkasi Hymn* is a poetic rendition of a beer-making recipe. In the half-century since then, two major advances have produced a far better interpretation of the text (and in doing so teach us something very important about how advances in understanding texts like this actually take place): Marten Stol produced a series of incredibly detailed studies of the realia involved in beer-making in the 1970s and 1980s (Stol 1971; Stol 1987–90), and more recently Zarnkow and coworkers were involved, as part of the excavations at Tell Bazi (northern Syria), in replicating the beer-making process in line with this and many other texts from early Mesopotamia (Zarnkow et al. 2006; Otto 2012).

These lines of research help us to see that *The Ninkasi Hymn* is not actually a recipe for producing beer, directly transmuted into a piece of Sumerian literature (thus quite different from an Akkadian parody like *At the Cleaners*); it is, instead, a hymnic description of the attributes of the goddess Ninkasi – what I have recently described as a type of ekphrastic description (Johnson 2019) – that lists each of the most important attributes of the goddess. Some of these attributes happen to correspond to steps in the production of sourdough, fermented mash, and a dried form of mash used for storage and long-distance journeys. The importance of these processes for our purposes here, however, is that they represent some of the most important metaphorical and symbolic meditations on chemical processes in all of Mesopotamian literature. The process can be summarized as the combination of an enzyme-rich "sourdough" (Sum. bappir = Akk. *bappiru* or *pappiru*), which takes the place of commercial yeast in modern recipes, with a sugar-rich mass of "malt" (Sum. $munu_4$ = Akk. *buqlu*) and other grains. This yielded a "fermented mash" (Sum. $sumun_2$ = *narṭabu*) that could be dried and stored in different ways (Sum. $titab_2$ = Akk. *titāpū*, or Sum. dida = Akk. *billatu*). Crucially, the key metaphors for the chemical processes involved in the making of beer are the movement of masses of insects ("undulating vermin" in lines 22 and 24) and the movement of waves ("the rising and sinking of waves" in lines 26 and 28). This is depicting nothing other than the normal operation of the enzymes, yeasts, and bacterial action that take place in a sourdough starter. And this type of kitchen chemistry was undoubtedly at the very center of Mesopotamian conceptualizations of other, less accessible chemical processes.

Yet nowhere in *The Ninkasi Hymn* do we find the term that has taken on a seemingly crucial role in the secondary literature for Mesopotamian contexts of transformation, namely Sum. $agarin_{3/4/5}$ = Akk. *agarinnu*. The danger with purely lexical studies is that they take the results of lexicographical study and generalize them into models of intellectual or cultural history. If, for example, we look at the entry for *agarinnu* in the standard Akkadian lexicon in English (CAD), we find three definitions: "beer mash," "mother," and "crucible"; it would be very easy to generalize *agarinnu* into a single term meant to cover all of the most important contexts of transformation in ancient Mesopotamia. If we make this term into a kind of symbolic representation for transformation in Mesopotamian thought, we are really imposing our own ideas on Mesopotamian society. As Sallaberger (2012) points out, the few lexical attestations of *agarinnu* take on a much clearer, if less dramatic, meaning in the context of the processes depicted in *The Ninkasi Hymn*. In the lexical list known as Hh XXIII, lines 1′–15′, the section moves from Sum. sa.hi.in.du "sourdough-yeast" through a number of different states or kinds of Sum. bappir "sourdough" and concludes with Sum. $sumun_2$ "mash." And it is near the start of this sequence that Sum. $agarin_4$ = Akk. *agarinnu* occurs, so it is clear from this context that it refers to the sourdough starter (containing lactic acid bacteria and wild yeast) before it was mixed into a new sourdough. Its extension to metalworking (Sum. $^{uruda}agarin_3$ = Akk. *agarinnu* in Diri IV 83), though with a distinct spelling and not used outside of lexical lists, is already found in the Old Babylonian period. The orthography of Sum. $agarin_4$, conventionally AMA.ŠIM but identical to $AMA.BAPPIR_2$, suggests that it was etymologized as the "source of the sourdough," with Akk.

ummu (= Sum. ama) in one of its perfectly regular senses ("source, origin") rather than a complex metaphor involving "mother" and hence "womb" (see generally Steinert 2017: 300).

GRECO-ROMAN WORLD

Matteo Martelli

Hephaestus (or Vulcan in Latin mythology) was the god of metalworking, a smith himself, expert in the working of a variety of metals, such as gold, silver, bronze, and iron. His house and workshop are described in the *Iliad*, when Thetis visits the smith god and asks him to forge new weapons for her son, Achilles (*Il.* XVIII 368–81; see Figure 8.5). Hephaestus' house was bright as a star and had been built in bronze by the god himself. He was also believed to have erected the houses of the gods, temples, and the palace of Alcinous, which had metallic walls and a bronze threshold:

FIGURE 8.5 Hephaestus working in his workshop, red-figure *kylix* (foundry painter, ca. 490–480 BCE), Altes Museum in Berlin. Public domain. Courtesy of Wikicommons.

> The brilliance over the high-roofed house of great-hearted Alkinoos / was like that of the sun or moon. There were bronze walls / that extended on both sides of the threshold into the interior, / and the topmost row of stones was made of lapis-lazuli (*kyanos*).
>
> (*Odyssey*, VII 84–7; transl. by Powell 2014)

After deciding to grant Thetis' wish, Hephaestus got down to work, and produced the famous shield of Achilles (*Il.* XVIII 470–608). Twenty self-moving bellows fed the fire below the crucibles, and Hephaestus put various metals on the fire. He worked bronze (or copper; *chalkos* in Greek), tin, precious gold, and silver on a great anvil, handling a heavy hammer with one hand, the tongs with the other. The metals were used to forge five different layers of the shield, as well as many details of the rich imagery that decorated it. The semiprecious blue stone *kyanos* (see Chapter 2, pp. 81–2) was also added to enrich the decoration. Hephaestus creates a scene of a vineyard full of large golden grapes, marked off by a dark blue trench (i.e. made of *kyanos*), and a fence of tin (*Il.* XVIII 564–5; for a recent and up-to-date discussion of the technology depicted in *Iliad* XVIII, see D'Acunto 2009).

As pointed out by Acton (2014: 7), since the preclassical period, craftsmen were well regarded in ancient societies, and their images in bronze and on vase paintings were common until at least the fifth century BCE. Some patterns may be identified in the way manual workers were represented: they usually have short hair, sometimes with bands or wreaths on their head; adult craftsmen usually have beards, while young workers appear beardless. Baldness has been interpreted either as a sign of old age or as an indication of coarseness. Bald bronze workers with long or snub noses, sunken jaws, and swollen lips, such as those depicted in a two-handled jar (*pelikē*) of the Museum of Fine Arts in Boston (inv. 13100; 440–430 BCE) or in a *lekythos* of the Museum of the Rhode Island School of Design in Providence (inv. 25109; 470–460 BCE), might seem to be caricatured figures (Chatzidimitriou 2014: 79–81). Indeed, a certain humor against workers has been detected in a *stamnos* (a type of jar) of the Metropolitan Museum in New York (inv. 17.230.37; 460–450 BCE), which depicts a carpenter who turns away from a rich woman; a slave between the two figures turns her face toward the mistress while pinching her nose, presumably because of the bad smell of the worker (Mitchell 2009: 70–1). On the other hand, bald old craftsmen were certainly more experienced, and sometimes baldness could simply mark workmen who had properly mastered their art (Birchler Emery 2008).

In vase paintings, craftsmen often wear a *himation*, a loose tunic that is either wrapped around their bodies, especially when they are seated, or tied around their waists when they are depicted working while standing. Along with semidraped workers, naked craftsmen are quite common, as recently pointed out by Roger Ulrich:

The Attic potter showed many of his subjects at work firing pottery, mining, woodworking, blacksmithing and plowing wearing nothing at all. ... It has been recognized that Greek athletes commonly exercised while nude, but one must wonder about the comfort and safety of working without any protective clothing.

(Ulrich 2008: 56)

If nudity may have been part of the artistic convention, blacksmiths and potters in particular, even when represented unclad, often wear a leather *pilos* ("skullcap"; see Figure 8.6), which was expected to protect artisans who work next to furnaces (Chatzidimitriou 2014: 65–71). Moreover, tools and instruments often accompany the depicted scenes (Ulrich 2008). For instance, a series of eight painted votive plaques found in the sanctuary of Poseidon at Penteskouphia (in the proximity of Corinth; sixth century BCE) display naked male potters next to kilns used for the firing process (see Figure 8.7; Hasaki 2019). In many plaques, beehive-shaped kilns stand at the center of the depicted scenes; they are usually slightly taller than the craftsmen, with the vent hole at the top and the loading door for inserting or removing the pottery depicted at the side. In only one case (Berlin, *Antikensammlung*, inv. 802b), a ladder is necessary to reach the top of a kiln.

FIGURE 8.6 Bronze workshop, red-figure *kylix* (foundry painter, ca. 490–480 BCE), Altes Museum in Berlin. Public domain. Courtesy of Wikicommons.

ART AND REPRESENTATION 229

FIGURE 8.7 A Corinthian black-figure *pinax* (ca. 575–550 BCE) found in Penteskouphia. Public domain. Courtesy of Wikicommons.

In some cases, ancient vases also feature gods and women at work. In scenes of weaving or spinning, women usually wear the *peplos* (a long garment with a broad overfold) or *chiton* (a sleeveless, rectangular garment); the same iconography appears in the Caputi *hydria* of the Torno collection in Milan (fifth century BCE), which depicts a woman decorating the handles of a volute-krater (i.e. a large vessel used to dilute wine with handles that curl in a volute). Hephaestus, on the other hand, is depicted in the interior scene (*tondo*) of the famous Berlin Foundry Cup (early fifth century BCE), a red-figure *kylix* (a wide-bowled drinking cup) kept at the Altes Museum in Berlin (see Figure 8.5); Thetis is standing in front of a bearded Hephaestus, who is seated on a stool cushioned with a fine pillow; he wears a loose tunic, according to the pattern that we have discussed above. The god holds a heavy hammer in one hand and Achilles' helmet in the other, while a hammer and an anvil are depicted behind the goddess Thetis. The scene continues in the exterior part, where we find one of the most detailed pictures of a bronze workshop from antiquity: a bearded male figure stokes coals in a furnace, while in the back a second young worker manipulates bellows to kindle the fire. Other craftsmen work to finish a statue, and various tools (hammers, a saw), along with various parts of statues (heads, feet, and hands), are depicted in the background (Mattusch 1980; see Figure 8.6).

In addition to potters and bronze-workers, other professions were also represented. For instance, three perfumers seated in an Athenian perfume shop are depicted in a *lekythos* (oil flask; late sixth century BCE) held by the Museum of Fine Arts in Boston (inv. 99.526). Oil flasks were the most common containers for perfumes and cosmetics in antiquity. The Boston black-figure vessel features three red-bearded men wearing elegant mantles and seated on sophisticated stools with animal legs. The man at the left holds a flower in his hand, perhaps one of the ingredients used to prepare the perfumes that were sold; the man at the center of the scene holds a siphon, probably used to permit the customers to taste the scented oils. On the other hand, a *pelikē* (a wine container; 520–490 BCE) kept at the National Archaeological Museum in Florence (inv. 72732) displays an old bald oil merchant who calls to a mantled woman and, at the same time, uses a stick to pick some oil (or wine) from an amphora in front of him: the painter recorded the words of the merchant, who takes hold of the woman's arm and says, "It is good!" The woman, however, eloquently pinches her nose (Mitchel 2009: 70–1).

As for the professional figures in the drug business, who are referred to with a variety of terms in ancient sources (root-cutters, root-sellers, drug-dealers, or perfume-makers), literary and scientific texts provide us with a more ambiguous representation. Scholars have often emphasized how these figures had a poor reputation in antiquity, at least according to the pictures depicted in medical sources (Boudon-Millot 2003; Samama 2006). After all, there was a certain degree of competition with the physicians, who also used *pharmaka* to cure their patients. As Laurence Totelin (2016b: 72–3) points out, drug-sellers knew about more than just plants, poisons, and aphrodisiacs; they also dealt with spiders and snakes. Aristotle, for instance, after observing the specimens of snakes and spiders kept alive in the shops of drug-sellers, infers that these animals can live for a long time even without food (*Hist. An.* VIII 4, 594a20–4). Aristotle also refers to the weak bites of the snakes of the druggists (*Hist. An.* IX 39, 622b34), who, in some cases, might have let themselves be bitten by harmless spiders in order to show the efficacy of the *pharmaka* they sold to their customers.

As for perfume-makers, Galen's representation is somewhat ambivalent. In some cases, he praises their expertise in the preparation of medicines; on the contrary, in a passage of his treatise *On the Capacities of Simple Drugs* (II 27 = XI 537 Kühn), he criticizes their method for making rose oil (*rhodinon*); perfume-makers, in fact, tended to adulterate its composition by adding ingredients that could strengthen the scent of the oil. On the other hand, Galen includes medicinal recipes taken from drug-sellers and root-cutters in his treatises on the composition of drugs and antidotes for poisons (Guardasole 2006), and Latin inscriptions found in Brescia (*CIL* V 4489, second to third centuries CE) preserve a record of the wealthy college of the *pharmacopolae* who were also in charge of estates (Totelin 2016b: 78, 85).

A certain opposition between *otium* ("leisure") and *negotium* ("daily business") is certainly evident in the work of Roman philosophers, such as the Stoic Seneca (first century CE). In *Epistle* 90, for instance, he criticized the earlier philosopher Posidonius, who attributed the invention of important technologies to ancient philosophers. Seneca contrasts wise philosophers, who must train their minds more than their hands, with industrious artisans, but other sources show the value that Roman craftsmen attributed to their own art. Roman reliefs on gravestones, in fact, often display craftsmen alone represented along with a few tools of their art. Likewise, many frescos discovered in Pompeii depict craftsmen or, in some cases, cupids who practice different arts. In the famous House of the Vettii, a room was decorated by thirteen panels depicting cupids and psyches at both work and play (62–79 CE). Ten of these panels remain, and three of them deal with arts. In one fresco, various cupids are depicted as perfume-makers extracting flagrances from flowers and stirring the scented oil in a cauldron (see Chapter 3, Figure 3.6). In another fresco, goldsmith cupids are depicted: a cupid works on a gold bowl at the right, a psyche melts the metal, while various objects for sale are depicted at the center of the scene. Finally, a third fresco depicts cupids at work in a cloth-treating shop or fullery (Clarke 2003: 100–5).

Scholars usually agree that fullers suffered from a poor reputation, probably due to the dirty nature of their work (see Chapter 5). While literary texts such as those of Roman historians and comic dramas (see Flohr 2013a: 324–8) seem to support this view, other sources, in particular various paintings in *fullonicae* excavated in Pompeii, seem to provide a different picture. For instance, *fullonica* VI 8.20 is decorated by four frescos that represent fullers at work (see Figure 8.8). As stressed by Clarke (2003: 117), the owner of the building, who certainly commissioned the frescos, "wanted to show his pride not only in his establishment but also in his workers. If he instructed the painters to make into portraits the vignettes of the workers … – as I believe he did – then he wanted to show not just what his employees did but also what they looked like."

The crafts were not only represented in reliefs and paintings – they were also praised by ancient scientists and physicians. In his *Exhortation to Study the Arts* (*Protrepticus*), Galen contrasts those who trust blind Fortune to those who cultivate their rational soul through the practice of the crafts (*technai*). Hermes is the patron of the second group, which is portrayed by Galen as a band marching in three circles (*Protr.* 5.1 = I 7 Kühn). Geometers, mathematicians, philosophers, doctors, and astronomers march closest to the god; painters, sculptors, carpenters, and experts in all the crafts are included in the other two bands. On the contrary, those who neglect to learn a craft (*technē*) are blind followers of the daimon Tychē ("Fortune"); unable to develop professional skills in any particular craft, they entrust themselves to the unpredictable Fate for their success. Indeed, learning a craft could be difficult and time-consuming,

FIGURE 8.8 Fullers at work. Fresco, Pompeii, *fullonica* VI 8,20. Public domain. Courtesy of Wikicommons.

and the easier path of Tychē could have tempted ancient craftsmen. Like Galen, the Greco-Egyptian alchemist Zosimos of Panopolis adopted a firmly critical stance against those who preferred to trust fortune rather than committing themselves to a study of the fundamental writings of the alchemical art. In his work *On the Letter Omega* (*Authentic Memoires*, I in Mertens 1995: 1–10), Zosimos directs a vitriolic attack against those alchemists who did not recognize the value of *On Furnaces*, an alchemical book ascribed to Maria the Jewess (see Chapter 3). In particular, those who, without being trained, are fortunate and succeed in dyeing techniques dismiss the teachings of *On Furnaces* as false or unnecessary and heap ridicule on the book. However, when their success comes to end, they reluctantly confess that there was some truth in the book. They are, in fact, escorts of Fate in its procession.

The Egyptian priest Neilos clearly exemplifies this attitude. In the chapter *On the Treatment of the Body of Magnēsia* (*CAAG* II 188–91), Zosimos fiercely criticizes Neilos and his entourage, who gave foolish and vain instructions about alchemical tinctures. Seeking false methods for obtaining gold, they did not base their practice on a solid theoretical foundation and, for this reason, easily made fools of themselves. This becomes evident at the end of the chapter, when Zosimos depicts the priest Neilos as a ridiculous novice. It is worth quoting the passage in full, since it provides us with one of the rare descriptions of an alchemist at work in Greco-Roman Egypt:

> For instance, once your priest Neilos provoked laughter, when he roasted *molybdochalkos* in a baking-oven: so that, if one adds some "bread" [i.e. slabs of *magnēsia*], he ends up kindling [the fire] with *kōbathia* [arsenic ores] all day long. Blind in his bodily eyes, he did not understand the failure he was doomed to, but he even puffed up with conceit, and he collected and showed the ashes, after they cooled down. When he was asked "where is the whitening?," he was at a loss and answered that it went deep into [the ashes]. Then he added copper, dyed its dross: in fact, there was nothing solid [to be dyed]. Then, he desisted and went away; and he fled deep into [those ashes], under which the same whitening of the *magnēsia* went. ... Pass my greetings to Neilos, the "*kōbathia*-burner".
>
> (*CAAG* II 191)

In this passage, Zosimos plays with the alchemical concept of "deep tincture," according to which the dyeing substances must sink into the metal in order to produce its complete transformation. On the contrary, Neilos' procedure does not bring about any transformation; the whitening penetrated so deeply, Zosimos says ironically, that it was not visible at all. From a technical point of view, Zosimos stresses that Neilos used a *klibanos* (Lat. *clibanus*), a portable oven used to bake bread, to make *magnēsia* white. Even though the *klibanos* could probably serve different purposes – it is, indeed, mentioned in pharmacological and alchemical texts as a tool for roasting a variety of ingredients (e.g. *CAAG* II 289, 346, 369, 391) – it is difficult to escape the impression that Zosimos disapproved of Neilos' alchemical experiment, which is presented as an extemporaneous performance. His approach to alchemical practice was as shallow and superficial as that of his followers. They were too self-confident and arrogant to listen to more skilled and experienced alchemists and, we can infer, to waste time reading the writings of the ancients – such as the treatise *On Furnaces* that Zosimos discusses in *On the Letter Omega*. Being more interested in making gold than in theories (*CAAG* II 190,20–1), these alchemists dared to prepare alchemical tinctures rashly and without understanding.

The tension between learned alchemists and untrained practitioners who approached alchemy with the sole purpose of making money (often by cheating their clients) is also evident in the few nonalchemical sources that deal with this craft in late antiquity. On the one hand, in his dialogue *Theophrastus*, the Neoplatonic philosopher Aeneas of Gaza (d. ca. 518 CE) claims that alchemists are able to turn matter into a superior state, as they do when transmuting metals into excellent gold or when making brilliant glass (Colonna 1958: 62–3; see Beretta 2009: 108–9). On the other hand, the Byzantine chronicler Malalas (sixth century CE) has preserved the following record of the misfortunes of an alchemist (actually a trickster) from Antioch, who tried to cheat the Byzantine emperor Anastasios I (reigned 491–518 CE; *Chronicle*, XVI 395 Dindorf):

> During [Anastasios'] reign a man named John Isthmeos, who came from the city of Amida, appeared in Antioch the Great. He was an alchemist and a tremendous imposter. He secretly went to the money dealers and showed them some hands and feet of statues made of gold, and also other figurines, saying that he had found a hoard of such figurines of pure gold. And so he tricked many of them and conned them out of a lot of money. The Antiochenes nicknamed him Bagoulas, which means a slick imposter. He slipped through everyone's fingers and fled to Constantinople, and there too he conned many money dealers and so came to the emperor's attention. When he was arrested and brought before the emperor, he offered him a horse's bridle of solid gold with the nose-piece inlaid with pearls. The emperor Anastasios took it, saying to him, "He you will not con," and banished him to Petra, where he died.
>
> (transl. by Jeffrey et al. 1986: 222)

CONCLUSIONS

Marco Beretta

Ancient representations of the chemical arts are both varied and numerous. In Egypt, craftsmen were richly represented in paintings, low-relief carvings, and in groups of small statuettes. However, these iconographic reconstructions did not play a didactic role and their main aim was to transmit the memory of their works. The most important achievements of the chemical arts were covered by secrecy and therefore their visual representations carefully avoided too detailed descriptions of the techniques involved. Despite the lack of technical details, Egyptian tombs offer rich iconographic views on beer- and wine-making, on metallurgical craftsmanship, on the different phases for the preparation of perfumes, and on other arts. In contrast to the richly imaginative world of the Egyptians, in Mesopotamia there were only few iconographic descriptions of the chemical arts, and none of them related directly to the technical processes. However, Mesopotamian texts introduced powerful metaphors such as that of the oven representing the womb that effectively described unseen domains within the body. Chemical processes too, such as a recipe for making beer, alluded to a description of the attributes of a goddess. The use of this kind of metaphor indicates the cultural importance achieved by the chemical arts and instruments.

Hephaestus, the Roman Vulcan, was depicted in his forge in Homer's *Iliad*, and this image became an exceedingly successful way to depict metallurgy within mythological contexts. However, both Greeks and Romans extensively represented craftsmen – with images and in words – that evoked the real workshops in which they operated. The frescos surviving in the House of the Vettii in Pompeii offer a vivid series of scenes of chemical workshops that are enriched with detailed descriptions of operations and instruments. At this time (79 CE), the chemical arts acquired an autonomous role in Roman society and culture.

BIBLIOGRAPHY

ABBREVIATIONS

Classical Greek and Latin authors are mostly cited according to the standard abbreviations of the *Oxford Classical Dictionary* (4th edition) with a few additions:
Diosc. = Dioscorides, *De materia medica*
DRN = Lucretius, *De rerum natura*
Lap. = Theophrastus, *On Stones*
NH = Pliny, *Naturalis historia*
CAAG = Berthelot, Marcelin, and Charles-Emile Ruelle. 1887–8. *Collection des anciens alchimistes grecs*. 3 vols. Paris: Georges Steinheil.

Abbri, Ferdinando. 2000. "Alchemy and Chemistry: Chemical Discourses in the Seventeenth Century." *Early Science and Medicine*, 5: 214–26.
Acton, Peter. 2014. *Poiesis: Manufacturing in Classical Athens*. Oxford: Oxford University Press.
Adams, Robert McC. 1974. "Anthropological Perspectives on Ancient Trade." *Current Anthropology*, 15: 239–58.
Agut-Labordère, Damien, and Michel Chauveau. 2011. *Héros, magiciens et sages oubliés de l'Égypte ancienne. Une anthologie de la littérature en égyptien démotique*. Paris: Belles Lettres.
Ailliaud, Georges-Julien. 1990. "Pastel et indigo ou les origines du bleu." *Revue d'Histoire de la Pharmacie*, 284: 13–20.
Al-Gailani, Lamia. 1965. "Tell edh Dhiba'i." *Sumer*, 21: 33–40.
Albayrak, Irfan. 2010. "The Understanding of Inheritance in Ancient Anatolia According to Testaments from Kültepe." In Fikri Kulakoğlu and Selmin Kangal (eds), *Anatolia's Prologue: Kultepe, Kanesh, Karum: Assyrians in Istanbul*, vol. 78. Kayseri: Kayseri Metropolitan Municipality.
Alizadeh, Abbas. 1985. "A Protoliterate Pottery Kiln from Chogha Mish." *Iran*, 23: 39–50.
Allam, Schafik, 1997. "La vie municipale à Deir el-Médineh: les supérieurs (ḥwtjw/ḥntjw) du village." *Bulletin de l'Institut français d'Archéologie orientale*, 97: 1–17.
Altenmüller, Hartwig. 2015. *Zwei Annalenfragmente aus dem frühen Mittleren Reich*. Hamburg: Buske.

Amigues, Susanne. 1990. "Un conte étymologique: Hélène et les serpents." *Journal des Savants* (3–4): 177–98. Available online: https://www.persee.fr/doc/jds_0021-8103_1990_num_3_1_1535.

Anthony, Flora Brooke. 2017. *Foreigners in Ancient Egypt: Theban Tomb Paintings from the Early Eighteenth Dynasty (1550–1372 BC)*. London: Bloombury.

Aperghis, Gerasimos G. 1998. "A Reassessment of the Laurion Mining Lease Records." *Bulletin of the Institute of Classical Studies*, 62: 1–20.

Archi, Alfonso. 1993. "Bronze Alloys in Ebla," pp. 615–25 in M. Frangipane et al. (eds.), *Between the Rivers and Over the Mountains: Archaeologica Anatolica et Mesopotamica A. Palmieri Dicata*. Rome: Università di Roma La Sapienza.

Archi, Alfonso. 2016. "Egypt or Iran in the Ebla Texts?" *Orientalia* NS, 85: 1–49.

Archi. Alfonso. 2017. "Metals in Third Millennium B.C." In Fikri Kulakoğlu and Gojko Barjamovic (eds), *Movement, Resources, Interaction: Proceedings of the 2nd Kültepe International Meeting Kültepe, 26–30 July 2015*. Turnhout: Brepols.

Arnold, Dieter, and Jeanine D. Bourriau (eds). 1993. *An Introduction to Ancient Egyptian Pottery*. Mainz: Philipp von Zabern.

Aston, Barbara, James A. Harrell, and Ian Shaw. 2000. "Stone." In Paul T. Nicholson and Ian Shaw (eds), *Ancient Egyptian Materials and Technology*. Cambridge: Cambridge University Press.

Aufrère, Sydney H. 1991. *L'Univers minéral dans la pensée égyptienne* (= *Bibliothèque d'Étude*, 105). 2 vols. Cairo: Institut Français d'Archéologie Orientale.

Aufrère, Sydney H. 1998a. "Bastet, l'albâtre, les parfums et les curiosités de la mer Rouge." In Rika Gyselen (ed.), *Parfums d'Orient* (= *Res orientales*, 11). Bures-sur-Yvette: Groupe pour l'Étude de la Civilisation du Moyen-Orient.

Aufrère, Sydney H. 1998b. "Évolution des idées concernant l'emploi des couleurs dans le mobilier et les scènes funéraires en Égypte jusqu'à l'époque tardive." In Sylvie Colinart and Michel Menu (eds), *La couleur dans la peinture et l'émaillage de l'Égypte Ancienne: Actes de la table ronde Ravello, 20–22 Marzo 1997* (= *Scienze e materiali del patrimonio culturale*, 4). Santo Spirito: Edipuglia.

Aufrère, Sydney H. 1998c. "Parfums et onguents liturgiques du Laboratoire d'Edfou: composition, codes végétaux et minéraux dans l'Égypte ancienne." In Rika Gyselen (ed.), *Parfums d'Orient* (= *Res orientales*, 11). Bures-sur-Yvette: Groupe pour l'Étude de la Civilisation du Moyen-Orient.

Aufrère, Sydney H. 1998d. "Religious Prospects of the Mine in the Eastern Desert in Ptolemaic and Roman Times (= Autour de *l'Univers minéral* VIII)." In Olaf E. Kaper (ed.), *Life on the Fringe, Living in the Southern Egyptian Deserts During the Roman and Early-Byzantine Periods: Proceedings of a Colloquium Held on the Occasion of the 25th Anniversary of the Netherlands Institute for Archaeology and Arabic Studies in Cairo, 9–12 December 1996* (= *CNWS publications*, 71). Leiden: Research School CNWS, School of Asian, African, and Amerindian Studies.

Aufrère, Sydney H. 1998e. "Les anciens Égyptiens et leur notion de l'Antiquité. Une quête archéologique et historiographique du passé." *Méditerranées*, 17: 11–56.

Aufrère, Sydney H. 1999. "L'importance des Trésors dans les temples égyptiens (Autour de *L'Univers minéral*, III)." In Annie Caubet (ed.), *Cornaline et Pierres précieuses. La Méditerranée de l'Antiquité à l'Islam: Actes du colloque du Musée du Louvre, 24 et 25 novembre 1995*. Paris: Documentation française.

Aufrère, Sydney H. 2001a. "Le rituel de cueillette des herbes médicinales du magicien égyptien traditionnel d'après le Papyrus Magique de Paris." In Sydney H. Aufrère (ed.), *Encyclopédie religieuse de l'Univers végétal. Croyances phytoreligieuses de l'Égypte ancienne*, vol. 2 (= *Orientalia Monspelliensia* 11). Montpellier: Presses universitaires de la Méditerranée.

Aufrère, Sydney H. 2001b. "Les encres magiques à composants végétaux." In Sydney H. Aufrère (ed.), *Encyclopédie religieuse de l'Univers végétal. Croyances phytoreligieuses de l'Égypte ancienne*, vol. 2 (= *Orientalia Monspelliensia*, 11). Montpellier: Presses universitaires de la Méditerranée.

Aufrère, Sydney H. 2001c. "Maladie et guérison dans les religions de l'Égypte ancienne. Au sujet du passage de Diodore Livre I, § LXXXII." In Jean-Marie Marconot (ed.), *"Représentation des maladies et de la guérison dans la Bible et ses traditions": Maladie et guérison dans les religions de l'Égypte ancienne*. Montpellier: Université de Montpellier.

Aufrère, Sydney H. 2001d, "Chapter 6 Technologie (Technology) § b Metallurgy, Chemistry, Alchemy, Glass-making." In Jan Assmann (ed.), *Storia della scienza*, vol. 1. Rome: Treccani.

Aufrère, Sydney H. 2001e. "Chapter 6 Technology § c Stone-quarrying and Mining, Architecture; Masonry, etc." In Jan Assmann (ed.), *Storia della scienza*, vol. 1. Rome: Treccani.

Aufrère, Sydney H. 2003. "La titulature de Djoser dans la stèle de la Famine. La redécouverte du vrai nom du constructeur de la pyramide à degrés." In Annie Gasse and Vincent Rondot (eds), *Séhel. Entre Égypte et Nubie. Inscriptions rupestres et graffiti de l'époque pharaonique: Actes du colloque international, 31 mai–1 juin 2002, Université Paul Valéry, Montpellier* (= *Orientalia Monspelliensia*, 13). Montpellier: Presses universitaires de la Méditerranée.

Aufrère, Sydney H. 2004a. "Imhotep et Djoser dans la région de la cataracte. De Memphis à Éléphantine." *Bulletin de l'Institut français d'Archéologie orientale*, 104: 1–20.

Aufrère, Sydney H. 2004b. "Le mythe de la Lointaine cananéenne et byblite et ses rapports avec les divinités égyptiennes." In Olivier Casabonne and Michel Mazoyer (eds), *Mélanges offerts à René Lebrun*. 2 vols. Paris: L'Harmattan.

Aufrère, Sydney H. 2005a. "Parfums et onguents liturgiques. Présentation des recettes d'Edfou." In Sydney H. Aufrère (ed.), *Encyclopédie religieuse de l'Univers végétal. Croyances phytoreligieuses de l'Égypte ancienne III.* (= *Orientalia Monspelliensia*, 16). Montpellier: Presses universitaires de la Méditerranée.

Aufrère, Sydney H. 2005b. "Threskiornis æthiopicus. Autour d'un mouvement migratoire de l'ibis dans l'Égypte ancienne." In Michel Mazoyer (ed.), *L'Oiseau. Entre ciel et terre*. Paris: L'Harmattan.

Aufrère, Sydney H. 2007. *Thot Hermès l'Égypte. De l'Infiniment grand à l'Infiniment petit*. Paris: L'Harmattan.

Aufrère, Sydney H. 2008. "Trouver et exploiter le filon: réalisme et perception divine au Pays-du-Dieu." *Égypte, Afrique & Orient*, 49: 3–18.

Aufrère, Sydney H. 2012a. "Manéthôn de Sebennytos, médiateur de la culture sacerdotale du Livre sacré? Questions diverses concernant l'origine, le contenu et la datation des *Aegyptiaca*." In Bernard Legras (ed.), *Transferts culturels et droits dans le monde grec et hellénistique. Actes du colloque international (Reims, 14–17 mai 2008)* (= *Histoire ancienne et médiévale*, 110). Paris: Sorbonne.

Aufrère, Sydney H. 2012b. "Symptomatologie des morsures d'ophidiens d'après le papyrus Brooklyn nos 47.218.48 et 85: aspects épistémologiques d'un texte égyptien ancien recopié au IVe siècle avant notre ère." In Sébastien Barbara and Jean Trinquier (eds), *Ophiaka, Anthropozoologica (Museum d'Histoire naturelle)* (= *Anthropozoologica*, 47 [1]). Paris: Musée d'Histoire Naturelle.

Aufrère, Sydney H. 2014. "Manéthon et la médiation du *Livre sacré*." In Charles Méla and Frédéric Möri (eds), *Alexandrie la Divine*, vol. 2. Geneva: Éditions de la Baconnière.

Aufrère, Sydney H. 2016a. "Sous le vêtement de lin du prêtre isiaque, le 'philosophe': le 'mythe' égyptien comme sagesse barbare chez Plutarque." In Sydney H. Aufrère and Frédéric Möri (eds), *Les Sagesses barbares. Échanges et réappropriations dans l'espace culturel gréco-romain*. Geneva: Éditions de la Baconnière.

Aufrère, Sydney H. 2016b. "Le savoir 'scientifique' des Égyptiens." In Alain Charron and Christophe Barbotin (eds), *Khâemouaset, le prince archéologue: Savoir et pouvoir à l'époque de Ramsès II*. Ghent: Schnoek.

Aufrère, Sydney H. 2016c. "La conception de l'Univers minéral chez les anciens Égyptiens." In Alain Charron and Christophe Barbotin (eds), *Khâemouaset, le prince archéologue: Savoir et pouvoir à l'époque de Ramsès II*. Ghent: Schnoek.

Aufrère, Sydney H. 2016d. "Recherches sur les interdits religieux des régions de l'Égypte ancienne d'après les encyclopédies sacerdotales." In Bernadette Menu (ed.), *Le Juste et le Sacré. Les territoires de la faute* (= Droits et Cultures. Revue internationale disciplinaire, 71). Paris: L'Harmattan. Available online: http://journals.openedition.org/droitcultures/3695.

Aufrère, Sydney H. 2017a. "Egyptian myths and trade of perfumes and spices from Punt and Africa." In Ilaria Incordino and Pearce Paul Creasman (eds), *Flora Trade between Egypt and Africa in Antiquity: Proceedings of a Conference held in Naples, Italy, 13 April 2015*. Oxford: Oxbow Books.

Aufrère, Sydney H. 2017b. "Osiris-Nil, Isis-Terre *versus* Typhon-Mer. Hypothèses sur les *boghaz* des lacs littoraux, l'*Ekrêgma* et les « Expirations de Typhon » du lac Sirbonis." In René Lebrun and Étienne Van Quickelberghe (eds), *Dieu de l'orage en Méditerranée antique* (= Homo religious, 17). Turnhout: Brepols.

Aufrère, Sydney H. 2018. "Ce que Typhon dissimule, Isis le révèle. Étymologies allégoriques des noms de Typhon et d'Osiris dans le *De Iside et Osiride* de Plutarque." In Françoise Graziani and Arnaud Zucker (eds), *Mythographie de l'étranger* (= Rencontres, 375). Paris: Classiques Garnier.

Aufrère, Sydney H. 2021a. "Phytopoïèse, mythes étiologiques et efficience des simples en tant que fluides divins en Égypte ancienne." In Antonio Ricciardetto, Nathan Carlig, Gabriel Nocchi Macedo, and Magali de Haro Sanchez (eds), *Le médecin et le livre. Hommages à Marie-Hélène Marganne*. Lecce: Pensa Multimedia.

Aufrère, Sydney H. 2021b. "Welches Ägypten-Bild zeigt das Corpus hermeticum?" In Dorothee Gall (ed.), *Die göttliche Weisheit des Hermes Trismegistos*. Pseudo-Apuleius, Asclepius. SAPERE-Band. Tübingen: Mohr-Siebeck. Forthcoming.

Aufrère, Sydney H., and Michel Menu. 1998. "Couleurs égyptiennes: de la chimie des matériaux et végétaux aux concepts religieux." In Sylvie Colinart and Michel Menu (eds), *La couleur dans la peinture et l'émaillage de l'Égypte Ancienne: Actes de la table ronde Ravello, 20–22 Marzo 1997* (= Scienze e materiali del patrimonio culturale, 4). Santo Spirito: Edipuglia.

Bailey, Kenneth C. 1929–32. *The Elder Pliny's Chapters on Chemical Subjects*. 2 vols. London: Edward Arnold.

Bardinet, Thierry. 1995. *Les papyrus médicaux de l'Égypte pharaonique*. Paris: Fayard.

Bardinet, Thierry. 2013. "Osiris et le gattilier." *Égypte nilotique et méditerranéenne*, 6: 33–78.

Bardinet, Thierry. 2017. *Médecins et magiciens à la cour du pharaon. Une étude du papyrus médical Louvre E 32847*. Paris: Éditions Khéops and Louvre Éditions.

Bardinet, Thierry. 2018. "Les couleurs du mal et de la maladie: pratiques magiques et examens cliniques chez les médecins de l'Égypte des pharaons." In Franck Collard and Évelyne Samama (eds), *Le corps polychrome, couleurs et santé. Antiquité, Moyen Âge, Époque moderne*. Paris: L'Harmattan.

Barguet, Paul. 1953. "Khnoum-Chou patron des arpenteurs." *Chronique d'Égypte*, 28: 223–7.

Baroni, Sandro, Giuseppe Pizzigoni, and Paola Travaglio. 2013. *Mappae Clavicula. Alle origini dell'alchimia in Occidente*. Saonara: Prato.

Barta, Winfried. 1963. *Die Altägyptische Opferliste von der Frühzeit bis zur Griechisch-römischen Epoque*. Berlin: Bruno Hessling.

Bartoš, Hynek. 2015. *Philosophy and Dietetics in the Hippocratic On Regimen. A Delicate Balance of Health*. Leiden: Brill.

Barucq, André, and François Daumas. 1980. *Hymnes et prières de l'Égypte ancienne* (= Littératures anciennes du Proche-Orient, 10). Paris: Cerf.

Bataille, André. 1952. *Les Memnonia. Recherches de papyrologie et d'épigraphie grecques sur la nécropole de la Thèbes d'Égypte aux époques hellénistique et romaines* (= Recherches d'Archéologie, de Philologie et d'Histoire, 23). Cairo: Institut Français d'Archéologie Orientale.

Baud, Michel. 1990. "La représentation de l'espace en Égypte ancienne: cartographie d'un itinéraire d'expédition." *Bulletin de l'Institut français d'Archéologie orientale*, 90: 51–63.

Baum, Nathalie. 1988. *Arbres et arbustes de l'Égypte ancienne. La liste de la tombe thébaine d'Inéni (n° 81)* (= Orientalia Lovaniensia Analecta, 31). Leuven: Departement Oriéntalistie.

Baum, Nathalie. 1994a. "La salle de Pount du temple de Repyt à Wennina." In Catherine Berger, Gisèle Clerc, and Nicolas Grimal (eds), *Hommages à J. Leclant*, vol. 4 (= Bibliothèque d'Étude, 106/2). Cairo: Institut Français d'Archéologie Orientale.

Baum, Nathalie. 1994b. "*Śnṯr*: une revision." *Revue d'Égyptologie*, 45: 17–39.

Baum, Nathalie. 1999. "L'organisation du règne végétal dans l'Égypte ancienne et l'identification des noms de végétaux." In Sydney H. Aufrère (ed.), *Encyclopédie religieuse de l'Univers végétal. Croyances phytoreligieuses de l'Égypte ancienne* (= Orientalia Monspelliensia, 10), vol. 1. Montpellier: Presses universitaires de la Méditerranée.

Beagon, Mary. 2005. *The Elder Pliny on the Human Animal. Natural History, Book 7, Translated with Introduction and Historical Commentary*. Oxford: Clarendon Press.

Beaulieu, Paul-Alain. 2010. "The Afterlife of Assyrian Scholarship in Hellenistic Babylonia." In Jeffrey Stackert, Barbara Nevling Porter, and David P. Wright (eds), *Gazing on the Deep: Ancient Near Eastern and Other Studies in Honor of Tzvi Abusch*. Bethesda, MD: CDL Press.

Beazot, Cinzia. 2017. "Ancient Ecology: Problems of Terminology." In Orietta Dora Cordovana and Gian Franco Chiai (eds), *Pollution and the Environment in Ancient Life and Thought*. Stuttgart: Franz Steiner.

Beck, Lily Y. 2011. *Pedanius Dioscorides of Anazarbus, De Materia Medica*, 2nd, rev. and enlarged ed. Hildesheim, Zürich, New York: Olms-Weidmann.

Belgiorno, Maria Rosaria. 2017. *The Perfume of Cyprus. From Pyrgos to François Coty, the Route of a Millenary Charm*, 3rd ed. Nicosia: De Strobel.

Belgiorno, Maria Rosaria. 2018. *Behind Distillation. Experimental Archaeology: Tepe Gawra and the spread of its channel jar to Slovakia, Sardinia and Cyprus*. Nicosia: De Strobel.

Bénédite, Georges. 1921. "Un thème nouveau de la décoration murale des tombes néo-memphites: La cueillette du lis et le 'lirinon', à propos d'un bas-relief et d'un fragment de bas-relief au Musée du Louvre." *Monuments et mémoires de la Fondation Eugène Piot*, 25: 1–28.

Beretta, Marco. 1991. "The Historiography of Chemistry during the 18th Century: A Preliminary Survey and Bibliography." *Ambix*, 39: 1–10.

Beretta, Marco. 2004. "Between Nature and Technology: Glass in the Ancient Chemical Philosophy." In Marco Beretta (ed.), *When Glass Matters. Studies in the History of Science and Art from Graeco-Roman Antiquity to Early Modern Era*. Florence: Olschki.

Beretta, Marco. 2009. *The Alchemy of Glass. Counterfeit, Imitation and Transmutation in Ancient Glassmaking*. Sagamore Beach, MA: Science History Publications.

Beretta, Marco. 2011. "The Changing Role of the Historiography of Chemistry in Continental Europe since 1800." *Ambix*, 58: 257–76.

Beretta, Marco. 2014. "Transmutations and Frauds in Enlightened Paris. Lavoisier and Alchemy." In Marco Beretta and Maria Conforti (eds), *Fakes!? Hoaxes, Counterfeits and Deception in Early Modern Science*. Sagamore Beach, MA: Science History Publications.

Beretta, Marco. 2015. *La rivoluzione culturale di Lucrezio. Filosofia e scienza nell'antica Roma*. Rome: Carocci.

Bergamasco, Marco. 1995. "Le διδασκαλικαί nella ricerca attuale." *Aegyptus*, 75: 95–167.

Berthelot, Marcellin, and Charles Emile Ruelle (eds). 1888. *Collections des anciens alchimistes grecs*. 3 vols. Paris: Georges Steinheil.

Besançon, Julien. 1954. "Les ressources du sous-sol en Égypte." *Annales de Géographie*, 63(338): 313–14.

Betz, Hans Dieter. 1986. *The Greek Magical Papyri in Translation, Including the Demotic Spells*. Chicago, IL: University of Chicago Press.

Bhayro, Siam. 2005. *The Shemihazah and Asael Narrative of 1 Enoch 6–11. Introduction, Text, Translation and Commentary with Reference to Ancient Near Eastern and Biblical Antecedents* (= Alter Orient und Altes Testament, 322). Münster: Ugarit-Verlag.

Bidez, Jean, and Franz Cumont. 1938. *Les mages hellénisés. Zoroastre, Ostanès et Hystaspe d'après la tradition grecque*. 2 vols. Paris: Belles Lettres.

Bietak, Manfred. 2009. "Perunefer: The Principal New Kingdom Naval Base." *Egyptian Archaeology*, 34: 15–17.

Biga, Maria Giovanna. 2010. "War and Peace in the Kingdom of Ebla (24th Century B.C.) in the First Years of Vizier Ibbi-Zikir under the Reign of the Last King Išar-Damu." In Maria Giovanna Biga and Mario Liverani (eds), *Ana Turri Gimilli. Studi Dedicati al Padre Werner R. Mayer, S.J. da Amici e Allievi*. Rome: Università di Roma La Sapienza.

Biga, Maria Giovanna. 2014. "Some Aspects of the Wool Economy at Ebla." In Catherine Breniquet and Cécile Michel (eds), *Wool Economy in the Ancient Near East: From the Beginnings of Sheep Husbandry to Institutional Textile Industry*. Oxford: Oxbow Books.

Biggs, Rogert. 1966. "Le lapis lazuli dans les textes sumériens archaïques." *Revue d'Assyriologie*, 60: 175–6.

Birchler Emery, Patrizia. 2008. "Du *banausos* au *technitês*. Contribution à l'étude du type iconographique de l'artisan en Grèce archaïque." In Martina Seifert (ed.), *Komplexe Bilder*. Berlin: Leonhard-Turneysser.

Bjorkman, Judith K. 1987. "Sargon II's Foundation Tablets." *Orienatlia Lovaniensia Periodica*, 18: 87–97.

Bjorkman, Judith K. 1993. "The Larsa Goldsmith's Hoards – New Interpretations." *Journal of Near Eastern Studies*, 52: 1–23.

Black, Matthew. 1970. *Apocalypsis Henochi Graece* (= *Pseudepigrapha Veteris Testamenti Graece*, 3). Leiden: Brill.

Blackman, M. James, Gil J. Stein, and Pamela B. Vandiver. 1993. "The Standardization Hypothesis and Ceramic Mass Production: Technological, Compositional, and Metric Indexes of Craft Specialization at Tell Leilan, Syria." *American Antiquity*, 58: 60–80.

Blet, Maryse, Bernard Guineau, and Bernard Bratuze. 1997. "Caractérisation de boules de bleu égyptien: analyse par absorption visible et par activation avec des neutrons rapides de cyclotron." *Archéosciences*, 21: 121–30.

Block, Yigal, and Wayne Horowitz. 2015. "Ura = hubullu XXII: The Standard Recension." *Journal of Cuneiform Studies*, 67: 71–125.

Bomhard, Anne-Sophie von. 2012. *The Decree of Sais. The Stelae of Thonis-Heracleion and Naukratis* (= *Oxford Centre for Maritime Archaeology monograph*, 7). Oxford: Oxford Centre for Maritime Archaeology.

Bondi, Jonas H. 1895. "Ägyptologisches aus der rabbinischen Litteratur. I. Ein Rezept für die Bereitung des *zuthos*." *Zeitschrift für Altägyptische Sprache und Altertumskunde*, 33: 62–4.

Borch, Ole. 1668. *De Ortu et progressu chemiae dissertatio*. Copenhagen: Sumptibus Petri Hauboldi.

Borch, Ole. 1674. *Hermetis, Ægyptiorum, et chemicorum sapientia ab Hermanni Conringii animadversionibus vindicate*. Copenhagen: Sumptibus Petri Hauboldi.

Borgard, Philippe, Jean-Pierre Brun, and Maurice Picon (eds). 2005. *L'alun de Méditerranée*. Naples: Centre Jean Bérard.

Bottéro, Jean. 1985. "The Cuisine of Ancient Mesopotamia." *The Biblical Archaeologist*, 48: 36–47.

Bottéro, Jean. 1987. "The Culinary Tablets at Yale." *Journal of the American Oriental Society*, 107: 11–19.

Bottéro, Jean. 1995. *Textes culinaires Mésopotamiens: Mesopotamian Culinary Texts* (= *Mesopotamian Civilizations*, 6). Winona Lake, IN: Eisenbrauns.

Bottéro, Jean. 2002. *La plus vieille cuisine du monde*. Paris: Louis Audibert.

Boudon-Millot, Véronique. 2000. *Galien. Œuvres, tome II. Exhortation à l'étude de la médecine, Art médical*. Paris: Belles Lettres.

Boudon-Millot, Véronique. 2002. "La thériaque selon Galien: poison salutaire ou remède empoissoné." In Frank Collard and Évelyne Samama (eds), *Le corps à l'épreuve. Poisons, remèdes et chirurgie: aspects des pratiques médicales dans l'Antiquité et au Moyen Âge*. Langrès: Dominique Guéniot.

Boudon-Millot, Véronique. 2003. "Au marges de la médecine rationnelle: médecins et charlatans à Rome au temps de Galien (IIe s. de notre ère)." *Revue des études grecques*, 116: 109–31.

Bourriau, Janine D., Paul T. Nicholson, and Pamela J. Rose. 2000. "Pottery." In Paul T. Nicholson and Ian Shaw (eds), *Ancient Egyptian Materials and Technology*. Cambridge: Cambridge University Press.

Bower, Bruce. 2005. "Ancient Glassmakers: Egyptians Crafted Ingots for Mediterranean Trade." *Science News*, 167: 388.

Brack-Bernsen, Lis, and John M. Steele. 2004. "Babylonian Mathemagics: Two Mathematical Astronomical-Astrological Texts." In Charles Burnett, Jan P.

Hogendijk, Kim Plofker, and Michio Yano (eds), *Studies in the History of Exact Sciences in Honour of David Pingree*. Leiden: Brill.

Brereton, Gareth (ed.). 2018. *I Am Ashurbanipal: King of the World, King of Assyria*. London: Thames and Hudson.

Bresson, Alain. 2017. "Anthropogenic Pollution in Greece and Rome." In Orietta Dora Cordovana and Gian Franco Chiai (eds), *Pollution and the Environment in Ancient Life and Thought*. Stuttgart: Franz Steiner.

Briant, Pierre, and Raymond Descat. 1998. "Un registre douanier de la satrapie d'Égypte à l'époque achéménide." In Nicolas Grimal and Bernadette Menu (eds), *Le commerce en Égypte ancienne* (= *Bibliothèque d'étude*, 121). Cairo: Institut Français d'Archéologie Orientale.

Brinkman, John A. 1988. "Textual Evidence for Bronze in Babylonia in the Early Iron Age, 1000–539 BC." In John Curtis (ed.), *Bronze-working Centres of Western Asia c.1000–539 BC*. London: Kegan Paul.

Brink, Martin, and Enoch G. Achgan-Dako (eds). 2012. *Ressources végétales de l'Afrique tropicale 16: Plantes à fibres*. Wageningen: Fondation PROTA/CTA.

Brisson, Luc. 1998. *Le même et l'autre dans la structure ontologique du Timée de Platon. Un commentaire systématique du Timée de Platon*. Sankt Augustin: Academia Verlag.

Broekhuis, Jan. 1971. *De godin Renenwetet*. Assen: Van Gorcum.

Broze, Michèle. 1989. *La princesse de Bakhtan. Essai d'analyse stylistique* (= *Monographies Reine Élisabeth*, 6). Brussels: Fondation Égyptologique Reine Élisabeth.

Brun, Jean-Pierre. 2000. "The Production of Perfumes in Antiquity: The Cases of Delos and Paestum." *American Journal of Archaeology*, 104: 277–308.

Brun, Jean-Pierre, Thomas Faucher, Bérangère Redon, and Florian Téreygeol. 2013. "L'or d'Égypte. L'exploitation des mines d'or dans le désert Oriental sous les Ptolémées." *L'Archéologue*, 126: 56–61.

Brunke, Hagan. 2011. "Feasts for the Living, the Dead, and the Gods." In Karen Radner and Eleanor Robson (eds), *The Oxford Handbook of Cuneiform Culture*. Oxford: Oxford University Press.

Brunke, Hagan, and Walther Sallaberger. 2010. "Aromata für Duftöl." In Alexandra Kleinerman and Jack M. Sasson (eds), *Why Should Someone Who Knows Something Conceal It? Cuneiform Studies in Honor of David I. Owen on His 70th Birthday*. Bethesda, MD: CDL Press.

Bryce, Trevor. 2003. *Letters of the Great Kings of the Ancient Near East: The Royal Correspondence of the Late Bronze Age*. London: Routledge.

Bull, Christian H. 2018. "Wicked Angels and the Good Demon: The Origins of Alchemy According to the *Physica* of Hermes." *Gnosis: Journal of Gnostic Studies*, 3: 3–33.

Burns, Dylan M. 2015 "Μίξεώς τινι τέχνῃ κρείττονι. Alchemical Metaphor in the *Paraphrase of Shem* (NHC VII, 1)." *Aries – Journal for the Study of Western Esotericism*, 15: 81–108.

Butler, Antony R., and Joseph Needham. 1980. "An Experimental Comparison of the East Asian, Hellenistic, and Indian (Gandhāran) Stills in Relation to the Distillation of Ethanol and Acetic Acid." *Ambix*, 27: 69–76.

Caley, Earle Radcliffe. 1926. "The Leiden Papyrus X: An English Translation with Brief Notes." *Journal of Chemical Education*, 3: 1149–66.

Caley, Earle Radcliffe. 1927. "The Stockholm Papyrus: An English Translation with Brief Notes." *Journal of Chemical Education*, 4: 979–1002.

Caley, Earle Radcliffe. 1928. "Mercury and Its Compounds in Ancient Times." *Journal of Chemical Education*, 5: 419–24.
Cambiano, Giuseppe. 1991. *Platone e le tecniche*. Bari: Laterza.
Campbell Thompson, Reginald. 1934. "An Assyrian Chemist's Vade-mecum." *Journal of the Royal Asiatic Society*, 66: 771–85.
Campbell Thompson, Reginald. 1936. *A Dictionary of Assyrian Chemistry and Geology*. Oxford: Clarendon Press.
Campbell Thompson, Reginald. 1949. *A Dictionary of Assyrian Botany*. London: British Academy.
Capasso, Luigi. 1995. "Archaeological Documentation of the Atmospheric Pollution in Antiquity." *Medicina nei Secoli*, 7: 435–44.
Caron, Chloé. 2014. "Des hommes de larmes, des hommes de tristesse? La conception anthropogonique dans les Textes des Sarcophages du Moyen Empire égyptien (2040–1785)." Ph.D. thesis, Université du Québec, Montreal.
Cassin, Elena. 1968. *La splendeur divine: introduction à l'étude de la mentalité mésopotamienne*. La Haye: Mouton.
Casson, Lionel. 1989. *The Periplus Maris Erythraei. Text with Introduction, Translation and Commentary*. Princeton, NJ: Princeton University Press.
Castel, Cécilia, Xavier Fernandez, Jean-Jacques Filippi, and Jean-Pierre Brun. 2012. "Les parfums antiques dans le bassin méditerranéen." *L'actualité chimique*, 359: 42–9.
Castel, Georges, and Georges Soukiassian. 1989. *Gebel el Zeit I: Les mines de galène (Égypte, IIe millénaire av. J.-C.) (= Fouilles de l'Institut français d'Archéologie orientale, 35)*. Cairo: Institut Français d'Archéologie Orientale.
Castel, Georges, Philippe Fluzin, and Pierre Tallet. 2008. "La métallurgie du cuivre au temps des pharaons." *Archéologia*, 460: 62–71.
Cauville, Sylvie. 2004. *Dendara V–VI: Traduction, Les Cryptes du Temple d'Hathor (= Orientalia Lovaniensia Analecta, 132)*. Leuven: Peeters.
Cauville, Sylvie. 2005. "Dendara: du sanatorium au tinctorium." *Bulletin de la Société française d'Égyptologie*, 160: 28–40.
Cauville, Sylvie. 2009. "Karnak et la quintessence de l'Égypte." *Bulletin de la Société française d'Égyptologie*, 172: 17–31.
Cauville, Sylvie, and Mohammed Ibrahim Ali. 2015. *Dendara. Itinéraire du visiteur*. Leuven: Peeters.
Cavassa, Laëtitia, François Delamare, and Monique Repoux. 2010. "La fabrication du bleu égyptien dans les Champs Phlégréens (Campanie, Italie) durant le Ier siècle de notre ère." In Pascale Chardron-Picault (ed.), *Aspects de l'artisanat en milieu urbain: Gaule et Occident romain*. Dijon: Revue Archéologique de l'Est.
Ceccarelli, Manuel. 2015. "Bemerkungen zur Entwicklung der Beschwörungen Marduk-Ea-Typs: Die Rolle Enlils." In Alfonso Archi (ed.), *Tradition and Innovation in the Ancient Near East: Proceedings of the 57th Rencontre Assyriologique Internationale at Rome 4–8 July 2011*. Winona Lake, IN: Eisenbrauns.
Charleston, Robert J. 1978. "Glass Furnaces through the Ages." *Journal of Glass Studies*, 20: 9–33.
Charron, Alain, and Christophe Barbotin (eds). 2016. *Khâemouaset, le prince archéologue: Savoir et pouvoir à l'époque de Ramsès II*. Ghent: Schnoek.
Charron, Régine. 2005. "The 'Apocryphon of John' (NHC II,1) and the Graeco-Egyptian Alchemical Literature." *Vigiliae Christianae*, 59: 438–56.

Charron, Régine, and Louis Painchaud. 2001. "'God is a Dyer.' The Background and Significance of a Puzzling Motif in the Coptic *Gospel According to Philip* (CG II,3)." *Le Muséon*, 114: 41–50.

Chartier-Raymond, Maryvonne. n.d. *Les habitats miniers du Sinaï et les conditions d'exploitation*. Thesis, University Lille III, Villeneuve-d'Ascq.

Chassinat, Émile. 1921. *Un papyrus médical copte* (= Mémoires de l'Institut français d'Archéologie orientale, 32). Cairo: Institut Français d'Archéologie Orientale.

Chassinat, Émile. 1928. "Une nouvelle mention du pseudo-architecte du temple d'Horus, à Edfou." *Bulletin de l'Institut français d'Archéologie orientale*, 28: 1–10.

Chassinat, Émile. 1966–8. *Le mystère d'Osiris au mois de Khoiak*. 2 vols. Cairo: Institut Français d'Archéologie Orientale.

Chatzidimitriou, Athina. 2014. "Craftsmen and Manual Workers in Attic Vase-Painting of the Archaic and Classical Period." In Anne-Catherine Gillis (ed.), *Corps, travail et status social. L'apport de la paléoanthropologie funéraire aux sciences historiques*. Villeneuve d'Ascq: Presses Universitaires du Septentrion.

Chermette, Michèle, and Jean-Claude Goyon. 1996. "Le catalogue raisonné des producteurs de styrax et d'oliban d'Edfou et d'Athribis de Haute-Égypte." *Studien zu Altägyptischen Kultur*, 23: 47–82.

Cherpion, Nadine, Jean-Pierre Corteggianni, and Jean-François Gout. 2007. *Le tombeau de Pétosiris à Touna el-Gebel* (= Bibliothèque générale, 27). Cairo: Institut Français d'Archéologie Orientale.

Cherpion, Nadine. 1994. "Le cône d'onguent, gage de survie." *Bulletin de l'Institut français d'Archéologie orientale*, 94: 79–106.

Christiansen, Thomas. 2017. "Manufacture of Black Ink in the Ancient Mediterranean." *The Bulletin of the American Society of Papyrologists*, 54: 167–95.

Cicarello, Mark. 1976. "Shesmu the Letopolite." In Janet H. Johnson and Edward F. Wente (eds), *Studies in Honor of George R. Hughes, January 12, 1977* (= Studies in Ancient Oriental Civilization, 39). Chicago, IL: Oriental Institute of the University of Chicago. Available online: https://oi.uchicago.edu/research/publications/saoc/saoc-39-studies-honor-george-r-hughes-january-12-1977.

Civil, Miguel. 1960. "Prescriptions médicales sumériennes." *Revue d'Assyriologie*, 54: 57–71.

Civil, Miguel. 1964. "A Hymn to the Beer Goddess and a Drinking Song." In Robert D. Biggs and John A. Brinkman (eds), *Studies Presented to A. Leo Oppenheim: June 7, 1964*. Chicago, IL: Oriental Institute of the University of Chicago. Available online: https://oi.uchicago.edu/research/publications/misc/studies-presented-leo-oppenheim-june-7-1964.

Civil, Miguel. 1987. "Feeding Dumuzi's Sheep: The Lexicon as a Source of Literary Inspiration." In Francesca Rochberg-Halton (ed.), *Language, Literature, and History: Philological and Historical Studies Presented to Erica Reiner* (= American Oriental Series, 67). New Haven, CT: American Oriental Society.

Civil, Miguel. 2008. *The Early Dynastic Practical Vocabulary A (Archaic HAR-ra A)* (= Archivi Reali di Ebla Studi, 4). Rome: Missione archeologica italiana in Siria.

Civil, Miguel. 2011. "The Law Collection of Ur-Namma." In Andrew R. George (ed.), *Cuneiform Royal Inscriptions and Related Texts in the Schøyen Collection* (= Cornell University Studies in Assyriology and Sumerology, 17). Bethesda, MD: CDL Press.

Clarke, John R. 2003. *Art in the Lives of Ordinary Romans: Visual Representation and Non-Elite Viewers in Italy, 100 B.C.–A.D. 315*. Berkeley: University of California Press.

Clarke, Mark. 2013. "The Earliest Technical Recipes. Assyrian Recipes, Greek Chemical Treatises and the *Mappae Clavicula* Text Family." In Ricardo Córdoba (ed.), *Craft Treatises and Handbooks: The Dissemination of Technical Knowledge in the Middle Ages*. Turnhout: Brepols.

Clerc, Gisèle, Vassos Karageorghis, Élisabeth Lagarce, and Jean Leclant (eds). 1976 *Fouilles de Kition II. Objets égyptiens et égyptisants: scarabées, amulettes et figurines en pâte de verre et en faïence, vase plastique en faïence. Sites I et II, 1959–1975*. Nicosia: Department of Antiquities, Cyprus.

Cline, Eric H., and David B. O'Connor. 2006. *Thutmose III: A New Biography*. Ann Arbor: University of Michigan Press.

Cochavi-Rainey, Zipora. 1999. *Royal Gifts in the Late Bronze Age, Fourteenth to Thirteenth Centuries B.C.E.; Selected Texts Recording Gifts to Royal Personages*. Beersheba: Ben-Gurion University of the Negev Press.

Cole, Thomas. 1967. *Democritus and the Sources of Greek Anthropology*. Cleveland, OH: Press of the Western Reserve University.

Colin, Frédéric. 2003. "Le parfumeur (pꜣ ꜥnt)." *Bulletin de l'Institut français d'Archéologie orientale*, 103: 73–109.

Colinart, Sylvie, and Michel Menu (eds). 1998. *La couleur dans la peinture et l'émaillage de l'Égypte Ancienne: Actes de la table ronde Ravello, 20–22 Marzo 1997* (= *Scienze e materiali del patrimonio culturale*, 4). Santo Spirito: Edipuglia.

Colinet, Andrée. 2010. *Les alchimistes grecs XI. Recettes alchimiques (Par. Gr. 2419; Holkhamicus 109), Cosmas le hiéromoine, chrysopée*. Paris: Belles Lettres.

Collar, Franck. 2007. *Pouvoir et poison. Histoire d'un crime politique de l'Antiquité à nos jours*. Paris: Seuil.

Colonna, Maria Elisabetta. 1958. *Enea di Gaza, Teofrasto*. Naples: Salvatore Iodice editore.

Conring, Hermann. 1669. *De Hermetica Medicina*. Helmstadt.

Couchoud, Sylvia. 1988. "La bière en Égypte pharaonique." *Bulletin du Cercle lyonnais d'Égyptologie*, 2: 7–11.

Cousin, Laura. 2013. "Female Perfume-makers in Neo-Assyrian and Neo-Babylonian Documents." *Carnet de Rôle Économique des Femmes en Mésopotamie Ancienne*. Available online: http://refema.hypotheses.org/806.

Couto-Ferreira, Érica. 2013. "The Circulation of Medical Practitioners in the Ancient Near East: The Mesopotamian Perspective." In Sergio Carro Martín, Arturo Echavarren, Esther Fernández Medina, Daniel Riaño Rufílanchas, Katja Šmid, Jesús Téllez Rubio, and David Torollo Sánchez (eds), *Mediterráneos: An Interdisciplinary Approach to the Cultures of the Mediterranean Sea*. Newcastle upon Tyne: Cambridge Scholars.

Couyat, Jules, and Pierre Montet. 1912. *Les inscriptions hiéroglyphiques et hiératiques du Ouâdi Hammâmât* (= *Mémoires de l'Institut français d'Archéologie orientale*, 24). Cairo: Institut Français d'Archéologie Orientale.

Craddock, Paul T. 2000. "From Hearth to Furnace: Evidences for the Earliest Metal Smelting Technologies in the Eastern Mediterranean." *Paléorient*, 26: 151–65.

Craddock, Paul T. 2016. "Classical Geology and the Mines of the Greeks and Romans." In Giorgia L. Irby (ed.), *A Companion to Science, Technology and Medicine in Ancient Greece and Rome*. Chichester: Wiley-Blackwell.

Crosby, Margaret. 1953. "The Leases of the Laurion Mines." *Hesperia*, 19: 189–312.
Cumont, Franz, 1937. *L'Égypte des astrologues*. Brussels: Fondation égyptologique Reine-Élisabeth.
Cunningham, Graham. 1997. *Deliver Me From Evil: Mesopotamian Incantations 2500–1500 BC* (= *Studia Pohl Series Maior*, 17). Rome: Pontifcio Istituto Biblico.
Cuvigny, Hélène. 1997. "Le crépuscule d'un dieu. Le déclin du culte de Pan dans le désert Oriental." *Bulletin de l'Institut français d'Archéologie orientale*, 97: 139–47.
D'Acunto, Matteo. 2009. "Efesto e le sue creazioni nel xviii libro dell'*Iliade*." *AION. Annali dell'Università degli studi di Napoli «L'Orientale»*, 31: 145–98.
Dachy, Tiphaine. 2014. "Réflexions sur le stockage alimentaire en Égypte, de la Préhistoire aux premières dynasties." *Archéo-Nil*, 24: 31–46.
Dahl, Jacob L. 2010. "A Babylonian Gang of Potters: Reconstructing the Social Organization of Crafts Production in the Late Third Millennium BC Southern Mesopotamia." In Leonid Kogan, Natalia Koslova, and Sergey Loesov (eds), *City Administration in the Ancient Near East. Proceedings of the 53rd Rencontre Assyriologique Internationale* (= *Babel und Bibel*, 5), vol. 2. Winona Lake, IN: Eisenbrauns.
Dalley, Stephanie. 1991. "Ancient Assyrian Textiles and the Origins of Carpet Design." *Iran*, 29: 117–35.
Darmstaedter, Ernst. 1925. "Vorläufige Bemerkungen zu den assyrischen chemisch-technischen Rezepten." *Zeitschrift für Assyriologie*, 36: 302–4.
Darmstaedter, Ernst. 1926. "Nochmals Babylonische 'Alchemie.'" *Zeitschrift für Assyriologie*, 37: 205–14.
Dasen, Véronique. 2013. *Dwarfs in Ancient Egypt and Greece*. Oxford: Clarendon Press.
Davey, Christopher. 1983. "The Metalworkers' Tools from Tell Edh Dhiba'i." *The Bulletin of the Institute of Archaeology University of London*, 20: 169–85.
Davey, Christopher J., and W. Ian Edwards. 2007. "Crucibles from the Bronze Age of Egypt and Mesopotamia." *Proceedings of the Royal Society of Victoria*, 120: 146–54.
David, A. Rosalie. 2000. "Mummification." In Paul T. Nicholson and Ian Shaw (eds), *Ancient Egyptian Materials and Technology*. Cambridge: Cambridge University Press.
Daumas, François. 1953. "Le trône d'une statuette de Pépi Ier trouvé à Dendara." *Bulletin de l'Institut français d'Archéologie orientale*, 52: 163–72.
Daumas, François. 1956. "La valeur de l'or dans la pensée égyptienne." *Revue de l'Histoire des Religions*, 149: 1–17.
Daumas, François. 1973. "Derechef Pépi Ier à Dendara." *Revue d'égyptologie*, 25: 7–20.
Daumas, François. 1977. "Le problème de la monnaie dans l'Égypte antique avant Alexandre." *Mélanges de l'École française de Rome et d'Athènes*, 89: 425–42.
Daumas, François. 1980. "Quelques textes de l'Atelier des Orfèvres." In Jean Vercoutter (ed.), *Livre du Centenaire, 1880–1980* (= *Mémoires de l'Institut français d'Archéologie orientale*, 104). Cairo: Institut Français d'Archéologie Orientale.
Davies, Norman de Garis. 1943. *The Tomb of Rekh-Mi-Ré at Thebes* (= *Publications of the Metropolitan Museum of Art Egyptian Expedition*, 11). 2 vols. New York: Plantin Press.
De Ryck, Ivan, Annemie Adriaens, and Freddy Adams. 2005. "An overview of Mesopotamian bronze metallurgy during the 3rd millennium BC." *Journal of Cultural Heritage*, 6: 261–8.

Delamare, François H. 1998a. "Le bleu égyptien, essai de bibliographie critique." In Sylvie Colinart and Michel Menu (eds), *La couleur dans la peinture et l'émaillage de l'Égypte Ancienne: Actes de la table ronde Ravello, 20–22 Marzo 1997* (= *Scienze e materiali del patrimonio culturale*, 4). Santo Spirito: Edipuglia.

Delamare, François H. 1998b. "De la composition du bleu égyptien utilisé en peinture murale gallo-romaine." In Sylvie Colinart and Michel Menu (eds), *La couleur dans la peinture et l'émaillage de l'Égypte Ancienne: Actes de la table ronde Ravello, 20–22 Marzo 1997* (= *Scienze e materiali del patrimonio culturale*, 4). Santo Spirito: Edipuglia.

Delamare, François. 2003. "La recette du *caeruleum* de Vitruve: le point de vue de la science des matériaux." *Archives Internationales d'Histoire des Sciences*, 53: 3–18.

Delamare, François H. 2007. *Bleus en poudres: de l'art à l'industrie: 5000 ans d'innovations*. Paris: ParisTech. (Available in English as *Blue Pigments: From Art to Industry*. London: Archetype Publications, 2013.)

Delatte, André. 1936. *Herbarius: recherches sur le cérémonial usité chez les anciens pour la cueillette des simples et des plantes magiques*. Paris: Les Belles Lettres.

Delitzsch, Friedrich. 1887. *Assyrisches Wörterbuch zur gesamten bisher veröffentlichten Keilschriftliteratur, Lieferung 1*. Leipzig: J.C. Hinrichs.

Dell'Acqua, Francesca. 2004. "Glassmakers in the West between Late Antiquity and the Middle Ages." In Marco Beretta (ed.), *When Glass Matters. Studies in the History of Science and Art from Graeco-Roman Antiquity to Early Modern Era*. Florence: Olschki.

Delougaz, Pinhas. 1940. *The Temple Oval at Khafajah* (= *University of Chicago Oriental Institute Publications*, 53), Chicago, IL: University of Chicago Press. Available online: https://oi.uchicago.edu/research/publications/oip/oip-53-temple-oval-khafajah.

Derchain, Philippe. 1964. *Le Papyrus Salt 825 (B.M. 10051). Rituel pour la conservation de la vie en Égypte* (= *Koninklijke Academie van België, Verhandelingen*, 58). Brussels: Palais des Académies.

Derchain, Philippe. 1975. "La Perruque et le Cristal." *Studien zur Altägyptischen Kultur*, 2: 55–74.

Derchain, Philippe. 1976. "La recette du kyphi." *Revue d'égyptologie*, 28: 61–5.

Derchain, Philippe. 1987. *Le dernier obélisque*. Brussels: Fondation égyptologique Reine Élisabeth.

Derchain, Philippe. 1990. "L'Atelier des Orfèvres à Dendara et les origines de l'Alchimie." *Chronique d'Égypte*, 65, fasc. 130: 219–42.

Dercksen, Jan Gerrit. 1999. "On the Financing of Old Assyrian Merchants." In Jan Gerrit Dercksen (ed.), *Trade and Finance in Ancient Mesopotamia: Proceedings of the First MOS Symposium (Leiden 1997)* (= *MOS Studies*, 1). Leiden: Nederlands Historisch-Archaeologisch Instituut te Istanbul.

Dercksen, Jan Gerrit. 2000. "Institutional and Private in the Old Assyrian Period." In A.C.V.M. Bongenaar (ed.), *Interdependency of Institutions and Private Entrepreneurs: Proceedings of the Second MOS Symposium (Leiden 1998)* (= *MOS Studies*, 2). Leiden: Nederlands Historisch-Archaeologisch Instituut te Istanbul.

Dercksen, Jan Gerrit. 2017. "Zinn (tin). A. I. Philologisch. In Mesopotamien." *Reallexikon der Assyriologie und Vorderasiatischen Archäologie*, 15: 301–4.

Desroches-Noblecourt, Christiane. 1975. "Les récentes acquisitions." *La Revue du Louvre*, 4: 249–51.

Dickson, Keith. 2005. "Enki and the Embodied World." *Journal of the American Oriental Society*, 125: 499–515.

Diels, Hermann. 1913. *Die Entdeckung des Alkohols*. Berlin: Verlag der königlichen Akademie der Wissenschaften.

Diels, Hermann, and Erwin Adelbert Schramm. 1920. *Exzerpte aus Philons Mechanik B. VII und VIII (vulgo fünftes Buch). Abhandlungen der preußischen Akademie der Wissenschaften, Philosophish-historische Klasse Nr. 12*. Berlin: Reimer.

Donbaz, Veysel. 1988. "Complementary Data on Some Assyrian Terms (ḫuruḫurātu, ḫuruḫuru, ḫurūtu, ḫuḫuru, akalu, kirrum)." *Journal of Cuneiform Studies*, 40: 69–80.

Dorandi, Tiziano. 2016. "Le *Divisiones quae vulgo dicuntur Aristoteleae*. Storia del testo e edizione delle *Recensiones* Marciana, Florentina e Leidensis." *Studia graeco-arabica*, 6: 1–58.

Douglas, Ronald W., and Susan Frank. 1972. *A History of Glass Making*. Henley-on-Thames: Foulis.

Doyen, Florence, and Eugène Warmenbol. 2004. *Pain et bière en Égypte ancienne: de la table à l'offrande: catalogue de l'exposition créée au Musée du Malgré-Tout à Treignes (Belgique) du 4 avril au 12 décembre 2004*. Treignes: Centre d'Étude et de Documentation Archéologiques (CEDARC).

Dufault, Olivier. 2015. "Transmutation Theory in the Greek Alchemical Corpus." *Ambix*, 62: 215–44.

Dufault, Olivier. 2019. *Early Greek Alchemy, Patronage and Innovation in Late Antiquity*. Berkeley: California Classical Studies.

Duistermaat, Kim. 2008. *The Pots and Potters of Assyria: Technology and organization of production, ceramic sequence and vessel function at Late Bronze Age Tell Sabi Abyad, Syria*. Turnhout: Brepols.

Dunand, Françoise. 2006. "La guérison dans les temples (Égypte, époque tardive)." *Archiv für Religionsgeschichte*, 8: 4–24.

Düring, Ingemar. 1944. *Aristotle's Chemical Treatise. Meteorologica, Book IV*. Gothenburg: Elanders.

Ebeling, Erich. 1950. *Parfümrezepte und kultische Texte aus Assur: Sonderdruck aus Orientalia 17-19* (= *Scripta Pontificii Instituti Biblici*, 101). Rome: Pontificium Institutum Biblicum.

Edel, Elmar. 1976. *Ägyptische Ärzte und ägyptische Medizin am hethitischen Königshof*. Opladen: VS Verlag für Sozialwissenschaften.

Edens, Christopher. 1992. "Dynamics of Trade in the Ancient Mesopotamian 'World System.'" *American Anthropologist*, 94: 118–39.

Edens, Christopher. 1999. "Khor Ile-Sud, Qatar: The Archaeology of Late Bronze Ago Purple-Dye Production in the Arabian Gulf." *Iraq*, 61: 71–88.

Edgerton, William. 1947. "The Nauri Decree of Seti." *Journal of Near Eastern Studies*, 6: 219–30.

Edzard, Dietz Otto. 2003. *Sumerian Grammar* (= *Handbook of Oriental Studies*. Section 1, *The Near and Middle East*, vol. 71). Leiden: Brill.

Egberts, Arnold. 1991. "The Chronology of The Report of Wenamun." *Journal of Egyptian Archæology*, 77: 57–67.

Eggert, Gerhard. 1995. "On the Origin of a Gilding Method of the Baghdad Silversmiths." *Gold Bulletin*, 28: 13–16.

Eichholz, David E. 1949. "Aristotle's Theory of Formation of Metals and Minerals." *The Classical Quarterly*, 43: 141–8.

Eichholz, David E. 1965. *Theophrastus, De lapidibus*. Oxford: Clarendon Press.

Eijk, Philip J. van der. 2000–1. *Diocles of Carystus: A Collection of the Fragments with Translation and Commentary* (= *Studies in Ancient Medicine*, 22). 2 vols. Leiden: Brill.

El-Shahawy, Abeer, 2010. *Recherche sur la décoration des tombes thébaines du Nouvel Empire. Originalités iconographiques et innovations* (= Internet-Beiträge zur Ägyptologie und Sudanarchäologie. Studies from the Internet on Egyptology and Sudanarchaeology, 13), London: Golden House. Available online: http://www2.rz.hu-berlin.de/nilus/net-publications/ibaes13/publikation/ibaes13_pub_text.pdf.

Eliade, Mircea. 1956. *Forgerons et alchimistes*. Paris: Flammarion. (Available in English as *The Forge and the Crucible: The Origins and Structures of Alchemy*. Chicago, IL: University of Chicago Press, 1978.)

Ellis, Richard S. 1968. *Foundation Deposits in Ancient Mesopotamia* (= Yale Near Eastern Researches, 2). New Haven, CT: Yale University Press.

Ellison, Rosemary. 1968. "The Uses of Pottery." *Iraq*, 46: 63–8.

Ellison, Rosemary. 1984. "Methods of Food Preparation in Mesopotamia (c. 3000–600 BC)." *Journal of the Economic and Social History of the Orient*, 27(1): 89–98.

Engelbach, Reginal. 1933. "The Quarries of the Western Nubian Desert. A Preliminary Report." *Annales du Service des Antiquités d'Égypte*, 33: 65–74.

Englund, Robert K. 1983. "Dilmun in the Archaic Uruk Corpus." In Daniel T. Potts (ed.), *Dilmun: New Studies in the Archaeology and Early History of Bahrain* (= Berliner Beiträge zum Vorderen Orient, 2). Berlin: Reimer. Available online: https://cdli.ucla.edu/?q=robert-k-englund.

Englund, Robert K. 1991. "Hard Work – Where Will It Get You? Labor Management in Ur III Mesopotamia." *Journal of Near Eastern Studies*, 50: 255–80. Available online: https://cdli.ucla.edu/?q=robert-k-englund.

Englund, Robert K. 1992. "Ur III Sundries." *Acta Sumerologica*, 14: 77–102. Available online: https://cdli.ucla.edu/?q=robert-k-englund.

Englund, Robert K. 1998. "Texts from the Late Uruk Period." In Josef Bauer, Robert K. Englund, and Manfred Krebernik (eds), *Mesopotamien: Späturuk-Zeit und Frühdynastische Zeit* (= Orbis Biblicus et Orientalis, 160/1). Freiburg: Universitätsverlag. Available online: https://cdli.ucla.edu/?q=robert-k-englund.

Englund, Robert K. 2006. "An Examination of the 'Textual' Witnesses to Late Uruk World Systems." In Yushu Gong and Yiyi Chen (eds), *A Collection of Papers on Ancient Civilizations of Western Asia, Asia Minor and North Africa* (= Oriental Studies Special Issue). Beijing: Beijing University. Available online: https://cdli.ucla.edu/?q=robert-k-englund.

Englund, Robert K. 2012. "Equivalency Values and the Command Economy of the Ur III Period in Mesopotamia." In John K. Papadopoulos and Gary Urton (eds), *The Construction of Value in the Ancient World*. Los Angeles, CA: Cotsen Institute of Archaeology Press. Available online: https://cdli.ucla.edu/?q=robert-k-englund.

Englund, Robert K., and Hans J. Nissen. 1993. *Archaische Texte aus Uruk. Bd.3 Die lexikalischen Listen der archaischen Texte aus Uruk* (= Ausgrabungen der Deutschen Forschungsgemeinschaft in Uruk-Warka, 13). Berlin: Gebr. Mann.

Escobar, Eduardo. 2017. "Technology as Knowledge: Cuneiform Technical Recipes and the Material World." Ph.D. thesis, University of California at Berkeley.

Espinel, A. Diego. 2011. *Abriendo los caminos de Punt. Contactos entre Egipto y el ambito afroarabe durante la Edad del Bronce (ca. 3000 a.C. – 1065 a.C.)*. Barcelona: Bellaterra Arqueologia.

Espinel, A. Diego. 2017. "The scents of Punt (and elsewhere): trade and functions of *snṯr* and *ꜥntw* during the Old Kingdom." In Ilaria Incordino and Pearce Paul Creasman (eds), *Flora Trade between Egypt and Africa in Antiquity. Proceedings of a Conference held in Naples, Italy, 13 April 2015*. Oxford: Oxbow Books.

Evrar-Derrick, Claire, and Jan Quaegebeur. 1979. "La situle décorée de Nesnakhetiou au Musée Royal de Mariemont." *Chronique d'Égypte*, 54(107): 26–56.

Ezer, Sabahattin. 2014. "Kültepe-Kanesh in the Early Bronze Age." In Levent Atici, Fikri Kulakoğlu, Gojko Barjamovic, and Andrew Fairbairn (eds), *Current Research at Kültepe-Kanesh. An Interdisciplinary and Integrative Approach to Trade Networks, Internationalism, and Identity* (= *Journal of Cuneiform Studies*, Supplemental Series, 4). Atlanta, GA: Lockwood Press.

Faist, Betina I. 2001. *Der Fernhandel des assyrischen Reiches zwischen dem 14. und 11. Jhr. v. Chr.* (= *Alter Orient und Altes Testament*, 265). Münster: Ugarit-Verlag.

Faivre, Xavier. 2009. "Récipients, ustensiles, et alimentation: fonctions et usages multiples ... mais lesquels?" *Cahier des thèmes transversaux ArScAn*, 9: 277–94. Available online: https://hal.archives-ouvertes.fr/hal-02264111/document.

Falkenstein, Adam. 1931. *Die Haupttypen der sumerischen Beschwörung, literarisch untersucht* (= *Leipziger semitistische Studien*, neue Folge, 1). Leipzig: August Pries.

Farber-Flügge, Gertrud. 1973. *Der Mythos "Inanna und Enki" unter besonderer Berücksichtigung der Liste der m e* (= *Studia Pohl: dissertationes scientificae de rebus Orientis antiqui*, 10). Rome: Biblical Institute Press.

Faucher, Thomas. 2018. "L'or des Ptolémées: l'exploitation de l'or dans le désert oriental." In Jean-Pierre Brun, Bérangère Redon, and Steven Sidebotham (eds), *Le désert oriental d'Égypte durant la période gréco-romaine: bilans archéologiques*. Paris: OpenEditions Books. Available online: https://books.openedition.org/cdf/5136?lang=en.

Feldman, Marian H. 2006. *Diplomacy by Design: Luxury Arts and an "International Style" in the Ancient Near East, 1400–1200 BCE*. Chicago, IL: University of Chicago Press.

Fermat, André. 2010. *Le rituel de la Maison de Vie. Papyrus Salt 825*. Paris: MdV.

Festugière, André Marie Jean. 1944. *La révélation d'Hermès Trismégiste. Vol. 1: L'astrologie et les sciences occultes*. Paris: Belles Lettres.

Festugière, André Marie Jean. 1967. "La création des âmes dans la Kore Kosmou." In André Marie Jean Festugière (ed.), *Hermétisme et mystique païenne*. Paris: Aubier-Montaigne.

Finkel, Irving L. 2000. "On Late Babylonian Medical Training." In Andrew R. George and Irving L. Finkel (eds), *Wisdom, Gods and Literature: Studies in Assyriology in Honour of W.G. Lambert*. Winona Lake, IN: Eisenbrauns.

Finkel, Irving L., Hero Granger-Taylor, and Dominique Cardon. 1998. "Un fragment de tablette inscrite en cunéiforme." In Dominique Cardon (ed.), *Teintures précisieuses de la Méditerranée: pourpre, kermes, pastel*. Carcassonne: Musée des Beaux-Arts de Carcassonne.

Firth, Richard. 2011. "A Discussion of the Use of im-babbar$_2$ by the Craft Workers of Ancient Mesopotamia." *Cuneiform Digital Library Journal*, (2). Available online: https://cdli.ucla.edu/pubs/cdlj/2011/cdlj2011_002.html.

Fleming, Stuart J. 1999. *Roman Glass. Reflections on Cultural Change*. Philadelphia: University of Pennsylvania Museum of Archaeology and Anthropology.

Flemming, Rebecca. 2003. "Empires of Knowledge: Medicine and Health in the Hellenistic World." In Andrew Erskine (ed.), *A Companion to the Hellenistic World*. Oxford: Wiley-Blackwell.

Flohr, Miko. 2003. "*Fullones* and Roman Society: A Reconsideration." *Journal of Roman Archaeology*, 16: 447–50.

Flohr, Miko. 2013a. *The World of the Fullo. Work, Economy, and Society in Roman Italy*. Oxford: Oxford University Press.

Flohr, Miko. 2013b. "The Textile Economy of Pompeii." *Journal of Roman Archaeology*, 26: 53–78.
Forbes, Robert J. 1950. *Metallurgy in Antiquity: A Notebook for Archaeologists and Technologists*. Leiden: Brill.
Forbes, Robert J. 1965. *Studies in Ancient Technology*, 2nd ed, vol. 3. Leiden: Brill
Forbes, Robert J. 1970. *A Short History of the Art of Distillation: From the Beginnings Up to the Death of Cellier Blumenthal*. Leiden: Brill.
Fournet, Jean-Luc. 2000. "'Un fragment de Néchepso' sur des formules apotropaïques pour chasser des maladies par les esprits du mal." In Henri Melaerts (ed.), *Papyri in honorem Johannis Bingen octogenarii*. Leuven: Peeters.
Frahm, Eckart. 2004. "Royal Hermeneutics: Observations on the Commentaries from Ashurbanipal's Libraries at Nineveh." *Iraq*, 66: 45–50.
Frame, Grant, and Andrew R. George. 2005. "The Royal Libraries at Nineveh: New Evidence for King Ashurbanipal's Tablet Collecting." *Iraq*, 67: 265–84.
Frankfort, Henri. 1944. "A Note on the Lady of Birth." *Journal of Near Eastern Studies*, 3: 198–200.
Frankfort, Henri, Henriette A. Frankfort, John A. Wilson, Thorkild Jacobsen, and William A. Irwin. 1946. *The Intellectual Adventure of Ancient Man: An Essay of Speculative Thought in the Ancient Near East*. Chicago, IL: University of Chicago Press. Available online: https://oi.uchicago.edu/research/publications/misc/intellectual-adventure-ancient-man-essay-speculative-thought-ancient-near.
Frayne, Douglas R. 1993. *Sargonic and Gutian Periods (2234–2113 BC)*. Toronto: University of Toronto Press.
Freu, Christel. 2011. "Apprendre et exercer un métier dans l'Égypte romaine (Ier–VIe siècles ap. J.-C." In Nicolas Monteix and Nicolas Tran (eds), *Les savoirs professionnels des gens de métier. Études sur le monde du travail dans les sociétés urbaines de l'empire romain*. Naples: Centre Jean Bérard.
Freu, Christel. 2016. "*Disciplina, patrocinium, nomen*: The Benefits of Apprenticeship in the Roman World". In Andrew Wilson and Miko Flohr (eds), *Urban Craftsmen and Traders in the Roman World*. Oxford: Oxford University Press.
Friberg, Jöran. 2007. *A Remarkable Collection of Babylonian Mathematical Texts: Manuscripts in the Schøyen Collection: Cuneiform Texts 1*. New York: Springer.
Froidefond, Christian (ed.). 1988. *Plutarque, Œuvres morales, tome V – 2^{e} partie, Isis et Osiris*. Paris: Belles Lettres.
Fronzaroli, Pelio. 1998. "A Pharmaceutical Text at Ebla (TM.75.G.1623)." *Zeitschrift für Assyriologie*, 88: 225–39.
Frymer-Kensky, Tikva. 1992. *In the Wake of the Goddesses: Women, Culture and the Biblical Transformation of Pagan Myth*. New York: Free Press.
Gabolde, Marc. 2009. "Égyptien *šdḥ*, grec *oinomeli* et *melitiès*, latin *mulsum*, grec d'Égypte *stagma*: la même ivresse?" In Isabelle Régen and Frédéric Servajean (eds), *Verba Manent. Recueil d'études dédiées à Dimitri Meeks* (= *Les Cahiers: Égypte Nilotique et Méditérranéenne*, 2). Montpellier: Université Paul Valéry.
Gadd, Cyril J., and R. Campbell Thompson. 1936. "A Middle-Babylonian Chemical Text." *Iraq*, 3: 87–96.
García-Alix, Antonio, Francisco J. Jiménez-Espejo, José Antonio Lozano, Gonzalo Jiménez-Moreno, Francisca Martínez-Ruiz, L. García Sanjuán, G. Aranda Jiménez, E. García Alfonso, Gerardo Ruiz-Puertas, and R. Scott Anderson. 2013. "Anthropogenic Impact and Lead Pollution throughout the Holocene in Southern Iberia." *Science of the Total Environment*, 449: 451–60.

Gardiner, Alan H. 1911. *Egyptian Hieratic Texts Transcribed, Translated and Annotated, Série I: Literary Texts of the New Kingdom. Part I: Papyrus Anastasi I and the Koller Papyrus Together with the Parallel Texts*. Leipzig: Hinrichs (new ed.: Hildesheim – Zürich – New York: Georg Holms Verlag, 2007).

Gardiner, Alan H. 1938. "The House of Life." *The Journal of Egyptian Archaeology*, 24: 157–79.

Gardiner, Alan H. 1947. *Ancient Egyptian Onomastica*. 3 vols. Oxford: Oxford University Press.

Gardiner, Alan H. 1959. *The Royal Canon of Turin*. Oxford: Oxford University Press.

Garenne-Marot, Laurence. 1984. "Le cuivre en Égypte pharaonique: sources et métallurgie." *Paléorient*, 10: 97–126.

Garenne-Marot, Laurence. 1985. "Le Travail du cuivre dans l'Égypte pharaonique d'après les peintures et les bas-reliefs." *Paléorient*, 11: 85–100.

Garstang, John. 1953. *Prehistoric Mersin, Yümük Tepe in Southern Turkey: The Neilson Expedition in Cilicia*. Oxford: Clarendon Press.

Gasse, Annie. 1987. "Une expédition au Ouadi Hammamat sous le règne de Sebekemsaf Ier." *Bulletin de l'Institut français d'Archéologie orientale*, 87: 207–18.

Gasse, Annie. 1988. "Amény, un porte-parole sous le règne de Sésostris Ier." *Bulletin de l'Institut français d'Archéologie orientale*, 88: 83–94.

Gauthier, Henri. 1918. "Un nouveau monument du dieu Imhotep." *Bulletin de l'Institut français d'Archéologie orientale*, 14: 33–49.

Gauthier, Patrick, and Béatrix Midant-Reynes. 1995. "La tête de massue du roi Scorpion." *Archéo-Nil*, 127: 87–127.

Geller, Markham J. 1985. *Forerunners to Udug-hul: Sumerian Exorcistic Incantations* (= *Freiburger altorientalische Studien*, 12). Stuttgart: Franz Steiner.

Geller, Markham J. 2016. *Healing Magic and Evil Demons: Canonical Udug-hul Incantations* (= Babylonisch-assyrische Medizin in Texten und Untersuchungen, 8). Boston, MA: De Gruyter.

Geller, Markham J. 2018. "A Babylonian Hippocrates." In Ulrike Steinert (ed.), *Assyrian and Babylonian Scholarly Text Catalogues: Medicine, Magic and Divination* (= *Babylonisch-assyrische Medizin in Texten und Untersuchungen*, 9). Boston, MA: De Gruyter.

George, Andrew R. 2016. *Mesopotamian Incantations and Related texts in the Schøyen Collection* (= *Cornell University Studies in Assyriology and Sumerology*, 32). Bethesda, MD: CDL Press.

Germer, Renate. 1992. *Die Textilfarberei und die Verwendung gefärbter Textilien im alten Ägypten* (= *Ägyptische Abhandlungen*, 53). Wiesbaden: Harrassowitz.

Giardina, Giovanna R. 2008. *La chimica fisica di Aristotele. Teoria degli elementi e delle loro proprietà. Analisi critica del De generatione et corruptione*. Rome: Aracne.

Gilan, Amir. 2013. "Once upon a Time in Kiškiluša: The Dragon-Slayer Myth in Central Anatolia." In JoAnn Scurlock and Richard H. Beal (eds), *Creation and Chaos: A Reconsideration of Hermann Gunkel's Chaoskampf Hypothesis*. Winona Lake, IN: Eisenbrauns.

Girard, Michel, and Jean Maley. 1987. "Autopsie d'une momie égyptienne du Muséum de Lyon: E-5, étude palynologique." *Nouvelles archives du Muséum d'histoire naturelle de Lyon*, 25: 103–10.

Gittinger, Mattiebelle. 1982. *Master Dyers to the World: Technique and Trade in Early Indian Dyed Cotton Textiles*. Washington, DC: Textile Museum.

Giumlia-Mair, Alessandra. 2002. "Zosimos the Alchemist – Manuscript 6.29, Cambridge, Metallurgical Interpretation." In Alessandra Giumlia-Mair (ed.), *I bronzi antichi: produzione e tecnologia: atti del XV Congresso internazionale sui bronzi antichi organizzato dall'Università di Udine, sede di Gorizia, Grado-Aquilieia, 22–26 maggio 2001*. Montagnac: Éditions Monique Mergoil.

Goddio, Franck (ed.) 2006. *Trésors engloutis d'Égypte*. Paris: Seuil.

Goedicke, Hans. 1975. *The Report of Wenamun*. Baltimore, MD: Johns Hopkins University Press.

Goedicke, Hans. 1984. "Abi-Sha(i)'s Representation in Beni Hasan." *Journal of the American Research in Egypt*, 21: 203–10.

Goltz, Dietlinde. 1972. *Studien zur Geschichte der Mineralnamen in Pharmazie, Chemie und Medizin von den Anfängen bis Paracelsus*. Wiesbaden: Franz Steiner.

Gouin, Philippe. 1993. "Bovins et laitages en Mésopotamie méridionale au 3ème millénaire: Quelques commentaires sur la 'frise à la laiterie' de el-'Obeid." *Iraq*, 55: 135–45.

Gowlett, John A.J., and Richard W. Wrangham. 2013. "Earliest Fire in Africa: Towards the Convergence of Archaeological Evidence and the Cooking Hypothesis." *Azania: Archaeological Research in Africa*, 48: 5–30.

Goyon, Georges. 1949. "Le papyrus deTurin dit 'des mines d'or' et le Wadi Hammamat." *Annales du Service des Antiquités de l'Égypte*, 49: 337–92.

Goyon, Georges. 1957. *Nouvelles inscriptions rupestres du Wadi Hammamat*. Paris: Imprimerie nationale.

Goyon, Jean-Claude. 1972. *Rituels funéraires de l'ancienne Égypte* (= Littératures anciennes du Proche-Orient, 4). Paris: Cerf.

Goyon, Jean-Claude. 1980. "Note pour servir à la connaissance des procédés tinctoriaux de l'ancienne Égypte." In Jean Vercoutter (ed.), *Le Livre du Centenaire de l'IFAO* (= Mémoires de l'Institut français d'Archéologie orientale, 104). Cairo: Institut Français d'Archéologie Orientale.

Goyon, Jean-Claude. 1982. "Ébauche d'un système étatique d'utilisation de l'eau: Égypte pharaonique de l'ancien au nouvel empire." *Publications de la Maison de l'Orient et de la Méditerranée*, 3: 61–68. In *L'homme et l'eau*, Lyon: MOM éditions. Available online: https://www.persee.fr/doc/mom_0766-0510_1982_sem_3_1_2018.

Goyon, Jean-Claude. 1996. "Le lin et sa teinture en Égypte. Des procédés ancestraux aux pratiques importées (VIIe siècle av. J.-C. à l'époque récente)." In *Aspects de l'artisanat du textile dans le monde méditerranéen (Égypte, Grèce, monde romain)* (= Collection de l'Institut d'archéologie et d'histoire de l'antiquité, 2). Lyon: Université Lumière Lyon 2.

Goyon, Jean-Claude. 2012. *Le Recueil de prophylaxie contre les agressions des animaux venimeux du Musée de Brooklyn* (= Studien zur spätägyptischen Religion, 5). Wiesbaden: Harrassovitz.

Goyon, Jean-Claude, Jean-Claude Golvin, Claire Simon-Boidot, and Gilles Martinet. 2004. *La construction pharaonique du Moyen Empire à l'époque gréco-romaine. Contexte et principes technologiques*. Paris: Picard.

Graham, Daniel W. 1999. "Empedocles and Anaxagoras: Responses to Parmenides." In Anthony A. Long (ed.), *The Cambridge Companion to Early Greek Philosophy*. Cambridge: Cambridge University Press.

Grandet, Pierre. 1994. *Le papyrus Harris I (BM 9999)*. 2 vols. (= Bibliothèque d'Étude, 109). Cairo: Institut Français d'Archéologie Orientale.

Grattan, John P., David D. Gilbertson, and Chris O. Hunt. 2007. "The local and global dimensions of metalliferous pollution derived form a reconstruction of a eight thousand year record of copper smelting and mining at a desert–mountain frontier in southern Jordan." *Journal of Archaeological Science*, 34: 83–110.

Grattan, John P., Steven Huxley, Lotus Abu Karaki, Harry Toland, David Gilbertson, Brian Pyatt, and Ziad Al Saad. 2002. "'Death more desirable than life …'? The Human Skeletal Record and Toxicological Implications of Ancient Copper Mining and Smelting in Wadi Faynan, South West Jordan." *Toxicology and Industrial Health*, 18: 297–307.

Grattan, John P., Lotus Abu Karaki, David Hine, Harry Toland, David Gilbertson, Ziad Al-Saad, and Brian Pyatt. 2005. "Analyses of patterns of copper and lead mineralisation in human skeletons excavated from an ancient mining and smelting centre in the Jordanian desert: a reconnaissance study." *Mineralogical Magazine*, 69: 653–66.

Greenaway, Frank. 1986. "Chemical Tests in Pliny." In Roger French and Frank Greenaway (eds), *Science in the Early Roman Empire: Pliny the Elder, His Sources and Influence*. London: Croom Helm.

Gremilliet, Jean-Paul, and Cyrille Delangle. 2017. "Le travail des roches dures dans l'Égypte ancienne." Available online: http://terraegenesis.org/wp-content/uploads/2019/10/CGTG_Etude_Egypte.pdf.

Grenier, Jean-Claude. 2002. "La stèle de la mère d'un Bouchis datée de Licinius et de Constantin." *Bulletin de l'Institut français d'Archéologie orientale*, 102: 247–58.

Grenier, Jean-Claude. 2009. "Les pérégrinations d'un Boukhis en Haute Thébaïde." In Christophe Thiers (ed.), *Documents de Théologies Thébaines Tardives (D3T 1)* (= *Cahiers: Égypte nilotique et méditerranéenne*, 3). Montpellier: Université Paul Valéry.

Griffith, Kenneth. 2006. "Images of the *rekhyt* from ancient Egypt." *Ancient Egypt* 7(2): 45–50.

Grimal, Nicolas. 2008–18. "Civilisation pharaonique: archéologie, philologie, histoire." In *L'Annuaire du Collège de France (Cours et travaux du Collège de France)*. Available online: https://journals.openedition.org/annuaire-cdf.

Grmek, Mirko. 1979. "Ruses de guerre biologiques dans l'Antiquité." *Revue des études grecques*, 92: 141–63.

Guardasole, Alessia. 2006. "Galien et le marché des simples aux Ier et IIe siècles de notre ère." In Franck Collard and Évelyne Samama (eds), *Pharmacolopes et apothicaires. Les "pharmaciens" de l'Antiquité au Grand Siècle*. Paris: L'Harmattan.

Guasch-Jané, Maria Rosa, Cristina Andrés-Lacueva, Olga Jáuregui, and Rosa M. Amuela-Raventós. 2006. "The Origin of the Ancient Egyptian Drink Shedeh Revealed using LC/MS/MS." *The Journal of Archaeological Science*, 33: 98–101.

Güell i Rous, Josep Maria. 2018. *The Tomb of Vizier Rekhmire (TT100). A Textual and Iconographic Study. Vol. 1: Rekhmire Receiving Foreign Tribute*. Barcelona: La Vocal de Lis.

Guilhou, Nadine. 1989. *La vieillesse des dieux*. Montpellier: Université Paul Valéry.

Gwinnett, A. John, and Leonard Gorelick. 1987. "The Change from Stone Drills to Copper Drills in Bronze Age Mesopotamia: An Experimental Perspective." *Expedition*, 29: 15–24.

Hadot, Pierre. 2004. *Le voile d'Isis: essai sur l'histoire de l'idée de nature*. Paris: Gallimard.

Haldon, John. 2006. "'Greek Fire' Revisited: Recent and Current Research." In Elizabeth Jeffreys (ed.), *Byzantine Style, Religion and Civilization. In Honour of Sir Steven Runciman*. Cambridge: Cambridge University Press.

Halleux, Robert. 1969. "Lapis-lazuli, azurite ou pâte de verre? À propos de *kuwano* et *kuwanowoko* dans les tablettes mycéniennes." *Studi micenei ed egeo-anatolici*, 9: 47–66.

Halleux, Robert. 1974. *Les problème des métaux dans la science antique*. Paris: Belles Lettres.

Halleux, Robert. 1975. "Les deux métallurgies du plomb argentifére dans *l'Histoire naturelle* de Pline." *Revue de philologie, de littérature et d'histoire anciennes*, 49: 72–88.

Halleux, Robert. 1977. "De stagnum 'étang' à stagnum 'étain.'" *L'Antiquité classique*, 46: 557–70.

Halleux, Robert. 1979. *Les textes alchimiques*. Turnhout: Brepols.

Halleux, Robert (ed.). 1981. *Les alchimistes grecs. Tome I: Papyrus de Leyde, Papyrus de Stockholm, recettes*. Paris: Belles Lettres.

Halleux, Robert. 1982. "Nouveaux textes sur la metallurgie antique." In *Mines et fonderies antiques de la Gaule: Université de Toulouse-Le Mirail, 21–22 novembre 1980: table ronde du CNRS*. Paris: Éditions du CNRS.

Halleux, Robert. 1985. "Méthodes d'essai et d'affinage des alliages auriféres dans l'Antiquité et au Moyen Age." In Cécile Morrisson, Claude Brenot, Jean-Pierre Callu, Jean-Noël Barrandon, Jacques Poirier, and Robert Halleux (eds), *L'or monnayé I. Purification et alterations de Rome à Byzance*. Paris: Éditions du CNRS.

Halleux, Robert. 1987. "La monnaie de fer de Lycurgue et le problème des acides en chimie antique." In Jean Servais, Tony Hackens, and Brigitte Servais-Soyez (eds), *Stemmata. Mélanges de philologie, d'histoire et d'archéologie grecques offerts à Jules Labarbe*. Liège: l'Antiquité Classique.

Halleux, Robert, and Anne Françoise Cannella. 1998. "Entre technologie et alchimie: de la teinture du verre à la fabrication des fausses pierres précieuses." In *Il colore nel Medioevo. Arte Simbolo Tecnica. Atti delle Giornate di Studi, Lucca, 2, 3, 4 maggio 1996*. Lucca: Istituto Storico Lucchese.

Halleux, Robert, and Paul Meyvaert. 1987. "Les origines de la *Mappae clavicula*." *Archives d'histoire doctrinale et littéraire du Moyen Âge*, 54: 7–58.

Halleux, Robert, and Jacques Schamp. 1985. *Les lapidaires grecs: Lapidaire orphique, Kerygmes, Lapidaires d'Orphée, Socrate et Denys, Lapidaire nautique, Damigéron, Evax*. Paris: Belles Lettres.

Hallum, Benjamin. 2008. "Theosebeia (ca. 250 – 300 CE)." In Paul Keyser and Georgia Irby-Massie (eds), *Encyclopedia of Ancient Natural Scientists. The Greek Tradition and Its Many Heirs*. London: Routledge.

Hampson, Michelle Theresa. 2012. "Men at Work. A Comparative Study of Workshop Scenes in Tombs of the Old Kingdom: Applying an Anatomisation Approach." Ph.D. thesis, Macquarie University, Macquarie Park.

Harrel, James Antony. 1990. "Misuse of the Term 'Alabaster' in Egyptology." *Göttinger Miszellen*, 119: 37–42.

Harrel, James Antony. 2004. "Archaeological Geology of the World's First Emerald Mine." *Geoscience Canada*, 31(2): 69–76. Available online: https://journals.lib.unb.ca/index.php/GC/issue/view/317.

Harris, William V. 2011. "Plato and the Deforestation of Attica." *Athenaeum*, 99: 479–82.

Harrison, George. 1987. "Martial 1.41: Sulphur and Glass." *Classical Quarterly*, 37: 203–7.

Hasaki, Eleni. 2013. "Craft Apprenticeship in Ancient Greece: Reaching beyond the Masters." In Willeke Wendrich (ed.), *Archaeology and Apprenticeship. Body Knowledge, Identity, and Communities of Practice*. Tucson: University of Arizona Press.

Hasaki, Eleni. 2019. *Potters at Work in Ancient Corinth: Industry, Religion, and the Penteskouphia Pinakes* (= *Hesperia Supplement*, 51). Princeton, NJ: American School of Classical Studies at Athens.

Hauptmann, Andreas, Sabine Klein, Paola Paoletti, Richard L. Zettler, and Moritz Jansen. 2018. "Types of Gold, Types of Silver: The Composition of Precious Metal Artifacts Found in the Royal Tombs of Ur, Mesopotamia." *Zeitschrift für Assyriologie*, 108: 100–31.

Healy, John F. 1999. *Pliny the Elder on Science and Technology*. Oxford: Oxford University Press.

Heessel, Nils P. 2000. *Babylonisch-assyrische Diagnostik* (= *Alter Orient und Altes Testament*, 43). Münster: Ugarit-Verlag.

Heessel, Nils P. 2005. "Stein, Pflanze und Holz: Ein neuer Text zur 'medizinischen Astrologie.'" *Orientalia NS*, 74: 1–22.

Heessel, Nils P. 2008. "Astrological Medicine in Babylonia." In Anna Akasoy, Charles Burnett, and Ronit Yoeli-Tlalim (eds), *Astro-Medicine: Astrology and Medicine, East and West*. Florence: Sismel.

Heessel, Nils P. 2009. "The Babylonian Physician Rabâ-ša-Marduk: Another Look at Physicians and Exorcists in the Ancient Near East." In Annie Attia and Gilles Buisson (eds), *Advances in Mesopotamian Medicine from Hammurabi to Hippocrates* (= *Cuneiform Monographs*, 37). Leiden: Brill.

Heessel, Nils P. 2010. "Neues von Esagil-kīn-apli: Die ältere Version der physiognomischen Omenserie alamdimmû." In Stefan M. Maul and Nils P. Heessel, (eds), *Assur-Forschungen: Arbeiten aus der Forschungsstelle "Edition literarischer Keilschrifttexte aus Assur" der Heidelberger Akademie der Wissenschaften*. Wiesbaden: Harrassowitz.

Heilen, Stephen. 2011. "Some Metrical Fragments from Nechepsos and Petosiris." In Isabelle Boehm and Wolfgang Hübner (eds), *La poésie astrologique dans l'Antiquité*: *Actes du colloque organisé les 7 et 8 décembre 2007* (= *Centre d'études et de recherches sur l'Occident romain*, 38). Paris: Université Jean-Moulin Lyon III.

Heimpel, Wolfgang. 1998. "The Industrial Park of Girsu in the Year 2042 B.C.: Interpretation of an Archive Assembled by P. Mander." *Journal of the American Oriental Society*, 118: 387–99.

Heimpel, Wolfgang. 2009. *Workers and Construction Work at Garšana* (= *Cornell University Studies in Assyriology and Sumerology*, 5). Bethesda, MD: CDL Press.

Heimpel, Wolfgang, Leonard Gorelick, and A. John Gwinnett. 1988. "Philological and Archaeological Evidence for the Use of Emery in the Bronze Age Near East." *Journal of Cuneiform Studies*, 40: 195–210.

Henrichs, Albert. 1984. "The Sophists and Hellenistic Religion: Prodicus as the Spiritual Father of the Isis Aretologies." *Harvard Studies in Classical Philology*, 88: 139–58.

Helck, Wolfgang. 1971. *Das Bier im Alten Ägypten*. Berlin: Gesellschaft für die Geschichte und Bibliographie des Brauwesens e.V.; Institut für Gärungsgewerbe und Biotechnologie.

Helwing, Barbara. 2011. "Conclusions: The Arismān Copper Production in a Wider Context." In Abdolrasool Vatandoust, Hermann Parzinger, and Barbara Helwing (eds), *Early Mining and Metallurgy on the Western Central Iranian Plateau*. Mainz: Philipp von Zabern.

Henderson, Julian. 2012. *Ancient Glass: An Interdisciplinary Exploration*. Cambridge: Cambridge University Press.

Herbin, François. 2010. "Les premières pages du Papyrus Salt 825." *Bulletin de l'Institut français d'Archéologie orientale*, 88: 95–112.

Hermann, Alfred. 1954. "Das Buch *kmj t* und die Chemie." *Zeitschrift für ägyptische Sprache*, 79: 99–105.

Herrmann, Georgina. 1968. "Lapis Lazuli: The Early Phases of Its Trade." *Iraq*, 30: 21–57.

Heseltine, Michael. 1913. *Petronius*. London: Heinemann.

Hett, Walter Stanley. 1936. *Aristotle, Minor Works* (= Loeb Classical Library, 307). London: Heinemann.

Hillman, Aubrey L., Mark B. Abbott, Blas Lorenzo Valero-Garcés, Mario Morellon, Fernando Barreiro-Lostres, and Daniel J. Bain. 2017. "Lead Pollution Resulting from Roman Gold Extraction in Northwestern Spain." *The Holocene*, 27: 1465–74.

Hinke, William J. 1907. *A New Boundary Stone of Nebuchadrezzar I from Nippur, with a Concordance of Proper Names and a Glossary of the Kudurru Inscriptions Thus Far Published*. Philadelphia: University of Pennsylvania Press.

Hirsch, Hans. 1963. "Die Inschriften der Könige von Agade." *Archiv für Orientforschung*, 20: 1–82.

Hoefer, Ferdinand. 1866. *Histoire de la chimie*. 2 vols. Paris: Firmin Didot Frères.

Hoffmeier, James K. 2006. "'The Walls of the Ruler' in Egyptian Literature and the Archaeological Record: Investigating Egypt's Eastern Frontier in the Bronze Age." *Bulletin of the American Schools of Oriental Research*, 343: 1–20.

Hoffner, Harry A. 1967. "Ugaritic *pwt*: A Term from the Early Canaanite Dyeing Industry." *Journal of the American Oriental Society*, 87: 300–3.

Holleran, Claire. 2018. "The Retail Trade." In Claire Holleran and Amanda Claridge (eds), *A Companion to the City of Rome*. Oxford: Wiley-Blackwell.

Hong, Sungmin, Jean-Pierre Candelone, Clair C. Patterson, and Claude F. Boutron. 1994. "Greenland Ice Evidence of Hemispheric Lead Pollution Two Millennia Ago by Greek and Roman Civilizations." *Science*, 265: 1841–3.

Hong, Sungmin, Jean-Pierre Candelone, Clair C. Patterson, and Claude F. Boutron. 1996a. "History of Ancient Copper Smelting Pollution During Roman and Medieval Times Recorded in Greenland Ice." *Science*, 272: 246–9.

Hong, Sungmin, Jean-Pierre Candelone, Michel Soutif, and Claude F. Boutron. 1996b. "A Reconstruction of Changes in Copper Production and Copper Emissions to the Atmosphere during the Past 7000 Years." *The Science of the Total Environment*, 188: 183–93.

Hopkins, Heather. 2008. "Using Experimental Archaeology to Answer the Unanswerable: A Case Study Using Roman Dyeing." In Penny Cunningham, Julia Heeb, and Roeland Paardekooper (eds), *Experiencing Archaeology by Experiment: Proceedings of the Experimental Archaeology Conference, Exeter 2007*. Oxford: Oxbow Books.

Hornung, Erik. 1997. *Der ägyptische Mythos von der Himmelskuh: Eine Ätiologie des Unvollkommenen* (= Orbis Biblicus Orientalis, 46). Freiburg: Universitätsverlag.

Høyrup, Jens. 2002. *Lengths, Widths, Surfaces: A Portrait of Old Babylonian Algebra and Its Kin*. New York: Springer.

Hughes, J. Donald. 2014. *Environmental Problems of the Greeks and Romans. Ecology in the Ancient Mediterranean*, 2nd ed. Baltimore, MD: Johns Hopkins University Press.

Hughes, J. Donald. 2017. "Deforestation and Forest Protection in the Ancient World." In Orietta Dora Cordovana and Gian Franco Chiai (eds), *Pollution and the Environment in Ancient Life and Thought*. Stuttgart: Franz Steiner.

Hunter, Erika C.D. 2002. "Beautiful Black Bronzes. Zosimos' Treatises in Cam. Mm. 6.29." In Alessandra Giumlia-Mair (ed.), *I bronzi antichi: produzione e tecnologia: atti del XV Congresso internazionale sui bronzi antichi organizzato dall'Università di Udine, sede di Gorizia, Grado-Aquilieia, 22–26 maggio 2001*. Montagnac: Éditions Monique Mergoil.

Huntingford, George W.B. 1980. *The Periplus of the Erythraean Sea by an Unknown Author. with Some Extracts from Agatharkhidēs "On the Erythraean Sea."* London: Hakluyt Society.

Hurry, Jamieson B. 1930. *The Woad Plant and Its Dye*. London: Oxford University Press.

Iamori, Marco. 2015. "The Eastern Palace of Qatna and the Middle Bronze Age Archiarchitectural Tradition of Western Syria." In P. Pfalzner and M. al-Maqdissi (eds), *Qatna and the Networks of Bronze Age Globalism: Proceedings of an International Conference in Stuttgart and Tübingen in October 2009 (= Qatna-Studien Supplementa, 2)*. Wiesbaden: Harrassowitz.

Ierodiakonou, Katerina. 2005. "Empedocles on Colour and Colour Vision." *Oxford Studies in Ancient Philosophy*, 29: 1–37.

Incordino, Ilaria. 2016. "Foreign Aromatic Products in the Cultural and Religious Identity of Ancient Egypt." In Ilaria Micheli (ed.), *Materiality and Identity. Selected Papers from the Proceedings of the ATrA Conferences of Naples and Turin 2015*. Trieste: EUT.

Incordino, Ilaria. 2017. "African *aromata* in Egypt: The 'ti-shepes.'" In Ilaria Incordino and Pearce Paul Creasman (eds), *Flora Trade between Egypt and Africa in Antiquity. Proceedings of a Conference Held in Naples, Italy, 13 April 2015*. Oxford: Oxbow Books.

Jacoby, Felix. 1923. *Die Fragmente der Griechischen Historiker*, Parts I–IV. Berlin: Weidmann.

Jacquet, Jean. 1994. *Le trésor de Thouthmôsis Ier. Installations antérieures ou postérieures au monument = Karnak-Nord VII. 2 (= Fouilles de l'Institut français d'Archéologie orientale, 36)*. Cairo: Institut Français d'Archéologie Orientale.

James, Matthew, Nicole Reifarth, and Richard Evershed. 2011. "Chemical Identification of Animal Dyestuffs from Mineralized Textile Fragments from the Royal Tomb." In Peter Pfälzner (ed.), *Interdisziplinäre Untersuchungen zur Königsgruft in Qaṭna (= Qaṭna Studien, 1)*. Wiesbaden: Harrassowitz.

James, Simon. 2011. "Stratagems, Combat and 'Chemical Warfare' in the Siege Mines of Dura-Europos." *American Journal of Archaeology*, 115: 69–101.

Jasnow, Richard, and Karl-Theodor Zauzich. 2005. *The Ancient Egyptian Book of Thot*. 2 vols. Wiesbaden: Harrassowitz.

Jeffreys, Elisabeth, Michael Jeffreys, and Roger Scott. 1986. *The Chronicle of John Malalas. A Translation*. Melbourne: Australian Association for Byzantine Studies.

Jelinková-Reymond, Eva. 1956. *Les inscriptions de la statue guérisseuse de Djer-ḥer-le-Sauveur (= Bibliothèque d'étude, 23)*, Cairo: Institut Français d'Archéologie Orientale.

Joachim, Harold H. 1903. "Aristotle's Conception of Chemical Combination." *Journal of Philology*, 29: 72–86.

Johnson, J. Cale. 2014. "The Cost of Cosmogony: Ethical Reflections on Resource Extraction, Monumental Architecture and Urbanism in the Sumerian Literary Tradition." In Natalie May and Ulrike Steinert (eds), *The Fabric of Cities: Aspects of Urbanism, Urban Topography and Society in Mesopotamia, Greece and Rome*. Leiden: Brill.

Johnson, J. Cale. 2015. "Depersonalized Case Histories in the Babylonian Therapeutic Compendia." In J. Cale Johnson (ed.), *In the Wake of the Compendia: Infrastructural Contexts and the Licensing of Empiricism in Ancient and Medieval Mesopotamia*. Berlin: De Gruyter.

Johnson, J. Cale. 2017. "The Stuff of Causation: Etiological Metaphor and Pathogenic Channeling in Babylonian Medicine." In John Z. Wee (ed.), *The Comparable Body: Analogy and Metaphor in Ancient Mesopotamian, Egyptian, and Greco-Roman Medicine* (*Studies in Ancient Medicine*, 49). Leiden: Brill.

Johnson, J. Cale. 2018. "Towards a New Perspective on Babylonian Medicine: The Continuum of Allegoresis and the Emergence of Secular Models in Mesopotamian Scientific Thought." In Ulrike Steinert (ed.), *Assyrian and Babylonian Scholarly Text Catalogues: Medicine, Magic and Divination* (= *Die babylonisch-assyrische Medizin in Texten und Untersuchungen*, 9). Berlin: De Gruyter.

Johnson, J. Cale. 2019. "Demarcating ekphrasis in Mesopotamia." In J. Cale Johnson and Alessandro Stavru (eds), *Visualizing the Invisible with the Human Body: Physiognomy and Ekphrasis in the Ancient World* (*Science, Technology, and Medicine in Ancient Cultures*, 10). Berlin: De Gruyter.

Johnson, J. Cale, and Markham J. Geller. 2015. *The Class Reunion – An Annotated Translation and Commentary on the Sumerian Dialogue Two Scribes* (= *Cuneiform Monographs*, 47). Leiden: Brill.

Jones, Horace Leonard. 1921. *Strabo, Geography, Books 3–5* (= *Loeb Classical Library*, 50). London: Heinemann.

Jones, Horace Leonard. 1930. *Strabo, Geography, Books 15–16* (= *Loeb Classical Library*, 241). London: Heinemann.

Jones, Tom B. 1961. "Review of *Le Travail du métal au pays de Sumer au temps de la IIIe dynastie d'Ur* by Henri Limet." *Journal of Cuneiform Studies*, 15: 114–16.

Jordan, David R. 2000. "A Personal Letter Found in the Athenian Agora." *Hesperia*, 69: 91–103.

Jouanna, Jacques. 1975. "Plutarque et la patine des statues à Delphes." *Revue de philologie, de littérature et d'histoire ancienne*, 49: 67–71.

Jung, Carl Gustav. 1944. *Psychologie und Alchemie*. Zurich: Rascher Verlag.

Jursa, Michael. 2005. "Parfüm(rezepte). A. In Mesopotamien." *Reallexikon der Assyriologie*, 10: 335–6.

Jursa, Michael. 2009. "Die Kralle des Meeres und andere Aromata." In Werner Arnold, Michael Jursa, Walter Müller, and Stephen Procházka (eds), *Philologisches und Historisches zwischen Anatolien und Sokotra: Analecta Semitica in Memoriam Alexander Sima*. Wiesbaden: Harrassowitz.

Jursa, Michael. 2010. "Business Companies in Babylonia in the First Millennium BC: Structure, Economic Strategies, Social Setting." In Myriam Wissa (ed.), *The Knowledge Economy and Technological Capabilities: Egypt, the Near East and the Mediterranean Second Millennium BC–First Millennium AD, Proceedings of a Conference Held at the Maison de la Chimie Paris, France* (= *Aula Orientalis Supplementa*, 26). Barcelona: Editorial Ausa.

Känel, Frédérique von. 1988. *Les prêtres ouâb de Sekhmet et les conjurateurs de Serket* (= *Bibliothèque de l'École des hautes études: Section des sciences religieuses*, 87). Paris: Presses Universitaires de France.

Kania, Katrin, Heather Hopkins, and Sabine Ringenberg. 2018. "The Influence of Metal Kettle Materials on the Mordanting and Dyeing Outcome." In Heater Hopkins and Katrin Kania (eds), *Ancient Textiles, Modern Science II* (= *Ancient Textiles Series*, 34). Oxford: Oxbow Books.

Karahashi, Fumi. 2004. "Lugal-e and the Song of Ullikummi: A Structural Comparison." *Jaarbericht "Ex Oriente Lux,"* 38: 77–82.

Kardos, Marie-José. 2002. "Quartiers et lieux de Rome dans les *Épigrammes* de Martial." *Bulletin de l'Association Guillame Budé*, 1: 119–35.

Kassianidou, Vasiliki, and A. Bernard Knapp. 2005. "Archaeometallurgy in the Mediterranean: The Social Context of Mining, Technology, and Trade." In Emma Blake and A. Bernard Knapp (eds), *The Archaeology of Mediterranean Prehistory*. Malden, MA: Wiley-Blackwell.

Katz, Esther. 2009. "Les larmes de la reine. Myrrhe et encens dans la Corne de l'Afrique." In Bernard Roussel and Bertrand Hirsch (eds), *Le Rift est-africain. Une singularité plurielle*. Paris: IRD.

Kaufman, David B. 1932. "Poisons and Poisoning among the Romans." *Classical Philology*, 27: 156–67.

Keimer, Louis. 1932. "Pendeloques en formes d'insectes faisant partie de colliers égyptiens." *Annales du Service des Antiquités de l'Égypte*, 32: 129–50.

Keimer, Louis. 1944. "L'horreur des Égyptiens pour les démons du desert." *Bulletin de l'Institut d'Égypte*, 31: 135–47.

Kemp, Barry J. 2000. "Soil (including Mud-brick Architecture)." In Paul T. Nicholson and Ian Shaw (eds), *Ancient Egyptian Materials and Technology*. Cambridge: Cambridge University Press.

Kemp, Barry J. 2005, *Ancient Egypt: Anatomy of a Civilization*, 2nd ed. Abingdon: Routledge.

Kerchove, Anna van der. 2012. *La voie d'Hermès: Pratiques rituelles et traités hermétiques* (= *Nag Hammadi and Manichaean Studies*, 77). Leiden: Brill.

Kessler, Dieter. 1987. "Die Asiatenkarawane von Beni Hassan." *Studien zu Altägyptischen Kultur*, 14: 147–66.

Kettel, Jeannot. 1994. "Canopes, *rdw.w* d'Osiris et Osiris-Canope." In *Hommages à Jean Leclant*, vol. 3: *Études isiaques* (= *Bibliothèque d'Étude*, 103). Cairo: Institut Français d'Archéologie Orientale.

Keyser, Paul T. 1993. "The Purpose of the Parthian Galvanic Cells: A First-Century A.D. Electric Battery Used for Analgesia." *Journal of Near Eastern Studies*, 52: 81–98.

Kidd, Ian G. (ed.). 1988. *Posidonius, II: The Commentary: (ii) Fragments 150–293*. Cambridge: Cambridge University Press.

Killen, Geoffrey, Nigel Hepper Peter Gasson, and Rowena Gale. 2000. "Wood." In Paul T. Nicholson and Ian Shaw (ed.), *Ancient Egyptian Materials and Technology*. Cambridge: Cambridge University Press.

Kingsley, Peter. 1995. *Ancient Philosophy, Mystery and Magic. Empedocles and the Pythagorean Tradition*. Oxford: Clarendon Press.

Kirk, Geoffrey S., John E. Raven, and Malcolm Schofield (eds). 1983. *The Presocratic Philosophers. A Critical History with a Selections of Texts*, 2nd ed. Cambridge: Cambridge University Press.

Kleingünther, Adolf. 1933. Πρῶτος εὑρετής. *Untersuchungen zur Geschichte einer Fragestellung.* Leipzig: Dieterich'sche Verlagsbuchhandlung.

Kleber, Kristin. 2016. "Arabian Gold in Babylonia." *Kaskal*, 13: 121–34.

Kleber, Kristin. 2019. "As Skillful as Croesus: Evidence for the Parting of Gold and Silver by Cementation from Second and First Millennium Mesopotamia." In Peter van Alfen and Ute Wartenberg (eds), *White Gold: Studies in Early Electrum Coinage.* New York: American Numismatic Society.

Klemm, Dietrich, and Klemm, Rosemerie. 1991. "Calcit-Alabaster oder Travertin? Bemerkungen zu Sinn und Unsinn petrographischer Bezeichnungen in der Ägyptologie." *Göttinger Miszellen*, 122: 57–70.

Klemm, Dietrich, and Klemm, Rosemarie. 1993. *Steine und Steinbrüche im Alten Ägypten.* Berlin: Springer. (Available in English as *Stone and Stone Quarries in Ancient Egypt.* London: British Museum Press, 2008.)

Köcher, Franz. 1953. "Der babylonische Göttertypentext." *Mitteilungen des Instituts für Orientforschung*, 1: 57–95.

Kockmann, Norbert. 2014. "History of Distillation." In Andrzej Górak and Eva Sorensen (eds), *Distillation. Fundamentals and Principles.* Amsterdam: Elsevier.

Koenig, Yvan. 1979–80. *Catalogue des étiquettes de jarres hiératiques de Deir el-Médineh* (= *Documents de fouilles de l'Institut français d'Archéologie orientale*, 21). Cairo: Institut Français d'Archéologie Orientale.

König, W. 1938. "Ein galvanisches Element aus der Partherzeit?" *Forschungen und Fortschritte*, 14: 8–9.

Korpela, Jukka. 1995. "*Aromatarii, pharmacopolae, thurarii et ceteri* zur Sozialgeschichte Roms." In Philip van der Eijk, Herman Frederik J. Horstmanshoff, and Piet H. Schrijvers (eds.), *Ancient Medicine in its Socio-Cultural Context. Papers Read at the Congress Held at Leiden University, 12–15 April 1992.* Vol. 1 (= Clio Medica, 27). Amsterdam and Atlanta, GA: Rodopi.

Krebernik, Manfred. 1984. *Die Beschwörungen aus Fara und Ebla: Untersuchungen zur ältesten keilschriftlichen Beschwörungsliteratur* (= *Texte und Studien zur Orientalistik*, 2). Hildesheim: Olms.

Kron, Geoffrey. 2016. "Classical Greek Trade in Comparative Perspective." In Edward M. Harris, David M. Lewis, and Mark Woolmer (eds), *The Ancient Greek Economy. Markets, Households and City-States.* Cambridge: Cambridge University Press.

Kroll, Wilhelm. 1899–1901. *Procli diadochi in Platonis rem publicam commentarii.* 2 vols. Leipzig: Teubner.

Kühn, Karl Gottlob. 1821–33. *Claudii Galeni opera omnia.* 20 vols. Leipzig: Knobloch.

Kurth, Dieter. 1996. "Der Erfolg des Harrurê in Serabit el-Chadim." *Göttinger Miszellen*, 154: 57–63.

Lafont, B. 1996. "L'extraction du minerai de cuivre en Iran à la fin du IIIe millénaire." In Önhan Tunca and D. Deheselle (eds), *Tablettes et images aux pays de Sumer et Akkad: Mélanges offerts à Monsieur H. Limet* (= *Memoirs (Association for the Promotion of Oriental History and Archeology)*, 1). Liège: Université de Liège.

Lagercrantz, Otto. 1938. *Das Wort Chemie* (= *Kungliga Vetenskaps-societetens Årsbok* 1937). Uppsala: AlmQvist & Wiksell.

Łajtar, Adam. 1991. "*Proskynema* Inscriptions of a Corporation of Iron-Workers from Hermonthis in the Temple of Hatshepsut in Deir el-Bahari: New Evidence for Pagan Cults in Egypt in the 4th Cent. A.D." *The Journal of Juristic Papyrology*, 21: 53–70.

Lalouette, Claire. 1979. "Le 'firmament de cuivre'. Contribution à l'étude du mot [*bja*]." *Bulletin de l'Institut français d'Archéologie orientale*, 79: 333–53.
Lambert, Wilfred G. 1957. "Ancestors, Authors, and Canonicity." *Journal of Cuneiform Studies*, 11: 1–14.
Lambert, Wilfred G. 1962. "A Catalogue of Texts and Authors." *Journal of Cuneiform Studies*, 16: 59–77.
Lambert, Wilfred G. 1967. "Enmeduranki and Related Matters." *Journal of Cuneiform Studies*, 21: 126–38.
Landsberger, Benno. 1957. *The Series HAR-ra* = hubullu: *Tablets I–IV* (= *Materialien zum Sumerischen Lexikon*, 5). Rome: Pontificium Institutum Biblicum.
Landsberger, Benno. 1958. *The Series HAR-ra* = hubullu: *Tablets V–VII* (*Materialien zum Sumerischen Lexikon*, 6). Rome: Pontificium Institutum Biblicum.
Landsberger, Benno. 1959. *The Series HAR-ra* = hubullu: *Tablets VIII–XII* (*Materialien zum Sumerischen Lexikon*, 7). Rome: Pontificium Institutum Biblicum.
Landsberger, Benno. 1965. "Tin and Lead: The Adventures of Two Vocables." *Journal of Near Eastern Studies*, 24: 285–96.
Landsberger, Benno. 1967. *The Series HAR-ra* = hubullu: *Tablet XV and Related Texts* (*Materialien zum Sumerischen Lexikon*, 9). Rome: Pontificium Institutum Biblicum.
Landsberger, Benno, and Erica Reiner. 1970. *The Series HAR-ra* = hubullu: *Tablets XVI, XVII and XIX and Related Texts* (*Materialien zum Sumerischen Lexikon*, 10). Rome: Pontificium Institutum Biblicum.
Larsen, Mogens Trolle. 1977. "Partnerships in the Old Assyrian trade." *Iraq*, 39: 119–45.
Larsson Lovén, Lena. 2013. "Female Work and Identity in Roman Textile Production and Trade: A Methodological Discussion." In Margarita Gleba and Judit Pásztókai-Szeöke (eds), *Making Textiles in Pre-Roman and Roman Times. People, Places, Identities*. Oxford: Oxbow Books.
Larsson Lovén, Lena. 2016. "Women, Trade, and Production in Urban Centres of Roman Italy." In Andrew Wilson and Miko Flohr (eds), *Urban Craftsmen and Traders in the Roman World*. Oxford: Oxford University Press.
Laskaris, Julie. 2016. "Metals in Medicine: From Telephus to Galen." In William V. Harris (ed.), *Popular Medicine in Graeco-Roman Antiquity: Explorations*. Leiden and Boston: Brill.
Lauer, Jean-Philippe. 1985. "À propos de l'invention de la pierre de taille par Imhotep pour la demeure d'éternité." In *Mélanges Gamal Eddin Mokhtar*, vol. 2 (= *Bibliothèque d'Étude*, 97). Cairo: Institut Français d'Archéologie Orientale.
Leach, Bridget, and John Tait. 2000. "Papyrus." In Paul T. Nicholson and Ian Shaw (eds), *Ancient Egyptian Materials and Technology*. Cambridge: Cambridge University Press.
Lechtman, Heather. 1979. "Issues in Andean Metallurgy." In Elizabeth P. Benson (ed.), *Pre-Columbian Metallurgy of South America: A Conference at Dumbarton Oaks*. Washington, DC: Dumbarton Oaks Research Library and Collection.
Leemans, Wilhelmus F. 1960. *Foreign Trade in the Old Babylonian Period*. Leiden: Brill.
Lefebvre, Gustave. 1923. *Le tombeau de Pétosiris II–III. Textes et planches*. Cairo: Institut Français d'Archéologie Orientale (new ed. 2007).
Lefebvre, Gustave. 1924. *Le tombeau de Pétosiris I. Description*. Cairo: Institut Français d'Archéologie Orientale (new ed. 2007).
Lefebvre, Gustave. 1949. *Romans et contes égyptiens de l'époque pharaonique*. Paris: Maisonneuve.

Lehner, Joseph W., and K. Aslıhan Yener. 2014. "Organization and Specialization of Early Mining and Metal Technologies in Anatolia." In Benjamin W. Roberts and Christopher P. Thornton (eds), *Archaeometallurgy in Global Perspective: Methods and Syntheses*. New York: Springer.

Leichty, Erle. 1979. "A Collection of Recipes for Dyeing." In *Studies in Honor of Tom B. Jones* (= *Alter Orient und Altes Testament*, 203). Kevelaer Neukirchen-Vluyn: Neukirchener Verlag.

Lenormant, François. 1878. "Les noms de l'airain et du cuivre dans les deux langues des inscriptions cunéiformes de la Chaldée et de l'Assyrie." *Transactions of the Society of Biblical Archaeology*, 6: 334–417.

Leon, Harry J. 1941. "Sulphur for Broken Glass (Martial 1.41.3–5)." *Transactions and Proceedings of the American Philological Association*, 73: 233–6.

Letrouit, Jean. 1995. "Chronologie des alchimistes grecs." In Didier Kahn and Sylvain Matton (eds), *Alchimie: art, histoire et mythes: Actes du 1er colloque international de la Société d'étude de l'histoire de l'alchimie (Paris, Collège de France, 14–15–16 mars 1991)*. Paris: S.É.H.A. and Milan: Arché.

Levey, Martin. 1955. "Evidences of Ancient Distillation, Sublimation and Extraction in Mesopotamia." *Centaurus*, 4: 23–33.

Levey, Martin. 1959. *Chemistry and Chemical Technology in Ancient Mesopotamia*. Amsterdam: Elsevier.

Levey, Martin. 1960. "The Earliest Stages in the Evolution of the Still." *Isis*, 51: 31–4.

Lewis, Naphtali. 1974. *Papyrus in Classical Antiquity*. Oxford: Clarendon Press.

Lilyquist, Christine, and Robert H. Brill (eds). 1993. *Studies in Early Egyptian Glass*. New York: Metropolitan Museum.

Limet, Henri. 1960. *Le travail du métal au pays de Sumer au temps de la IIIe dynastie d'Ur*. Paris: Belle Lettres.

Limet, Henri. 1986. *Textes administratifs relatifs aux métaux* (= *Archives royales de Mari*, 25). Paris.

Lindsay, Jack. 1970. *The Origins of Alchemy in Graeco-Roman Egypt*. London: Frederik Muller.

Lipìnski, Edward (ed.). 1992. *Dictionnaire de la civilisation phénicienne et punique*. Turnhout: Brepols.

Lippmann, Edmund O. von. 1913. *Abhandlungen und Vorträge zur Geschichte der Naturwissenschaften*, vol. 2. Leipzig: Veit.

Lippmann, Edmund O. von. 1919. *Entstehung und Ausbreitung der Alchemie*. Berlin: Springer.

Lippmann, Edmund O. von. 1923. *Beiträge zur Geschichte der Naturwissenschaften und der Technik*. Berlin: Springer.

Littré, Émile. 1839–61. *Oeuvres completes d'Hippocrate*. 10 vols. Paris: J.B. Baillière.

Liu, Jinyu. 2017. "Group Membership, Trust Networks, and Social Capital: A Critical Analysis". In Koenraad Verboven and Christian Laes (eds), *Work, Labour, and Professions in the Roman World*. Leiden: Brill.

Liverani, Mario (ed.). 1998. *Le lettere di el-Amarna*, 2 vols. Brescia: Paideia.

Liverani, Mario. 2000. "The Great Powers' Club." In Raymond Cohen and Raymond Westbrook (eds), *Amarna Diplomacy: The Beginnings of International Relations*. Baltimore, MD: Johns Hopkins University Press.

Liverani, Mario. 2001. "Mesopotamian Historiography and the Amarna Letters." In Tzvi Abusch, Carol Noyes, William W. Hallo, and Irene Winter (eds), *Proceedings* of the XLV Rencontre assyriologique internationale. Part 1 Historiography in the Cuneiform World: Harvard University. Bethesda, MD: CDL Press.

Lloyd, Alan B. 1982. "The Inscription of Udjahorresnet: A Collaborator's Testament." *Journal of Egyptian Archaeology*, 68: 166–80.

Longo, Oddone. 1998. "La zoologia delle porpore nell'antichità Greco-romana." In Oddone Longo (ed.), *La porpora. Realtà ed immaginario di un colore simbolico*. Venice: Istituto veneto di scienze lettere ed arti.

Loret, Victor. 1892. *Flore pharaonique*, 2nd ed. Paris: Leroux.

Loret, Victor. 1928. "La Turquoise chez les anciens Égyptiens." *Kêmi*, 1: 99–114.

Loret, Victor. 1930. "Deux racines tinctoriales de l'Égypte ancienne: orcanette et garance." *Kêmi*, 3: 23–32.

Loret, Victor. 1949. *La Résine de térébinthe ("sonter") chez les anciens Égyptiens* (= Recherches d'Archéologie, de Philologie et d'Histoire, 19). Cairo: Institut Français d'Archéologie Orientale.

Lowe, Benedict. 2016. "The Dye Shops of Pompeii." In Jónatan Ortiz García, Carmen Alfaro Giner, Luis G. Turell Coll, and María Julia Martínez García (eds), *Textiles, Basketry and Dyes in the Ancient Mediterranean World: Proceedings of the Vth International Symposium on Textiles and Dyes in the Ancient Mediterranean World (Montserrat, 19–22 March, 2014)* (= Purpureae Vestes, 5). Valencia: Universitat de València.

Lucas, Alfred. 1962. *Ancient Egyptian Materials and Industries*, 4th ed. rev. and enlarged by James R. Harris. London: Edward Arnold.

Lüchtrath, Agnes. 1999. "Das Kyphirezept." In Dieter Kurth (ed.), *Edfu: Bericht über drei Surveys. Materialien und Studien* (= Die Inschriften des Tempels von Edfu Begleitheft, 5). Wiesbaden: Harrassowitz.

Luthy, Christoph. 2000. "The Fourfold Democritus on the Stage of Early Modern Science." *Isis*, 91: 443–79.

Maekawa, Kazuya. 1980. "Female Weavers and Their Children." *Acta Sumerologica* 2: 81–125.

Makra, Lázló. 2015. "Anthropogenic Air Pollution in Ancient Times." In Philip Wexler (ed.), *History of Toxicology and Environmental Health. Toxicology in Antiquity*, vol. 2. Waltham, MA: Academic Press.

Mallowan, Max E.L. 1966. *Nimrud and its Remains*. 2 vols. London: Collins.

Mamane, Yaacov. 1987. "Air Pollution Control in Israel during the First and Second Century." *Atmospheric Environment*, 21: 1861–3.

Manning, Joseph G. 2002. "Irrigation et État en Égypte antique." *Annales*, 57: 611–23.

Marasco, Gabriele. 1995. "Cleopatra e gli esperimenti su cavie umane." *Historia: Zeitschrift für Alte Geschichte*, 44: 317–25.

Marganne, Marie-Hélène, and Sydney H. Aufrère. 2014. "La question de l'interface entre les sciences égyptiennes et grecques." In Charles Méla and Frédéric Möri (eds), *Alexandrie la Divine*, vol. 1. Geneva: Baconnière.

Marriott, John, and Karen Radner. 2015. "Sustaining the Assyrian Army among Friends and Enemies in 714 BCE." *Journal of Cuneiform Studies*, 67: 127–43.

Martelli, Matteo. 2009. "Divine Water in the Alchemical Writings of Pseudo-Democritus." *Ambix*, 56: 5–22.

Martelli, Matteo. 2011. "Greek Alchemists at Work: 'Alchemical Laboratory' in the Greco-Roman Egypt." *Nuncius*, 26: 271–311.

Martelli, Matteo. 2013. *The Four Books of Pseudo-Democritus* (= *Sources of Alchemy and Chemistry*, 1). Leeds: Society for the History of Alchemy and Chemistry.
Martelli, Matteo. 2014a. "The Alchemical Art of Dyeing: The Fourfold Division of Alchemy and the Enochian Tradition." In Sven Dupré (ed.), *Laboratories of Art. Alchemy and Art Technology from Antiquity to the 18th Century*. Cham: Springer.
Martelli, Matteo. 2014b. "Properties and Classification of Mercury between Natural Philosophy, Medicine and Alchemy." *AION. Annali dell'Università degli studi di Napoli "L'Orientale,"* 36: 17–47.
Martelli, Matteo. 2014c. "Alchemical Textiles: Colourful Garments, Recipes and Dyeing Techniques in Graeco-Roman Egypt." In Mary Harlow and Marie-Louise Nosch (eds), *Greek and Roman Textiles and Dress. An Interdisciplinary Anthology*. Oxford: Oxbow Books.
Martelli, Matteo. 2017. "Alchemy, Medicine and Religion: Zosimus of Panopolis and the Egyptian Priests." *Religion in the Roman Empire*, 3: 202–20.
Martelli, Matteo. 2019. *L'alchimista antico. Dall'Egitto greco-romano a Bisanzio*. Milan: Editrice Bibliografica.
Martelli, Matteo, and Maddalena Rumor. 2014. "Near Eastern Origins of Graeco-Egyptian Alchemy." In Klaus Geus and Markham Geller (eds), *Esoteric Knowledge in Antiquity* (= *Preprint Max-Planck-Institut für Wissenschaftsgeschichte*, 454). Berlin: Max-Planck-Institut für Wissenschaftsgeschichte.
Marzano, Annalisa. 2013. *Harvesting the Sea: The Exploitation of Marine Resources in the Roman Mediterranean*. Oxford: Oxford University Press.
Mathieu, Bernard. 2000. "L'énigme du recrutement des 'enfants du kap' dans l'Égypte pharaonique: une solution?" *Göttinger Miszellen*, 177: 41–8.
Mathieu, Bernard. 2009. "Les couleurs dans les Textes des Pyramides: approche des systèmes chromatiques." *Égypte nilotique et méditerranéenne*, 2: 25–52.
Matoïan, Valérie, and Anne Bouquillon. 2000. "Le 'bleu égyptien' à Ras Shamra-Ougarit (Syrie)." In Paolo Matthiae (ed.), *Proceedings of the First International Congress on the Archaeology of the Near East, Rome, May 18th–23rd 1998*, vol. 2. Rome: Università degli studi di Roma "La Sapienza."
Mattern, Susan P. 2013. *The Prince of Medicine: Galen in the Roman Empire*. Oxford: Oxford University Press.
Mattingly, David J. 1990. "Paintings, Presses and Perfume Production at Pompeii." *Oxford Journal of Archaeology*, 9: 71–90.
Mattusch, Carol C. 1980. "The Berlin Foundry Cup: The Casting of Greek Bronze Statuary in the Early Fifth Century B.C." *American Journal of Archaeology*, 84: 435–44.
Mayhew, Robert. 2011. *Prodicus the Sophist. Texts, Translations and Commentary*. Oxford: Oxford University Press.
Mayor, Adrienne. 2003. *Greek Fire, Poison Arrows, and Scorpion Bombs: Biological and Chemical Warfare in the Ancient World*. London: Duckworth.
McConnell, Joseph R., Andrew I. Wilson, Andreas Stohl, Monica M. Arienzo, Nathan J. Chellman, Sabine Eckhardt, Elisabeth M. Thompson, A. Mark Pollard, and Jørgen Peder Steffensen. 2018. "Lead Pollution Recorded in Greenland Ice Indicates European Emissions Tracked Plagues, Wars, and Imperial Expansion during Antiquity," *Proceedings of the National Academy of Sciences*, 115: 5726–31.
McCorriston, Joy. 1997. "The Fiber Revolution: Textile Extensification, Alienation and Social ratification in Ancient Mesopotamia." *Current Anthropology*, 38: 517–35.

McKerrell, Hugh, and Ronald F. Tylecote. 1972. "The Working of Copper–Arsenic Alloys in the Early Bronze Age and the Effect on the Determination of Provenance." *Proceedings of the Prehistory Society*, 38: 209–18.

Meeks, Dimitri. 1971. "Génies, anges, démons en Égypte." In *Génies, anges et démons: Égypte, Babylone, Israël, Islam, peuples Altaïques, Inde, Birmanie, Asie du Sud-est, Tibet, Chine*. (= *Sources orientales*, 8). Paris: Seuil.

Meeks, Dimitri. 1991. "Oiseaux des carrières et des cavernes." In Ursula Verhoeven and Erhart Graefe (eds), *Religion und Philosophie in alten Ägypten. Festgabe für Philippe Derchain zu seinem 65. Geburtstag am 24. Juli 1991* (= *Orientalia Lovaniensia Analecta*, 38). Leuven: Peeters.

Meeks, Dimitri. 2003. "Locating Punt." In David O'Connor and Stephen Quirke (eds), *Mysterious Lands*. London: UCL Press. Available online: https://archive.org/stream/MysteriousLandsEncountersWithAncientEgypt_201711/Mysterious%20Lands%20Encounters%20With%20Ancient%20Egypt_djvu.txt.

Meeks, Dimitri. 2006. *Mythes et légendes du Delta selon le papyrus Brooklyn 47.218.84* (= *Mémoires de l'Institut français d'Archéologie orientale*, 125). Cairo: Institut Français d'Archéologie Orientale.

Meeks, Dimitri. 2012. "La hiérarchie des êtres vivants selon la conception égyptienne." In Annie Gasse, Frédéric Servajean, and Christophe Thiers (eds), *Et in Aegypto et ad Aegyptum. Recueil d'études dédiées à Jean-Claude Grenier* (= *Cahiers: Égypte nilotique et méditerranéenne*, 5), vol. 3. Montpellier: Université Paul Valéry.

Menu, Bernadette (ed.). 1994. *Les problèmes institutionnels de l'eau en Égypte ancienne et dans l'antiquité méditerranéenne (2nd Actes du Colloque AIDEA, Vogüé 1992)* (= *Bibliothèque d'étude*, 110). Cairo: Institut Français d'Archéologie Orientale.

Mertens, Michèle. 1984. "Un traité gréco-égyptien d'alchimie: la *Lettre d'Isis à Horus*." Master's thesis, Université de Liège.

Mertens, Michèle. 1988. "Une scène d'initiation alchimique: la *Lettre d'Isis à Horus*." *Revue de l'histoire des religions*, 205: 3–23.

Mertens, Michèle. 1995. *Les alchimistes grecs*, vol. 4, part I. *Zosime de Panopolis, Mémoires authentiques*. Paris: Belles Lettres.

Mewaldt, Johannes. 1914. *Galeni In Hippocratis: De natura hominis commentaria tria; In Hippocratis de victu acutorum; De diaeta Hippocratis in morbis acutis*. Leipzig: Teubner.

Michalowski, Piotr. 1993. "The Torch and the Censer." In Mark E. Cohen, Daniel C. Snell, and David B. Weisberg (eds), *The Tablet and the Scroll: Near Eastern Studies in Honor of William W. Hallo*. Bethesda, MD: CDL Press.

Michalowski, Piotr. 2005. "Iddin-Dagan and His Family." *Zeitschrift für Assyriologie*, 95: 65–76.

Michalowski, Piotr. 2011. *Correspondence of the Kings of Ur: Epistolary History of an Ancient Kingdom*. Winona Lake, IN: Eisenbrauns.

Michel, Cécile. 2012. "L'alimentation au Proche-Orient ancien: les sources et leur exploitation." *Dialogues d'histoire ancienne supplément*, 7: 17–45.

Michel, Marianne. 2010. "Le *pefsou*, un élément de recette du boulanger et du brasseur égyptiens." In *Décrire, nommer ou rêver les lieux en Orient. Géographie et toponymie entre réalité et fiction. XLVIIIe session des journées des orientalistes belges* (= *Acta Orientalia Belgica*, 24).

Middeke-Conlin, Robert. 2014. "The Scents of Larsa: A Study of the Aromatics Industry in an Old Babylonian Kingdom." *Cuneiform Digital Library Journal*, (1). Available online: https://cdli.ucla.edu/pubs/cdlj/2014/cdlj2014_001.html.

Miller, Dana R. 2003. *The Third Kind in Plato's* Timaeus. Göttingen: Vandenhoeck and Ruprecht.
Mirelman, Sam. 2015. "Birds, Balag̃s and Snakes (K.4206+)." *Journal of Cuneiform Studies*, 67: 169–86.
Mitchel, Alexandre G. 2009. *Greek Vase-Painting and the Origins of Visual Humour*. Cambridge: Cambridge University Press.
Mittermayer, Catherine. 2009. *Enmerkara und der Herr von Aratta: Ein ungleicher Wettstreit*. Freiburg: Academic Press.
Mondolfo, Rodolfo. 1982. *Polis lavoro e tecnica. Introduzione e cura di Massimo Venturi Ferriolo, con un saggio di André Aymard*. Milan: Feltrinelli.
Monteix, Nicolas. 2015. "Baking and Cooking." In John Wilkins and Robin Nadeau (eds), *A Companion to Food in the Ancient World*. Oxford: Wiley-Blackwell.
Monteix, Nicolas, and Jean-Pierre Brun. 2009. "Les parfumeries en Campanie antique." In Jean-Pierre Brun (ed.), *Artisanats antiques d'Italie et de Gaule. Mélanges offerts à Maria Francesca Buonaiuto*. Naples: Centre Jean Bérard.
Montet, Pierre. 1952a. "Ptah-Patèque et les orfèvres." *Revue Archéologique*, 40: 8–12.
Montet, Pierre. 1952b. "Ptah patèque et les orfèvres nains." *Bulletin de la Société française d'égyptologie*, 11: 73–4.
Moran, William L. (ed.). 1992. *The Amarna Letters*. Baltimore, MD: Johns Hopkins University Press.
Morandi Bonacossi, Daniele. 2016. "Werkstatt." *Reallexikon der Assyriologie*, 15: 62–4.
Moreno Garcia, Juan Carlos. 2013. *Ancient Egyptian Administration* (= Handbook of Oriental Studies. Section 1 The Near and Middle East, 104). Leiden: Brill.
Moorey, Peter R.S. 1985. *Materials and Manufacture in Ancient Mesopotamia: The Evidence of Art and Archaeology – Metals and Metalwork, Glazed Materials and Glass* (= BAR International Series, 237). Oxford: British Archaeological Reports.
Moorey, Peter R.S. 1994. *Ancient Mesopotamian Materials and Industries: The Archaeological Evidence*. Oxford: Oxford University Press (also published in 1999 by Eisenbrauns, Winona Lake, IN).
Morley, Neville. 2007. *Trade in Classical Antiquity*. Cambridge: Cambridge University Press.
Mosshammer, Alden A. 1984. *Georgii Syncelli Ecloga chronographica*. Leipzig: Teubner.
Muhs, Brian. 2016. *The Ancient Egyptian Economy*. Cambridge: Cambridge University Press.
Müller, Marcus. 2015. "The Repit Temple at Athribis after the Worship of Repit." In Alejandro Jiménez-Serrano and Cornelius von Pilgrim (eds), *From the Delta to the Cataract: Studies Dedicated to Mohamed el-Bialy*. Leiden: Brill.
Multhauf, Robert P. 1982. *The Origins of Chemistry*. Canton, MA: Science History Publications/USA.
Murray, Mary Anne. 2000. "Cereal production and Processing." In Paul T. Nicholson and Ian Shaw (eds), *Ancient Egyptian Materials and Technology*. Cambridge: Cambridge University Press.
Murray, Mary Anne, Neil Boulton, and Carl Heron. 2000. "Viticulture and Wine Production." In Paul T. Nicholson and Ian Shaw (eds), *Ancient Egyptian Materials and Technology*. Cambridge: Cambridge University Press.
Musco, Stefano, Paola Catalano, Angela Caspio, Walter Pantano, Kristina Killgrove, and Emilie Formoso. 2008. "Le complexe archéologique de Casal Bertone." *Les Dossiers d'Archéologie*, 330: 32–9.

Mynářová, Jana. 2014. "Egyptian State Correspondence of the New Kingdom: The Letters of the Levantine Client Kings in the Amarna Correspondence and Contemporary Evidence." In Karen Radner (ed.), *State Correspondence in the Ancient World: From New Kingdom Egypt to the Roman Empire*. New York: Oxford University Press.

Na'aman, Nadav. 2007. "The Contribution of the Suhu Inscriptions to the Historical Research on the Kingdoms of Israel and Judah." *Journal of Near Eastern Studies*, 66: 107–22.

Nardelli, Jean-Fabrice. 2017. *L'Osiris de Plutarque: un commentaire de "De Iside et Osiride" chapitres 12–19* (= *Exemplaria Classicas*, Anejo 9). Huelva: Universidad de Huelva.

Needham, Joseph. 1983. *Science and Civilization in China. Spagyrical Discovery and Invention: Physiological Alchemy*. Cambridge: Cambridge University Press.

Nelson, Max. 2001. "Beer in Greco-Roman Antiquity." Ph.D. thesis, University of British Columbia, Vancouver. Available online: https://open.library.ubc.ca/cIRcle/collections/ubctheses/831/items/1.0090870.

Neumann, Hans. 1993. *Handwerk in Mesopotamien: Untersuchungen zu seiner Organisation in der Zeit der III. Dynastie von Ur*, 2nd ed. Berlin: Akademie-Verlag.

Newberry, Percy E. 1920. "A Glass Chalice of Thoutmosis III." *The Journal of Egyptian Archaeology*, 1: 155–60.

Newton, Claire, Thomas Whitbread, Damien Agut-Labordère, and Michel Wuttmann. 2013. "L'agriculture oasienne à l'époque perse dans le sud de l'oasis de Kharga (Égypte, V^e–IVe s. AEC)." *Revue d'ethnoécologie*, 4: 1–20.

Nezafati, Nima, Ernst Pernicka, and Morteza Momenzadeh. 2009. "Introduction of the Deh Hosein Ancient Tin-Copper Mine, Western Iran: Evidence from Geology, Archaeology, Geochemistry and Lead Isotope Data." *Tuba-ar*, 12: 223–36.

Nicholson, Paul T. 2006. "Glass Vessels from the Reign of Thutmose III and a Hitherto Unknown Glass Chalice." *Journal of Glass Studies*, 48: 11–21.

Nicholson, Paul T. 2012. "'Stone … That Flows': Faience and Glass as Man-Made Stones in Egypt." *Journal of Glass Studies*, 54: 11–23.

Nicholson, Paul T., and Ian Shaw (ed.). 2000. *Ancient Egyptian Materials and Technology*. Cambridge: Cambridge University Press.

Nissen, Hans J. 1988. *The Early History of the Ancient Near East, 9000–2000 B.C.* Chicago, IL: University of Chicago Press.

Nissen, Hans J. 1989. "The 'Ubaid Period in the Context of the Early History of the Ancient Near East." In Elizabeth F. Henrickson and Ingolf Thuesen (eds), *Upon This Foundation: The 'Ubaid Reconsidered: Proceedings from the 'Ubaid Symposium, Elsinore, May 30th–June 1st 1988*. Copenhagen: Museum Tusculanum Press.

Nissen, Hans J., Peter Damerow, and Robert K. Englund. 1993. *Archaic Bookkeeping: Early Writing and Techniques of the Economic Administration in the Ancient Near East*. Chicago, IL: University of Chicago Press.

Nix, Ludwig, and Wilhelm Schmidt (eds). 1900. *Herons von Alexandria Mechanik und Katoptrik*. Leipzig: Teubner.

Nolte, Birgit. 1968. *Die Glasgefässe im alten Ägypten*. Berlin: Bruno Hessling.

Nolte, Birgit. 1971. "An Egyptian Glass Vessel in The Metropolitan Museum of Art." *The Metropolitan Museum Journal*, 4: 167–71. Available online: http://resources.metmuseum.org/resources/metpublications/pdf/An_Egyptian_Glass_Vessel_The_Metropolitan_Museum_Journal_v_4_1971.pdf.

Norris, John A. 2006. "The Mineral Exhalation Theory of Metallogenesis in Pre-Modern Mineral Science." *Ambix*, 53: 43–65.

Nougayrol, Jean. 1970. *Le palais royal d'Ugarit, 6: Textes en cunéiformes babyloniens des archives du grand palais et du palais sud d'Ugarit* (= *Mission de Ras Shamra*, 12). Paris: Klincksieck.

Nutton, Vivian. 1988. "The Drug Trade in Antiquity." In Vivian Nutton (ed.), *From Democedes to Harvey: Studies in the History of Medicine*. London: Variorum Reprints.

Oates, David, Joan Oates, and Helen MacDonald. 1997. *Excavations at Tell Brak. Vol. 1. The Mitanni and Old Babylonian periods*. London: British School of Archeology in Iraq.

Ogden, Jack. 2000. "Metals." In Paul Nicholson and Ian Shaw (eds), *Ancient Egyptian Materials and Technology*. Cambridge: Cambridge University Press.

Oldfather, Charles Henry. 1935. *Diodorus Siculus, Library of History. Vol. 2. Books 2.35–4.58*. (= *Loeb Classical Library*, 303). London: Heinemann.

Olivieri, Alessandro. 1935. *Aetii Amideni Libri medicinales I–IV* (= *Corpus Medicorum Graecorum*, VIII.1). Leipzig: Teubner.

Oppenheim, A. Leo. 1966. "Mesopotamia in the Early History of Alchemy." *Revue d'Assyriologie et d'Archéologie Orientale*, 60: 29–45.

Oppenheim, A. Leo. 1967. "Essay on Overland Trade in the First Millennium B.C." *Journal of Cuneiform Studies*, 21: 236–54.

Oppenheim, A. Leo. 1970. "The Cuneiform Texts." In A. Leo Oppenheim, Robert H. Brill, Dan P. Barag, and Axel von Saldern (eds), *Glass and Glassmaking in Ancient Mesopotamia. An Edition of the Cuneiform Texts which Contain Instructions for Glassmakers with a Catalogue of Surviving Objects*. Corning, NY: Corning Museum of Glass.

Oppenheim, A. Leo. 1973. "Towards a History of Glass in the Ancient Near East." *Journal of the American Oriental Society*, 93: 259–66.

Oren, Eliezer D. (ed.). 2000. *The Sea Peoples and Their World: A Reassement*. Philadelphia: University of Pennsylvania Museum.

Osing, Jürgen. 1998. *Hieratische Papyri aus Tebtunis I: Text* (= *CNI Publications*, 17). Copenhagen: Museum Tusculanum Press.

Otto, Adelheid. 2012. "Defining and Transgressing the Boundaries between Ritual Commensality and Daily Commensal Practices: The Case of Late Bronze Age Tell Bazi." In Susan Pollock (ed.), *Between Feasts and Daily Meals: Towards an Archaeology of Commensal Spaces* (= *Berlin Studies of the Ancient World*, 30). Berlin: Edition Topoi. Available online: http://www.edition-topoi.org/books/details/809.

Ouyang, Xiaoli. 2013. *Monetary Role of Silver and its Administration in Mesopotamia during the Ur III Period (c. 2112–2004 BCE): A Case Study of the Umma Province*. Madrid: Consejo Superior de Investigaciones Científicas.

Pagès-Camagna, Sandrine. 1998. "Pigments bleu et vert égyptiens en question: vocabulaire et analyses." In Sylvie Colinart and Michel Menu (eds), *La couleur dans la peinture et l'émaillage de l'Égypte ancienne: Actes de la table ronde, Ravello, 20–22 Mars 1997* (= *Scienze e materiali del patrimonio culturale*, 4). Santo Spirito: Edipuglia.

Pantalacci, Laure. 1982. "Une conception originale de la survie osirienne d'après les listes de Basse Époque?" *Göttinger Miszellen*, 52: 57–66.

Pantalacci, Laure. 1996. "Un été à Sérabit el-Khadim (encore sur l'inscription de Horourrê, Sinaï n° 90." *Göttinger Miszellen*, 150: 87–91.

Paoletti, Paola. 2016a. "Werkstatt. A. I. Philologisch. In Mesopotamien." *Reallexikon der Assyriologie und Vorderasiatischen Archäologie*, 15: 56–61.

Paoletti, Paola. 2016b. "'Raffiniertes' Gold? Gold und seine Qualität(sbezeichnung) en in den altbabylonischen Quellen aus Mesopotamien im 20.–18. Jahrhundert v. Chr." In Rupert Gebhard and Rüdinger Krause (eds), *Bernstorf: Archäologisch-naturwissenschaftliche Analysen der Gold- und Bernsteinfunde vom Bernstorfer Berg bei Kranzberg, Oberbayern* (= Abhandelungen und Bestandskataloge der Archäologischen Staatssammlung, 3). Munich: Archäologische Staatssammlung München.

Papathanassiou, Maria K. 2017. *Stephanos von Alexandria und sein alchemistisches Werk. Die kritische Edition des griechischen Textes eingeschlossen*. Athens: COSMOSWARE.

Pare, Christopher F.E. 2000. "Bronze and the Bronze Age." In Christopher F.E. Pare (ed.), *Metals Make the World Go Round: The Supply and Circulation of Metals in Bronze Age Europe*. Oxford: Oxbow Books.

Partington, James Riddick. 1935. *Origins and Development of Applied Chemistry*. London: Longmans.

Partington, James Riddick. 1970. *A History of Chemistry. Vol. 1, part 1, Theoretical Background*. London: Macmillan.

Partington, James Riddick. 1999. *A History of Greek Fire and Gunpowder. Introduction by Bert S. Hall*. Baltimore, MD: Johns Hopkins University Press.

Payne, Elizabeth. 2007. "The Craftsmen of the Neo-Babylonian Period: A Study of Textile and Metal Workers of the Eanna Temple." Ph.D. thesis, Yale University, New Haven, CT.

Peremanns, Willie. 1967. "Diodore de Sicile et Agatharchide de Cnide." *Historia: Zeitschrift für Alte Geschichte*, 16: 432–55.

Pernicka, Ernst. 2004. "Copper and Silver in Arisman and Tappeh Sialk and the Early Metallurgy in Iran." In Thomas Stöllner, Rainer Slotta, and Abdolrasool Vatandoust (eds), *Persiens antike Pracht: Bergbau, Handwerk, Archäologie: Katalog der Ausstellung des Deutschen Bergbau-Museums Bochum vom 28. November 2004 bis 29. Mai 2005* (= Veröffentlichungen aus dem Deutschen Bergbau-Museum Bochum, 128). Bochum: Deutsches Bergbau-Museum.

Pernicka, Ernst, Thilo Rehren, and Sigrid Schmitt-Strecker. 1998. "Late Uruk Silver Production by Cupellation at Habuba Kabira, Syria." In Thilo Rehren, Andreas Hauptmann, and James D. Muhly (eds), *Metallurgica Antiqua: In Honour of Hans-Gert Bachmann and Robert Maddin* (= Anschnitt, Beiheft, 8). Bochum: Deutsches Bergbau-Museum.

Petrie, Cameron A. 2012. "Ceramic Production." In Daniel T. Potts (ed.), *A Companion to the Archaeology of the Ancient Near East*, vol. 1. Oxford: Wiley-Blackwell.

Petrie, W.M. Flinders. 1889. *Two Hieroglyphic Papyri from Tanis. II. The Geographical Papyrus (an Almanack)* (= Extra Memoir [9th] of the Egypt Exploration Fund). London: Trübner.

Petrie, W.M. Flinders. 1894. *Tell el Amarna*. London: Methuen.

Petrie, W.M. Flinders. 1908. *Athribis* (= Publications, British School of Archaeology in Egypt, 14). London: British School of Archaeology in Egypt.

Petrie, W.M. Flinders. 1924–5. "Glass Found in Egypt." *Transactions of the British Newcomen Society*, 5: 72–6.

Pfister, René. 1935. *Teinture et alchimie dans l'Orient hellénistique*. Prague: Institut Kondakov.

Photos-Jones, Effie. 2018. "From Mine to Apothecary: An Archaeo-biomedical Approach to the Study of Greco-Roman Lithotherapeutics Industry." *World Archaeology*, 50: 418–33.

Photos-Jones, Effie, and Allan J. Hall. 2011. *Lemnian Earth and the Earths of the Aegean: An Archaeological Guide to Medicines, Pigments and Washing Powders*. Glasgow: Potingair Press

Pleše, Zlatko. 2005. "Platonist Orientalism." In Aurelio Pérez Jiménez and Frances Titchener (eds), *Historical and Biographical Values of Plutarch's Works. Studies Devoted to Professor Philip A. Stadler by the International Plutarch Society*. Málaga: Universidad de Málaga.

Plouvier, Liliane. 2010. "À la table du roi Hammurabi de Babylone – d'après les tablettes de la Yale Babylonian Collection." In *Proceedings of the XV UISPP World Congress (Lisbon, 4–9 September 2006)* (= BAR International Series, 1804). Oxford: Archaeopress.

Pollock, Susan. 1999. *Ancient Mesopotamia: The Eden That Never Was*. Cambridge: Cambridge University Press.

Pongratz-Leisten, Beate. 2015. "Imperial Allegories: Divine Agency and Monstrous Bodies in Mesopotamia's Body Description Texts." In Beate Pongratz-Leisten and Karen Sonik (eds), *The Materiality of Divine Agency* (= Studies in Ancient Near Eastern Records, 8). Boston, MA: De Gruyter.

Pons Mellado, Esther. 2005. "Los enanos orfebres en los talleres matalúrgico del antiguo Egipto." In Josep Cervelló Autuori, Juan Montserrat Díaz de Crio, and David Rull Ribó (eds), *Actas del Segundo Congreso ibérico de egiptología. Bellaterra, 12–15 de Marzo de 2001* (= Aula Ægyptiaca Studia, 5). Bellaterra: Universitat Autónoma de Barcelona.

Pons Mellado, Esther, 2006. "Trade of Metals between Egypt and Other Countries from the Old until the New Kingdom." *Chronique d'Égypte*, 81(161): 7–16.

Porcier, Stéphanie. 2012. "Apis, Mnévis, l'Occident et l'Orient." In Annie Gasse, Frédéric Servajean, and Christophe Thiers (eds), Et in Ægypto et ad Ægyptum. *Recueil d'études dédiées à Jean-Claude Grenier* (= Cahiers Égypte nilotique et méditerranéenne, 5), vol. 3. Montpellier: Université Paul Valéry.

Porcier, Stéphanie. 2014. "Le Mnévis d'Héliopolis: Bos primigenius ou Bos taurus?" In Armelle Gardeisen and Christophe Chandezon (eds), *Équidés et bovidés de la Méditerranée antique. Rites et combats. Jeux et savoirs*. Actes du colloque organisé par l'axe Animal et sociétés méditerranéennes. Lattes: Édition de l'Association pour le Développement de l'Archéologie en Languedoc-Roussillon.

Posener, Georges. 1936. *La première domination perse en Égypte* (Bibliothèque d'Étude, 11). Cairo: Institut Français d'Archéologie Orientale.

Post, Edwin. 1908. *Selected Epigrams of Martial*. Boston, MA: Ginn & Co.

Postgate, Nicholas. 1994. *Early Mesopotamia: Society and Economy at the Dawn of History*. London: Routledge.

Postgate, Nicholas. 1997. "Mesopotamian Petrology: Stages in the Classification of the Material World." *Cambridge Archaeological Journal*, 7: 205–24.

Potter, Paul. 1980. *Hippokrates: Über die Krankheiten III, herausgegeben, übersetzt und erläutert von Paul Potter* (= Corpus Medicorum Graecorum, 1, 2, 3). Berlin: Akademie.

Potts, Daniel T. 1986. "Eastern Arabia and the Oman Peninsula during the Late Fourth and Early Third Millennium B.C." In Uwe Finkbeiner and Wolfgang Röllig (eds), *Gamdat Nasr: Period or Regional Style?* Wiesbaden: Reichert.

Potts, Daniel T. 1997. *Mesopotamian Civilization: The Material Foundations.* London: Athlone.
Potts, Daniel T. 2007. "Babylonian Sources of Exotic Raw Materials." In Gwendolyn Leick (ed.), *The Babylonian World.* New York: Routledge.
Potts, Daniel T. 2010. "Adamšah, Kimaš and the Miners of Lagaš." In Heather Baker, Eleanor Robson, and Gábor Zólyomi (eds), *Your Praise is Sweet: A Memorial Volume for Jeremy Black from Students, Colleagues and Friends.* London: British Institute for the Study of Iraq.
Potts, Daniel T. 2017. "Resource Origins and Resource Movement in and around the Persian Gulf." In Anke K. Scholz, Martin Bartlesheim, Roland Hardenberg, and Jörn Staecker (eds), *ResourceCultures: Sociocultural Dynamics and the Use of Resources: Theories, Methods, Perspectives.* Tübingen: Universität Tübingen. Available online: https://publikationen.uni-tuebingen.de/xmlui/handle/10900/74124.
Powell, Barry B. 2014. *Homer. The Odyssey, Translation, Introduction and Notes.* Oxford: Oxford University Press.
Principe, Lawrence M. 2013. *The Secrets of Alchemy.* Chicago, IL: University of Chicago Press.
Principe, Lawrence M. 2018. "Texts and Practices: The Promises and Problems of Laboratory Replication and the Chemical Explanation of Early Alchemical Processes." In Efthymios Nicolaidis (ed.), *Greek Alchemy from Late Antiquity to Early Modernity.* Turnhout: Brepols.
Pusch, Edgar B. 1994. "Divergierende Verfahren der Metallverarbeitung in Theben und Qantir?" *Ägypten und Levante,* 4: 145–70.
Putter, Thierry de, and Karlshausen, Christina. 1992. *Les pierres utilisées dans la sculpture et l'architecture* (= *Connaissance de l'Égypte ancienne,* Etude, 4). Brussels: Connaissance de l'Égypte ancienne.
Puybaret, Marie-Pierre, Philippe Borgard, and Roger Zérubia. 2008. "Teindre comme à Pompéi: Approche expérimentale." In Carmen Alfaro Giner and Lilian Karalē (eds), *Vestidos, textiles y tintes. Estudios sobre la producción de bienes de consumo en la Antigüedad: Actas del II Symposium Internacional sobre Textiles y Tintes del Mediterráno en el Mundo Antiguo (Atenas, 24 al 26 de noviembre, 2005)* (= *Purpureae Vestes,* 2). Valencia: Universitat de València.
Quack, Joachim F. 2014. "Imhotep – der Weise, der zum Gott wurde." In Verena M. Lepper (ed.), *Persönlichkeiten aus dem Alten Ägypten im Neuen Museum: Für das Ägyptische Museum und Papyrussammlung, Staatliche Museen zu Berlin.* Petersberg: Michael Imhof.
Quaegebeur, Jan. 1995. "Diodore I, 20 et les mystères d'Osiris." In Terence Duquesne (ed.), *Hermes Ægyptiacus: Egyptological studies for B.H. Stricker on his 85th Birthday* (*Discussions in Egyptology* Special Number, 2). Oxford: DE Publications.
Rabot, Alexandre, and Isabelle Goncalves. 2015. "Photo-interprétation du site pharaonique de Samut el-Beda." Available online: https://desorient.hypotheses.org/421.
Rackham, Harris. 1952. *Pliny, Natural History, Volume IX: Books 33–35* (= *Loeb Classical Library,* 394). Cambridge, MA: Harvard University Press.
Rainey, Anson F. 2015. *A New Edition of the Cuneiform Letters from the Site of El-Amarna base don Collations of all Extant Tablets.* 2 vols (= *Handbook of Oriental Studies,* 110). Leiden: Brill.

Rasmussen, Seth C. 2014. *The Quest for Aqua Vitae. The History and Chemistry of Alcohol from Antiquity to the Middle Ages*. Cham: Springer.

Ray, Pafulla Chandra. 1903–9. *A History of Hindu Chemistry from the Earliest Times to the Middle of the Sixteenth Century*, A.D.: *with Sanskrit Texts, Variants, Translation and Illustrations*, 2nd ed., 2 vols. Calcutta: Bengal Chemical and Pharmaceutical Works.

Ray, Pafulla Chandra, and Priyadaranjan Ray. 1956. *History of Chemistry in Ancient and Medieval India*. Calcutta: Indian Chemical Society.

Rayor, Diane J. 2004. *The Homeric Hymns. A Translation with Introduction and Notes*. Berkeley: University of California Press.

Reade, Julian E. 1998–2001. "Ninive (Nineveh)." *Reallexikon der Assyriologie und Vorderasiatischen Archäologie*, 9: 388–433.

Reddé, Michel. 2018. *Fortins routiers du désert Oriental d'Égypte*. In Jean-Pierre Brun, Thomas Faucher, Bérangère Redon, and Steven Sidebotham (eds), *Le désert oriental d'Égypte durant la période gréco-romaine: bilans archéologiques*. Paris: Collège de France. Available online: https://books.openedition.org/cdf/5163?lang=en.

Redon, Bérangère. 2016. "Travailler dans les mines d'or ptolémaïques du désert Oriental égyptien." *Les Nouvelles de l'archéologie*, 143: 5–7. Available online: https://journals.openedition.org/nda/3291.

Redon, Bérangère, Matthieu Vanpeene, and Mikaël Pesenti. 2016. "'La vigne a été inventée dans la ville égyptienne de Plinthine': À propos de la découverte d'un fouloir saïte à Kôm el-Nogous (Maréotide)." *Bulletin de l'Institut français d'archéologie orientale, Institut français d'archéologie orientale*, 116: 303–23.

Reger, Gary. 2005. "The Manufacture and Distribution of Perfume." In Zofia H. Archibald, John K. Davies, and Vincent Gabrielsen (eds), *Making, Moving and Managing: The New World of Ancient Economies, 323–31 BC*. Oxford: Oxbow Books.

Rehder, John E. 2000. *The Mastery and Uses of Fire in Antiquity*. Montreal: McGill-Quenn's University Press.

Rehren, Thilo. 1997. "Ramesside Glass-colouring Crucibles." *Archaeometry*, 39: 355–68.

Reiner, Erica. 1956. "Lipšur Litanies." *Journal of Near Eastern Studies*, 15: 129–49.

Reiner, Erica. 1995. *Astral Magic in Babylonia*. Philadelphia, PA: American Philosophical Society.

Renger, Johannes. 2002. "Royal Edicts of the Old Babylonian Period: Structural Background." In Michael Hudson and Marc Van de Mieroop (eds), *Debt and Economic Renewal in the Ancient Near East*. Bethesda, MD: CDL Press.

Reynolds, Frances. 2002. "Describing the Body of a God." In Cornelia Wunsch (ed.), *Mining the Archives: Festschrift for Christopher Walker on the Occasion of his 60th Birthday, 4 October 2002*. Dresden: Islet.

Reynolds, Frances. 2010. "A Divine Body: New Joins in the Sippar Collection." In Heather D. Baker, Eleanor Robson, and Gábor Zólyomi (eds), *Your Praise Is Sweet. A Memorial Volume for Jeremy Black from Students, Colleagues and Friends*. London: British Institute for the Study of Iraq.

Riccardelli, Carolyn. 2017. "Egyptian Faience: Technology and Production." In *Heilbrunn Timeline of Art History*. New York: The Metropolitan Museum of Art. Available online: https://www.metmuseum.org/toah/hd/egfc/hd_egfc.htm.

Richardson, John S. 1976. "The Spanish Mines and the Development of Provincial Taxation in the Second Century B.C." *Journal of Roman Studies*, 66: 139–52.

Richter, Thomas. 2012. *Bibliographisches Glossar des Hurritischen*. Wiesbaden: Harrassowitz.

Riggs, Christina. 2014. *Unwrapping Ancient Egypt*. London: Bloomsbury.
Rihll, Tracey E. 2001. "Making Money in Classical Athens." In David J. Mattingly and John Salmon (eds), *Economies Beyond Agriculture in the Classical World*. London: Routledge.
Rihll, Tracey E., and John V. Tucker. 2002. "Practice Makes Perfect: Knowledge of Materials in Classical Athens." In Christopher J. Tuplin and Tracey Rihll (eds), *Science and Mathematics in Ancient Greek Culture*. Oxford: Oxford University Press.
Robson, Eleanor. 2001. "Technology in Society: Three Textual Case Studies from Late Bronze Age Mesopotamia." In Andrew J. Shortland (ed.), *The Social Context of Technological Change: Egypt and the Near East, 1650–1550 BC*. Oxford: Oxbow Books.
Robson, Eleanor. 2008. *Mathematics in Ancient Iraq: A Social History*. Princeton, NJ: Princeton University Press.
Robson, Eleanor. 2013. "Reading the libraries of Assyria and Babylonia." In Jason König, Katerina Oikonomopolou, and Greg Woolf (eds), *Ancient Libraries*. Cambridge: Cambridge University Press.
Rochberg-Halton, Francesca. 1984. "Canonicity in Cuneiform Texts." *Journal of Cuneiform Studies*, 36: 127–44.
Romano, Elisa. 1998. "I colori artificiali e le origini della chimica." In Gilbert Argoud and Jean-Yves Guillaumin (eds), *Sciences exactes et sciences appliquées à Alexandrie (IIIe siècle av. J.-C. – Ier siècle apr. J.-C.): Actes du colloque international de Saint-Etienne (6–8 juin 1996) (= Mémoires [Centre Jean Palerne]*, 16). Saint-Étienne: Publication de l'Université de Saint-Étienne.
Rossetti, Livio. 2002. "Il più antico decreto ecologico a noi noto e il suo contesto." In Thomas M. Robinson and Laura Westra (eds), *Thinking about the Environment: Our Debt to the Classical and Medieval Past*. Lanham, MD: Lexington Books.
Rouchon, Olivier, Jane Fabre, Marie-Pierre Etcheverry, and Max Schvoerer. 1990. "Pigments d'Égypte, étude physique de matières colorantes bleue, rouge, blanche, verte et jaune, provenant de Karnak." *Revue d'Archéométrie*, 14: 87–97.
Rumor, Maddalena. 2015. *Babylonian Pharmacology in Graeco-Roman Dreckapotheke (with an Edition of Uruanna III 1–143)*. Ph.D. thesis, Freie Universität Berlin.
Rumor, Maddalena. 2018. "At the Dawn of Plant Taxonomy: Shared Structural Design of Herbal Descriptions in Šammu šikinšu and Theophrastus' *Historia plantarum* IX." In Strahil V. Panayotov and Luděk Vacín (eds), *Mesopotamian Medicine and Magic: Studies in Honor of Markham J. Geller*. Leiden: Brill.
Russo, Simona. 1999. *I gioielli nei papiri di età greco-romana*. Florence: Istituto Papirologico "G. Vitelli."
Ryholt, Kim. 2011. "New Light on the Legendary King Nechepsos of Egypt." *Journal of Egyptian Archaeology*, 97: 61–72.
Saffrey, Henry D. 1995. "Historique et description du manuscrit alchimique de Venise *Marcianus Graecus* 299." In Didier Kahn and Sylvain Matton (eds), *Alchimie: art, histoire et mythes: Actes du 1er colloque international de la Société d'étude de l'histoire de l'alchimie (Paris, Collège de France, 14–15–16 mars 1991)*. Paris: S.É.H.A. and Milan: Arché.
Salavert, Aurélie, and Tengberg, Margareta. 2005. "Les préparations alimentaires dans les cuisines et les boulangeries du Ramsesseum. Premiers résultats de l'étude carpologique du secteur D." *Memnonia*, 16: 121–31.
Saldern, Axel von. 1970. "Other Mesopotamian glass vessels (1500–600 BC)." In A. Leo Oppenheim, Robert. H. Brill, Dan P. Barag, and Axel von Saldern (eds), *Glass*

and Glassmaking in Ancient Mesopotamia. An Edition of the Cuneiform Texts which Contain Instructions for Glassmakers with a Catalogue of Surviving Objects. Corning, NY: Corning Museum of Glass.

Saliou, Catherine. 1996. *Le traité d'urbanisme de Julien d'Ascalon (VIe siècle). Droit et architecture en Palestine au VIe siècle.* Paris:: De Boccard.

Saliou, Catherine. 2012. "Artisanats et espace urbain dans le monde romain: droit et projets urbains (Ier siècle av. J.-C.–VIe siècle ap. J.-C.)." In Arianna Esposito and Giorgios M. Sanidas (eds), *"Quartiers" artisanaux en Grèce ancienne, une perspective méditerranéenne.* Villeneuve d'Ascq: Presses Universitaires du Septentrion.

Sall, Babacar. 2005–6. "L'Égypte était-elle un don du Nil?" *Ankh,* 14–15: 34–51. Available online: http://www.ankhonline.com/ankh_num_14_15/ankh_14_15_b_sall_egypte_etait-elle_un_don_du_nil_.pdf.

Sallaberger, Walther. 1999. "Ur III-Zeit." In Walther Sallaberger and Aage Westenholz (eds), *Mesopotamien: Akkade-Zeit und Ur III-Zeit.* Göttingen: Vendenhoeck and Ruprecht.

Sallaberger, Walther. 2012. "Bierbrauen in Versen: Eine neue Edition und Interpretation der Ninkasi-Hymne." In Catherine Mittermayer and Sabine Ecklin (eds), *Altorientalische Studien zu Ehren von Pascal Attinger: mu-ni u$_4$ ul-li$_2$-a-aš ĝa$_2$-ĝa$_2$-de$_3$.* Göttingen: Vandenhoeck and Ruprecht.

Sallaberger, Walther. 2014. "The Value of Wool in Early Bronze Age Mesopotamia: On the Control of Sheep and the Handling of Wool in the Presargonic to the Ur III Periods (c. 2400–2000 BC)." In Catherine Breniquet and Cécile Michel (eds), *Wool Economy in the Ancient Near East: From the Beginnings of Sheep Husbandry to Institutional Textile Industry.* Oxford: Oxbow Books.

Salonen, Armas. 1964. "Die Öfen der alten Mesopotamier." *Baghdader Mitteilungen*, 3: 100–24.

Samama, Évelyne. 2006. "*Thaumatopoioi pharmakopôlai*: la singulière image des préparateurs et vendeurs de remèdes dans les textes grecs." In Franck Collard and Évelyne Samama (eds), *Pharmacolopes et apothicaires. Les "pharmaciens" de l'Antiquité au Grand Siècle.* Paris: L'Harmattan.

Samuel, Delwen. 2000. "Brewing and baking." In Paul T. Nicholson and Ian Shaw (eds), *Ancient Egyptian Materials and Technology.* Cambridge: Cambridge University Press.

Santorelli, Biagio. 2013. *Giovenale, Satira V. Introduzione, Traduzione e Commento.* Berlin: De Gruyter.

Sauneron, Serge. 1954. "La manufacture d'armes de Memphis." *Bulletin de l'Institut français d'Archéologie orientale,* 54: 7–12.

Sauneron, Serge. 1961. "Remarques de philologie et d'étymologie (en marge des texts d'Esna): 5. Une étymologie égyptienne du nom de Thot." In *Mélanges Mariette (= Bibliothèque d'Étude,* 32). Cairo: Institut Français d'Archéologie Orientale.

Sauneron, Serge. 1962a. "Une allusion inattendue à la 'matière divine.'" *Kêmi,* 16: 38–9.

Sauneron, Serge (ed.). 1962b. *Le rituel de l'Embaumement, pap. Boulaq III, Pap. Louvre 5.158.* Cairo: Imprimerie nationale.

Sauneron, Serge. 1963. "Un hymne à Imouthès." *Bulletin de l'Institut français d'Archéologie orientale,* 63: 73–87.

Sauneron, Serge. 1964. "Villes et légendes d'Égypte. I. Khnoum de Chashotep, crateur des animaux." *Bulletin de l'Institut français d'Archéologie orientale,* 62: 22–57.

Sauneron, Serge. 1967. *Les prêtres de l'ancienne Égypte* (= *Le Temps qui court*, 6). Paris: Seuil. (Available in English as *The Priests of Ancient Egypt*. Ithaca, NY: Cornell University Press, 2000.)

Sauneron, Serge. 1968. *Le temple d'Esna* (*Esna*, 3). Cairo: Institut Français d'Archéologie Orientale.

Sauneron, Serge. 1983. *Villes et légendes d'Égypte*, 2nd ed., rev. and corr. (= *Bibliothèque d'Étude*, 90). Cairo: Institut Français d'Archéologie Orientale.

Sauneron, Serge. 1989. *Un traité égyptien d'ophiologie: Papyrus du Brooklyn Museum N°s 47.218.48 et 85* (= *Bibliothèque générale*, 11). Cairo: Institut Français d'Archéologie Orientale.

Scarborough, John. 1983. "Theoretical Assumptions in Hippocratic Pharmacology." In François Lasserre and Philippe Mudry (eds), *Formes de pensée dans la collection hippocratique*. Geneva: Droz.

Schmidtchen, Eric. 2018a. "Esagil-kīn-apli's Catalogue of Sakikkû and Alamdimmû." In Ulrike Steinert (ed.), *Assyrian and Babylonian Scholarly Text Catalogues: Medicine, Magic and Divination* (= *Die babylonisch-assyrische Medizin in Texten und Untersuchungen*, 9). Berlin: De Gruyter.

Schmidtchen, Eric. 2018b. "The Edition of Esagil-kīn-apli's Catalogue of the Series *Sakikkû* and *Alamdimmû*." In Ulrike Steinert (ed.), *Assyrian and Babylonian Scholarly Text Catalogues: Medicine, Magic and Divination* (= *Die babylonisch-assyrische Medizin in Texten und Untersuchungen*, 9). Berlin: De Gruyter.

Schmitz, Rudolf, and Helmut Conradi. 1968. "Glasgefässe im Dienste der Körperpflege und Heilkunde zur Zeit der altagyptischen Hochkulturen (1500 v. Chr. bis Zeitenwende)." *Deutsche Apotheker-Zeitung*, 108: 773–8.

Schreiber, Marvin. 2018. "Die astrologische Medizin der spätbabylonischen Zeit." Ph.D. thesis, Humboldt University of Berlin.

Schuster-Brandis, Anais. 2008. *Steine als Schutz- und Heilmittel: Untersuchung zu ihrer Verwendung in der Beschwörungskunst Mesopotamiens im 1. Jt. v. Chr.* (= *Alter Orient und Altes Testament*, 46). Münster: Ugarit-Verlag.

Schwechler, Coralie Charlène. 2017. "Les noms des pains en Égypte ancienne: étude lexicologique." Ph.D. thesis, University of Geneva.

Scurlock, JoAnn. 1995. "*pizzer* or *upinsir*: Creepy Medicine." *NABU: Nouvelles Assyriologiques Brèves et Utilitaires*, (4): 95–6. Available online: https://sepoa.fr/?page_id=172.

Selz, Gebhard, Colette Grinevald, and Orly Goldwasser. 2017. "The Question of Sumerian 'Determinatives': Inventory, Classifer Analysis, and Comparison to Egyptian Classifiers from the Linguistic Perspective of Noun Classification." In Daniel A. Werning (ed.), *Proceedings of the Fifth International Conference on Egyptian-Coptic Linguistics: (Crossroads V), Berlin, February 17–20, 2016* (= *Lingua Aegyptia*, 25). Hamburg: Widmaier.

Serpico, Margaret, and Raymond White. 2000. "Resins, Amber and Bitumen". In Paul T. Nicholson and Ian Shaw (eds), *Ancient Egyptian Materials and Technology*. Cambridge: Cambridge University Press.

Shimy, Mohammed Abdel-Hamid. 1998. *Parfums et parfumerie dans l'ancienne Égypte: De l'Ancien Empire à la fin du Nouvel Empire*. Villeneuve d'Ascq: Presses Universitaires du Septentrion.

Šichan, Daniel. 2011. "Harbours in Ancient Egypt." Master's thesis, Charles University, Prague. Available online: https://www.academia.edu/5608915/Harbours_in_ancient_Egypt.

Simkó, Krisztián. 2015. "Emery Abrasive in the Lapidary Craft of the Ur III Period? Some Further Remarks on the Stone ú-na₄-gug and its Old Babylonian Counterpart." *Aula Orientalis*, 33: 141–55.

Singer, Peter N. 1997. *Galen, Selected Works. Translated with an Introduction and Notes*. Oxford: Oxford University Press.

Smirniou, Melina, and Thilo Rehren. 2016. "The Use of Technical Ceramics in Early Egyptian Glass-making." *The Journal of Archaeological Science*, 67: 52–63.

Smith, Cyril Stanley, and John G. Hawthorne 1974. "*Mappae clavicula*. A Little Key to the Word of Medieval Techniques." *Transactions of the American Philosophical Society*, 64(4): 1–128.

Smith, Wesley D. 1990. *Hippocrates, Pseudepigraphic Writings. Letters–Embassy–Speech from the Altar–Decree, Edited and Translated with an Introduction*. Leiden: Brill.

Soldt, Wilfred H. van. 1990. "Fabrics and Dyes at Ugarit." *Ugarit Forschungen*, 22: 321–57.

Soukiassian, Georges, Michel Wuttmann, and Laure Pantalacci. *Les ateliers de potiers d'ʿAyn-Aṣīl: Fin de l'Ancien Empire, première période intermédiaire* (= Fouilles de l'Institut français d'Archéologie orientale du Caire, 34), Cairo: Institut Français d'Archéologie Orientale.

Sprengel, Kurt (ed.). 1830. *Pedanii Dioscoridis Anazarbei De materia medica libri quinque, tomus secundus*. Leipzig: Cnobloch.

Stannard, Jerry. 1961. "Hippocratic Pharmacology." *Bulletin of the History of Medicine*, 35: 497–518.

Stadhouders, Henry. 2011. "The Pharmacopoeial Handbook *Šammu šikinšu*: An Edition." *Le Journal des Médecines Cunéiformes*, 18: 3–51.

Stadhouders, Henry. 2012. "The Pharmacopoeial Handbook *Šammu šikinšu* – A Translation." *Le Journal des Médecines Cunéiformes*, 19: 1–21.

Stadhouders, Henry, and J. Cale Johnson. 2018. "A Time to Extract and a Time to Compile: The Therapeutic Compendium BM 78963." In Strahil V. Panayotov and Luděk Vacín (eds), *Mesopotamian Medicine and Magic. Studies in Honor of Markham J. Geller*. Leiden: Brill.

Staubli, Thomas. 1991. *Das Image der Nomaden im alten Israel und in der Ikonographie seiner sesshaften Nachbarn* (= Orbis Biblicus et Orientalis, 107). Göttingen: Vandenhoeck and Ruprecht.

Stauder, Andreas. 2018a; "Staging Restricted Knowledge: The Sculptor Irtysen's 239 Self-presentation (ca. 2000 BC)." In Gianluca Miniaci, Juan Carlos Moreno García, Stephen Quirke, and Andreas Stauder (eds), *The Arts of Making in Ancient Egypt. Voices, Images, and Objects of Material Producers 2000–1550 BC*. Leiden: Sidestone Press.

Steele, John M. 2011. "Astronomy and Culture in Late Babylonian Uruk." In Clive L.N. Ruggles (ed.), *Archaeoastronomy and Ethnoastronomy: Building Bridges between Cultures: Proceedings of the 278th Symposium of the International Astronomical Union and "Oxford IX" International Symposium on Archaeoastronomy, Held in Lima*, Peru, *January 5–14, 2011* (= International Astronomical Union Symposium, 278). Cambridge: Cambridge University Press.

Steinert, Ulrike. 2013. "Fluids, Rivers, and Vessels: Metaphors and Body Concepts in Mesopotamian Gynaecological Texts." *Le Journal des Médecines Cunéiformes*, 22: 1–23.

Steinert, Ulrike. 2017. "Concepts of the Female Body in Mesopotamian Gynaecological Texts." In John Z. Wee (ed.), *The Comparable Body: Analogy and*

Metaphor in Ancient Mesopotamian, Egyptian, and Greco-Roman Medicine (= Studies in Ancient Medicine, 49). Leiden: Brill.

Steinert, Ulrike (ed.). 2018. *Assyrian and Babylonian Scholarly Text Catalogues* (= Babylonisch-assyrische Medizin in Texten und Untersuchungen, 9). Boston, MA: De Gruyter.

Steiniger, Daniel. 2011. "Excavations of Slagheaps in Arisman." In Abdolrasool Vatandoust, Hermann Parzinger, and Barbara Helwing (eds), *Early Mining and Metallurgy on the Western Central Iranian Plateau: The First Five Years of Work* (= Archäologie in Iran und Turan, 9). Mainz: Philipp von Zabern.

Steinkeller, Piotr. 1996. "The Organization of Crafts in Third Millennium Babylonia: The Case of Potters." *Altorientalische Forschungen*, 23: 232–53.

Steinkeller, Piotr. 2016. "The role of Iran in the Inter-Regional Exchange of Metals: Tin, Copper, Silver and Gold in the Second Half of the Third Millennium BC." In Kazuya Maekawa (ed.), *Ancient Iran: New Perspectives from Archaeology and Cuneiform Studies: Proceedings of the International Colloquium Held at the Center for Eurasian Cultural Studies, Kyoto University, December 6–7, 2014* (= Ancient Text Studies in the National Museum, 2). Kyoto: Nakanishi Printing Company.

Stern, E. Marianne. 1999. "Roman Glassblowing in a Cultural Context." *American Journal of Archaeology*, 103: 441–84.

Stierlin, Henri. 2007. *The Gold of the Pharaohs*. Paris: Terrail.

Stol, Marten. 1971. "Zur altmesopotamischen Bierbereitung." *Bibliotheca Orientalis*, 28(3/4): 167–71.

Stol, Marten. 1983. "Leder(industrie)." *Reallexikon der Assyriologie und Vorderasiatischen Archäologie*, 6: 527–43.

Stol, Marten. 1987–90. "Malz." *Reallexikon der Assyriologie und Vorderasiatischen Archäologie*, 7: 322–9.

Stol, Marten. 2000. *Birth in Babylonia and the Bible: Its Mediterranean Setting*. Groningen: Styx.

Stol, Marten. 2004. "Wirtschaft und Gesellschaft in Altbabylonischer Zeit." In Pascal Attinger, Walther Sallaberger, and Markus Wäfler (eds), *Mesopotamien. Die altbabylonische Zeit* (= Orbis biblicus et orientalis, 160/4). Freiburg: Academic Press.

Stol, Marten. 2006. "The Digestion of Food According to Babylonian Sources." In Laura Battini and Pierre Villard (eds), *Médecine et médecins au Proche-Orient ancien* (= BAR International Series, 1528). Oxford: British Archaeological Reports.

Streily, Hansen. 2000. "Early Pottery Kilns in the Middle East." *Paléorient*, 26: 69–81.

Suda. 2014. New Digital Edition (*Suda On Line: Byzantine Lexicography*). Available online: http://www.stoa.org/sol/.

Takacs, Laszo. 2000. "Quicksilver from Cinnabar: The First Documented Mechanochemical Reaction?" *JOM: Journal of Minerals, Metals and Materials Society*, 52(1): 12–13.

Tallet, Pierre. 1995. "Le shedeh, étude d'un procédé de vinification en Égypte ancienne." *Bulletin de l'Institut français d'Archéologie orientale*, 95: 459–92.

Tallet, Pierre. 2000. "La zone minière du Sud-Sinaï à l'époque pharaonique." *Égypte, Afrique & Orient*, 59: 17–32.

Tallet, Pierre. 2009. "Les Égyptiens et le littoral de la mer Rouge à l'époque pharaonique." *Comptes rendus de l'Académie des Inscriptions et Belles Lettres*, 153(2): 687–719. Available online: https://www.persee.fr/doc/crai_0065-0536_2009_num_153_2_92529.

Tallet, Pierre. 2013–18. *La zone minière pharaonique du Sud-Sinaï (I–III)*. Cairo: Institut Français d'Archéologie Orientale.

Tallet, Pierre. 2015. "Les 'ports intermittents' de La mer rouge à L'époque pharaonique: caractéristiques et chronologie." In Bruno Argémi and Pierre Tallet (eds), *Entre nil et mers la navigation en égypte ancienne. Actes des rencontres de Provence Égyptologie. Musée Départemental Arles Antique le 12 avril 2014.* (= *NeHet*, 3). Available online: http://www.nehet.fr/NEHET3/03-NeHeT%20 3-01-POMEY-p.1-29.pdf

Tallet, Pierre. 2017. *Les papyrus de la mer Rouge I. Le « Journal » de Merer (Papyrus Jarf A et B)* (= *Mémoires de l'Institut français d'Archéologie orientale*, 136). Cairo: Institut Français d'Archéologie Orientale.

Tallet, Pierre, Georges Castel, and Philippe Fluzin. 2011. "Metallurgical Sites of South Sinai (Egypt) in the Pharaonic Era: New Discoveries." *Paléorient*, 37: 79–89.

Taylor, Frank Sherwood. 1930. "A Survey of Greek Alchemy." *The Journal of Hellenic Studies*, 50: 109–39.

Taylor, Frank Sherwood. 1945. "The Evolution of the Still." *Annals of Science*, 5: 185–202.

Thavapalan, Shiyanthi. 2020. *The Meaning of Color in Ancient Mesopotamia*. Leiden: Brill.

Thavapalan, Shiyanthi, Jens Stenger, and Carol Snow. 2016. "Color and Meaning in Ancient Mesopotamia: The Case of Egyptian Blue." *Zeitschrift für Assyriologie*, 106: 198–214.

Thornton, Christopher Peter. 2009. "The Emergence of Complex Metallurgy on the Iranian Plateau: Escaping the Levantine Paradigm." *Journal of World Prehistory*, 22: 301–27.

Tite, Michael S., and Mavis Bimson. 1989. "Glazed Steatite: An Investigation of the Methods of Glazing Used in Ancient Egypt." *World Archaeology*, 21(1): 87–100.

Tite, Michael S., Andrew J. Shortland, Paul T. Nicholson, and Caroline Jackson. 1998. "The Use of Copper and Cobalt Colorants in Vitreous Materials in Ancient Egypt." In Sylvie Colinart and Michel Menu (eds), *La couleur dans la peinture et l'émaillage de l'Égypte ancienne: Actes de la table ronde Ravello, 20–22 Marzo 1997* (= *Scienze e materiali del patrimonio culturale*, 4). Santo Spirito: Edipuglia.

Tolaini, Francesca. 2004. "*De tinctio omnium musivorum*. Technical Recipes for Glass in the So-Called *Mappae Clavicula*." In Marco Beretta (ed.), *When Glass Matters. Studies in the History of Science and Art from Graeco-Roman Antiquity to Early Modern Era*. Florence: Olschki.

Tosi, Maurizio. 1984. "The Notion of Craft Specialization and Its Representation in the Archaeological Record of Early States in the Turanian Basin." In Matthew J.T. Spriggs (ed.), *Marxist Perspectives in Archaeology*. Cambridge: Cambridge University Press.

Totelin, Laurence M.V. 2004. "Mithridates' Antidote: A Pharmacological Ghost." *Early Science and Medicine*, 9: 1–19.

Totelin, Laurence M.V. 2009. *Hippocratic Recipes. Oral and Written Transmission of Pharmacological Knowledge in the Fifth- and Fourth-Century Greece* (= *Studies in Ancient Medicine*, 34). Leiden: Brill.

Totelin, Laurence M.V. 2016a. "The World in a Pill. Local Specialties and Global Remedies in the Graeco-Roman World." In Rebecca Futo Kennedy and Molly Jones-Lewis (eds), *The Routledge Handbook of Identity and the Environment in the Classical and Medieval Worlds*. London and New York: Routledge.

Totelin, Laurence M.V. 2016b. "Pharmakopōlai: A Re-Evaluation of the Sources." In William V. Harris (ed.) *Popular Medicine in Graeco-Roman Antiquity: Explorations*. Leiden and Boston: Brill.

Touwaide, Alain. 1983. "L'authenticité et l'origine de deux traités de toxicologie attribués à Dioscoride. I. Historique de la question. II. Apport de l'histoire du texte grec." *Janus*, 70: 1–53.

Touwaide, Alain. 1992. "Les deux traités de toxicologie attribués à Dioscoride: tradition manuscrite, établissement du texte et critique d'authenticité." In Antonio Garzya (ed.), *Tradizione e ecdotica dei testi medici tardoantichi e bizantini. Convegno internazionale: Papers*. Naples: M. D'Auria.

Touwaide, Alain. 1998a. "Le médicament en Alexadrie: de la pratique à l'épistemologie." In Gilbert Argoud and Jean-Yves Guillaumin (eds), *Sciences exactes et sciences appliquées à Alexandrie (IIIe siècle av. J.-C. – Ier siècle apr. J.-C.)*. Saint-Étienne: Publication de l'Université de Saint-Étienne.

Touwaide, Alain. 1998b. "Therapeutic Strategies: Drugs." In Mirko D. Grmek (ed.), *Western Medical Thought from Antiquity to the Middle Ages*. Cambridge, MA: Harvard University Press.

Traunecker, Claude. 1989. "Le 'Château de l'Or' de Thoutmosis III et les magasins nord du temple d'Amon." *Cahiers de Recherche de l'Institut de Papyrologie et d'Egyptologie de Lille*, 11: 89–111.

Traunecker, Claude. 2004. "L'anticipation dans la pensée de l'Égypte antique. À propos du texte de la Théologie Memphite." In Rudolph Sock and Béatrice Vaxelaire (eds), *L'anticipation. À l'horizon du Présent*. Liège: Mardaga.

Treister, Michail Yu. 1996. *The Role of Metals in Ancient Greek History* (= *Mnemosyne, Supplements*, 156). Leiden: Brill.

Treister, Michail Yu. 1998. "Ionia and the North Pontic Area. Archaic Metalworking. Tradition and Innovation." In Gocha R. Tsetskhladze (ed.), *The Greek Colonisation of the Black Sea Area*. Stuttgart: Franz Steiner Verlag.

Trinquier, Jean. 2013. "*Cinnabaris* et 'sang-dragon': le 'cinabre' des anciens entre minéral, végétal et animal." *Revue archéologique*, 56: 305–46.

Ulrich, Roger. 2008. "Representations of Technical Processes." In John Peter Oleson (ed.), *The Oxford Handbook of Engineering and Technology in the Classical World*. Oxford: Oxford University Press.

Ursin, Johann Heinrich. 1661. *De Zoroastre bactriano, Hermete Trismegisto, Sanchoniathone phoenicio eorumque scriptis et aliis contra Mosaicae Scripturae antiquitatem*. Nuremberg.

Vaiman, Aizik A. 1982. "Eisen in Sumer." *Archiv für Orientforschung Beiheft*, 19: 33–8.

Valbelle, Dominique. 1982. *Les ouvriers de la Tombe: Deir el-Médineh à l'époque ramesside* (= *Bibliothèque d'Étude*, 96). Cairo: Institut Français d'Archéologie Orientale.

Valbelle, Dominique, and Charles Bonnet. 1996. *Le sanctuaire d'Hathor, maîtresse de la turquoise*. Paris: Picard.

Van De Mieroop, Marc. 1987. *Crafts in the Early Isin Period: A Study of the Isin Craft Archive the Reigns of Ishbi-Erra and Shu-ilishu* (= *Orientalia Lovaniensia analecta*, 24). Leuven: Departement Orientalistiek.

Van De Mieroop, Marc. 2009. *Eastern Mediterranean in the Age of Ramesses II*. Malden, MA: Wiley-Blackwell.

Van Minnen, Peter. 1987. "Urban Craftsmen in Roman Egypt." *Münstersche Beiträge zur antike Handelsgeschichte*, 6: 31–88.

Vandersleyen, Claude. 2013. *Le Rapport d'Ounamon (vers 1065 avant Jésus-Christ): Analyse d'une mission manquée* (= Connaissance de l'Égypte Ancienne, 15). Brussels: Safran.

Vandier, Jacques. 1936. *La Famine dans l'Égypte Ancienne* (= Recherches d'Archéologie, de Philologie et d'Histoire, 7). Cairo: Institut Français d'Archéologie Orientale.

Vandier, Jacques. 1963. *Le papyrus Jumilhac*. Paris: CNRS.

Vandier, Jacques. 1969. *Manuel d'archéologie égyptienne V: Bas-reliefs et peintures. Scènes de la vie quotidienne*. Paris: Picard.

Vandiver, Pamela B. 1998. "A Review and Proposal of new Criteria for Production Technologie of Egyptian Faience." In Sylvie Colinart and Michel Menu (eds), *La couleur dans la peinture et l'émaillage de l'Égypte Ancienne: Actes de la table ronde Ravello, 20–22 Marzo 1997* (= Scienze e materiali del patrimonio culturale, 4). Santo Spirito: Edipuglia.

Vanstiphout, Herman L.J. 1984. "On the Sumerian Disputation between the How and the Plough." *Aula Orientalis*, 2: 239–51.

Vanstiphout, Herman L.J. 1991. "Lore, Learning and Levity in the Sumerian Disputations: A Matter of Form, or Substance?" In Gerrit J. Reinink and Herman L.J. Vanstiphout (eds), *Dispute Poems and Dialogues in the Ancient and Mediaeval Near East: Forms and Types of Literary Debates in Semitic and Related Literatures* (= Orientalia Lovaniensia analecta, 42). Leuven: Department Oriëntalistiek.

Vanstiphout, Herman L.J. 2003. *Epics of the Sumerian Kings: The Matter of Aratta*. Atlanta, GA: Society of Biblical Literature.

Varberg, Jeanette, Bernard Gratuze, and Flemming Kaul. 2015. "Between Egypt, Mesopotamia and Scandinavia: Late Bronze Age Glass Beads Found in Denmark." *The Journal of Archaeological Science*, 54: 168–81.

Varille, Alexandre. 1938. *La tombe de Ny-Ânkh-Pépi à Zâouyet el-Mayetîn* (Mémoires de l'Institut français d'Archéologie orientale, 70). Cairo: Institut Français d'Archéologie Orientale.

Vartavan, Christian T. de. 1998. "Sources végétales possibles ou connues pour les colorants, liants et vernis de l'Égypte ancienne." In Sylvie Colinart and Michel Menu (eds), *La couleur dans la peinture et l'émaillage de l'Égypte Ancienne: Actes de la table ronde Ravello, 20–22 Marzo 1997* (= Scienze e materiali del patrimonio culturale, 4). Santo Spirito: Edipuglia.

Veldhuis, Niek. 2014. *History of the Cuneiform Lexical Tradition* (= Guides to the Mesopotamian Textual Record, 6). Münster: Ugarit Verlag.

Verner, Miroslav. 1989. "La tombe d'Oudjahorresnet et le cimetière saïto-perse d'Abousir." *Bulletin de l'Institut français d'Archéologie orientale*, 89: 283–90.

Vernier, Émile. 1907. *Bijoux et orfèvrerie* (= Catalogue général du musée du Caire). 2 vols. Cairo: Institut Français d'Archéologie Orientale.

Venticinque, Philip F. 2016. *Honor Among Thieves. Craftsmen, Merchants, and Associations in Roman and Late Roman Egypt*. Ann Arbor, MI: University of Michigan Press.

Viano, Cristina. 1995. "Olympiodore l'alchimiste et les Présocratiques. Une doxografie de l'unité (De arte sacra, § 18–27)." In Didier Kahn and Sylvain Matton (eds), *Alchimie. Art, histoire et mythes: Actes du 1er colloque international de la Société d'étude de l'histoire de l'alchimie (Paris, Collège de France, 14–15–16 mars 1991)*. Paris: S.É.H.A. and Milan: Arché.

Viano, Cristina. 1996. "Aristote et l'alchimie grecque: la trasmutation et le modèle aristotélicien entre théorie et pratique." *Revue d'histoire des sciences*, 49:189–213.

Viano, Cristina (ed.). 2002. *Aristoteles chemicus: il IV libro dei "Metereologica" nella tradizione antica e medievale*. Sankt Augustin: Academia Verlag.
Viano, Cristina. 2005. "Les alchimistes gréco-alexandrins et le *Timée* de Platon." In Cristina Viano (ed.), *L'alchimie et ses racines philosophiques. La tradition grecque et la tradition arabe*. Paris: Vrin.
Viano, Cristina, 2006. *La matière des choses. Le livre IV des Météorologiques d'Aristote et son interprétation par Olympiodore*. Paris: Vrin.
Viano, Cristina. 2015. "*Mixis* and *Diagnôsis*: Aristotle and the 'Chemistry' of the Sublunary World." *Ambix*, 62: 203–14.
Viano, Cristina. 2018. "Olympiodore l'alchimiste et la *taricheia*. La transformation du minerai d'or: technê, nature, histoire et archéologie." In Efthymios Nicolaidis (ed.), *Greek Alchemy from Late Antiquity to Early Modernity*. Turhout: Brepols.
Vogelsang-Eastwood, Gillian. 2000. "Textiles." In Paul T. Nicholson and Ian Shaw (eds), *Ancient Egyptian Materials and Technology*. Cambridge: Cambridge University Press.
Vogler, Herbert. 2013. "The Craft of Dyeing in Ancient Egypt." *Textile History*, 13: 159–63.
Von Staden, Heinrich. 1989. *Herophilus: The Art of Medicine in Early Alexandria*. Cambridge: Cambridge University Press.
Vos, René L. 1993. *The Apis Embalming Ritual. P. Vindob 3873* (= Orientalia Lovaniensia Analecta, 50). Leuven: Peeters.
Waddell, W. Gillian. 1980. *Manetho* (= Loeb Classical Library, 350). London: Heinemann.
Waetzoldt, Hartmut. 1972. *Untersuchungen zur neusumerischen Textilindustrie* (= Studi Economici e technologici, 1). Rome: Centro per le antichità e la storia dell'arte del Vicino Oriente.
Waetzoldt, Hartmut. 1981. "Zur Terminologie der Metalle in den Texten aus Ebla." In L. Cagni (ed.), *La Lingua di Ebla: Atti del convegno internazionale (Napoli, 21–23 aprile 1980)* (= Seminario di Studi Asiatici Series Minor, 14). Naples: Istituto Universitario Orientale.
Waetzoldt, Hartmut. 1985. "Ölpflanzen und Planzenöle im 3. Jahrtausend." *Bulletin on Sumerian Agriculture*, 2: 77–96.
Waetzoldt, Hartmut, and Hans-Gert Bachmann. 1984. "Zinn-und Arsenbronzen in den Texten aus Ebla und aus dem Mesopotamien des 3. Jahrtausends." *Oriens Antiquus*, 23: 1–18.
Wagensonner, Klaus. 2010. "Early Lexical Lists Revisited: Structures and Classification as a Mnemonic Device." In *Language in the Ancient Near East: Proceedings of the 53ᵉ Rencontre assyriologique internationale* (= Babel und Bibel: Annual of ancient near eastern old testament, and semitic studies, 4/1–4/2). Winona Lake, IN: Eisenbrauns.
Wagner, Günther A., and Önder Öztunalı. 2000. "Prehistoric Copper Sources in Turkey." In Ünsal Yalçın (ed.), *Anatolian Metals I*. Bochum: Deutsches Bergbau-Museum.
Waldron, Harry A. 1973. "Lead Poisoning in the Ancient World." *Medical History*, 17: 391–9.
Wallraff, Martin, Carlo Scardino, Laura Mecella, and Christophe Guignard (eds). 2012. *Iulius Africanus: Cesti*. Berlin: De Gruyter.

Wasserman, Nathan. 2013. "Treating Garments in the Old Babylonian Period: 'At the Cleaners' in a Comparative View." *Iraq*, 75: 255–77.
Waterfield, Robin. 2008. *Plato. Timaeus and Critias*. Oxford: Oxford University Press.
Watkins, Trevor. 2005. "Obsidian. B. Archäologischer Befund." *Reallexikon der Assyriologie*, 10: 9–13.
Watkins, Trevor. 2008. "Supra-Regional Networks in the Neolithic of Southwest Asia." *Journal of World Prehistory*, 21: 139–71.
Wee, John Z. 2016. "Virtual Moons over Babylonia: The Calendar Text System, Its Micro-Zodiac of 13, and the Making of Medical Zodiology." In John M. Steele (ed.), *The Circulation of Astronomical Knowledge in the Ancient World*. Leiden: Brill.
Wee, John Z. (ed.). 2017. *The Comparable Body: Analogy and Metaphor in Ancient Mesopotamian, Egyptian and Greco-Roman Medicine* (= Studies in Ancient Medicine, 49). Leiden: Brill.
Weeks, Lloyd. 2012. "Metallurgy." In Daniel T. Potts (ed.), *A Companion to the Archaeology of the Ancient Near East*, vol. 1. Wiley-Blackwell.
Weeks, Lloyd. 2013. "Iranian Metallurgy of the Fourth Millennium BC in Its Wider Technological and Cultural Context." In Cameron A. Petrie (ed.), *Ancient Iran and Its Neighbours: Local Developments and Long-Range Interactions in the Fourth Millennium BC*. Oxford: Oxbow Books.
Wenkebach, Ernst (ed.). 1956. *Galeni in Hippocratis Epidemiarum librum VI* (= Corpus Medicorum Graecorum, V.10.2.2). Berlin: Akademie Verlag.
Wertime, Theodore A. 1973. "The Beginnings of Metallurgy: A New Look." *Science*, 182(4115): 875–87.
Westbrook, Raymond. 2000. "International Law in the Amarna Age." In Raymond Cohen and Raymond Westbrook (eds), *Amarna Diplomacy: The Beginnings of International Relations*. Baltimore, MD: Johns Hopkins University Press
Widmer, Ghislaine. 2003. "Les fêtes en l'honneur de Sobek dans le Fayoum à l'époque gréco-romaine." *Égypte. Afrique & Orient*, 32: 23–32.
Wildung, Dietrich. 1977. *Imhotep und Amenhotep. Gottwerdung im Alten Ägypten* (= Münchner Ägyptologische Studien, 36). Munich: Deutscher Kunstverlag.
Williams, Michael. 2006. *Deforesting the Earth. From Prehistory to Global Crisis. An Abridgment*. Chicago, IL: University of Chicago Press.
Wilson, C. Anne. 1984. *Philosophers, Iōsis and the Water of Life*. Leeds: Leeds Philosophical and Literary Society.
Wilson, Malcolm. 2013. *Structure and Method in Aristotle's* Meteorologica. *A More Disorderly Nature*. Cambridge: Cambridge University Press.
Winter, Irene J. 1983. "The Program of the Throneroom of Assurnasirpal II." In Prudence O. Harper and Holly Pittman (eds), *Essays on Near Eastern Art and Archaeology in Honor of Charles Kyrle Wilkinson*. New York: The Metropolitan Museum of Art.
Wipszycka, Ewa. 1965. *L'industrie textile dans l'Égypte Romaine*. Wroclaw: Zakład Narodowy im. Ossolińskich.
Woolley, C. Leonard. 1956. *Ur Excavations IV. The Early Periods*: A Report on the Sites and Objects Prior in Date to the Third Dynasty of Ur Discovered in the Course of the Excavations. Oxford: Oxford University Press.
Wright, M. Rosemary. 1981. *Empedocles: The Extant Fragments*. New Haven, CT: Yale University Press.
Yoffee, Norman. 1981. *Explaining Trade in Ancient Western Asia*. Malibu, CA: Undena Publications.

Yoyotte, Jean. 1975. "Les Sementiou et l'exploration des régions minières à l'Ancien Empire." *Bulletin de la Société française d'Égyptologie*, 73: 44–55.

Yoyotte, Jean. 1977. "Une notice biographique du roi Osiris." *Bulletin de l'Institut français d'Archéologie orientale*, 77: 145–49.

Zaccagnini, Carlo. 1983. "Patterns of Mobility among Ancient Near Eastern Craftsmen." *Journal of Near Eastern Studies*, 42: 245–64.

Zarnkow, Martin, et al. 2006. "Interdisziplinäre Untersuchungen zum altorientalischen Bierbrauen in der Siedlung von Tall Bazi/Nordsyrien vor rund 3200 Jahren." *Technikgeschichte*, 73: 3–25.

Zhmud, Leonid. 2006. *The Origin of the History of Science in Classical Antiquity*. Berlin: De Gruyter.

Ziegler, Christiane. 1981. "Une famille de 'grands des djebels de l'or' d'Amon." *Revue d'Égyptologie*, 33: 125–32.

Zimmern, Heinrich. 1925. "Assyrische Chemisch-technische Rezepte, Insbesondere für Herstellung Farbiger Glasierter Ziegel, im Umschrift und Übersetzung." *Zeitschrift für Assyriologie*, 36: 177–208.

Zimmern, Heinrich. 1926. "Vorläufiger Nachtrag zu den assyrischen chemisch-technischen Rezepten." *Zeitschrift für Assyriologie*, 37: 213–14.

LIST OF CONTRIBUTORS

Sydney H. Aufrère, Directeur de recherche emeritus at the CNRS

Marco Beretta, Professor of History of Science at the University of Bologna

Cale Johnson, Professur für Wissensgeschichte des Altertums at the Freie Universität Berlin

Matteo Martelli, Professor of History of Ancient Science at the University of Bologna

INDEX

Acton, Peter 227
Aeneas of Gaza 133, 234
Aeschylus 130
Africanus, Julius 77, 155–6
Agatharchides 56, 154
Agathodaimon 43
Agricola, Georgius 2, 147–8
Akkadian step-by-step procedure 199–201
alchemy 1–3, 10–11, 20, 75–6, 121–2, 131–3, 159, 233–4
   derivation of the term 207
Alcohol 110
d'Alembert, Jean le Rond 3
alembics 105, 108–10
Alexander the Great 181
alkali 67
alloys 144, 160, 198
alum 13, 166
Amarna 175
amber 73
Amenhotep 120
Amenhotep II 163
Anastasios I 234
Anaximander 9, 43
Anazimenes 9
Andromakos the Elder 136
Annals of Amenemhat II 166
antiquity of authorship 128
Anubis 115
Apollonios 159, 184

apprenticeship 205–9
Arachne 130
Aristophanes 179
Aristotle and Aristotelianism 42, 45–9, 76, 83, 129, 133–4, 230
aromatics 74–5, 216
artisans 7–8, 19, 135, 185, 205, 210
   autonomy of 138
Assurbanipal Library 202–3, 239
Atomism (ancient) 43–44
At the Cleaners 199–200, 224
Aufrère, Sydney H. 21

Bedouin tribes 162
beer-brewing 77, 212, 223–4
Berthelot, Marcellin 3, 15–16
beveled-rim bowl 94
beverages 77
biological evolution in man 5
Biringuccio, Vannoccio 2
blowpipes 14, 102
Body Description texts 41
Boncossi, Morandi 97
book-burning 11
book-keeping techniques 169
Books of Breathings 30
Borch, Ole 2–3
Bortéro, Jean 198, 216, 218
Boyle, Robert 43–4
bread-making 212
British Museum 3

bronze 58, 80, 64–5, 144
  Corinthian bronze 80

cadmia 80, 107, 181–2
Cairo Papyrus 131
camels 170
case histories, depersonalized 196
Catalogue of Texts and Authors 128
cementation 66
ceramics 58, 89, 142
cereal grains 61
chemical apparatus 16–17
chemical arts 2–12, 15, 19–21, 43, 48,
    83–4, 112, 129, 134, 137, 160, 210,
    234–5
  creation of gems by use of 19
  cultural importance of 21
  polluting effects of 21
  role in economic systems 2
  specialization in 83
  technology 7–8
chemistry
  academic status of 1, 5
  emergence as a science 50
  etymology of the word 10–11
  origins 2–5, 11–12
  practical 2, 12–17
  public image of 1–2
Cheops 120–1
Chymes 42
China, chemistry in 4
cinnabar 80–1, 155
Civil, Miguel 37, 223–4
Clarke, John R. 231
Class Reunion (Sumerian dialogue) 152
clay and clay tablets 32–4
Clement of Alexandria 29–30, 120,
    192
Cleopatra 13, 121, 133, 142
clothing 229
Codex Urnamma 196
cognitive functions 114
Colin, Frédéric 190
colors 44–5
  primary 44
compendia, of recipes 195
conceptual metaphor theory 222–3
contexts of transformation 220
Conring, Hermann 2
cooking 6

copper 46, 57–8, 63–5, 87, 102, 142–3,
    153, 214
Copper and Silver (Sumerian dialogue) 126
cosmology 9
Couto-Ferreira, Erica 220
Craddock, Paul T. 101
crafts 203–5, 209–10, 231–2
  gods associated with 115, 129
  listing of 132
  specialist and *non-specialist* 127
craftsmanship 139–40, 163
craftsmen 15, 104, 165, 185, 227, 231, 234
Critias 179–80
cryptic orthography 202
cuneiform writing 33, 36, 39–40, 94
cupellation 148
currency, lack of 86–4
Curse of Agade 146
Cydias 77, 179
cylinder seal 34
Cyprus 105, 107, 153, 165, 178, 181–2

dairy fats, processing 218
Davey, Christopher 102, 217
Deckname 41, 202
Deforestation 152–3
Delos 105, 180
Del Purpurarius 183
Democritus 43–4, 82, 131, 153
Demosthenes 178, 204
Dendara 26–9, 92–4, 117, 194
dentistry 75
deposits foundation 38
Dercksen, Jan Gerrit 175
Diagnostic Handbook 200–1
Dickson, Keith 124
Diderot, Denis 3
Diels, Hermann 110
Digital Corpus of Cuneiform Lexical Texts
    (DCCLT) 40
Diocles of Caristus 135
Diocletian, Emperor 11, 159
Diodorus Siculus 49, 114, 130, 134, 154,
    161, 167
Diogenes of Apollonia 42–3
Diogenes Laertius 44
Dionysius 130
Dioscorides 15, 20–1, 76, 79, 81, 107,
    135, 154, 181, 208
directives, second-person in recipes 198

distillation 75, 105, 108–10
divinity, associations with 113–14, 137
domestic production 34
drugs and drug-dealers 167, 230
Dura Europos, city of 156
Düring, Ingemar 48
dyeing 60–1, 69–71, 92–3, 133, 135, 142, 158–9, 184

Early Dynastic Practical Vocabularies (EDPV) 37
Ebla archive 65–6
economy, silver-based 146
Edens, Christopher 171
Edfu laboratory 187–91, 216
Edwin Smith Surgical Papyrus 193
Eggert, Gerhard 103
Egypt 2, 11–15, 23–32, 50, 85–94, 112–22, 137–8, 150, 161–8, 187–93, 234
    seen as the land of chemistry and alchemy 121
    seen as the land of poisoners 142–3
'Egyptian blue' (or *kyanos*) 13, 59, 81, 167, 179
Eisler, Robert 129
elements, theories of 44–6
Eliade, Mircea 129
Ellis, Richard S. 38
Ellison, Rosemary 103
embalming 142, 226
Embalming Ritual 30
emery particles 34–5, 235
Empedocles 8–9, 44–5, 50
Englund, Robert K. 36, 64, 145
Enki and Ninhursag 122–5, 222
Enki and the World Order 125
Enmerkara and the Lord of Aratta 71
Enochic myth 131–2, 159
equivalent names 193
Eratosthenes of Cyrene 153
ergastērion, use of 104
Esagil-kin-apli 128, 200–1
Evil Demons 200
exchange, of technical specialists 177
exhalations theory 49, 134
exotic products 164
experimental practice 21
Exploits of Nimurta 125

fabrics 70–1; *see also* textiles
fast wheel 102
fermentation process 223
Festugière, André Marie Jean 43
fiber revolution 66
Finkel, Irving L. 195
fire, power and controlled use of 5–7
fish, preservation of 213
Fleming, Stuart J. 14, 184
Flohr, Miko 184–5
foodstuffs
    collective production of 86
    distribution of 89
    elite, notation system 95
Forbes, Robert J. 147–8
forest degradation 152
Frankfort, Henry 222
freed slaves 204
Frymer-Kensky, Tikva 122, 125, 220
fuller's earth 67
fulling 231
funerary architecture 141
furnaces 14, 99–101, 106–8

Gadd, Cyril J. 202
Galen 44–5, 78, 107, 134–5, 153–4, 180–1, 208, 230–2
Gardiner, Alan H. 55
gas emissions 155
Gassendi, Pierre 43–4
Geller, Markham J. 128
geological phenomena 134
glassmaking and glass technology 8–9, 12–20, 60, 71–3, 81–4, 99, 106, 157, 166, 181–4, 202–3, 213
glosses 193
Gnosticism 133
Gods
    creators 114
    associated with craftsmanship in general 115, 129
    associated with particular kilns 219–20
    invention of 114
gold and goldsmithing 46–7, 56–7, 65–6, 87, 130, 144–5, 149–50, 214
    debasement 150
    grades / quality of 66
Gorelick, Leonard 34
Gospel of Philip 133

Göttertypentext 41
Grattan, John P. 148–9
Great Harris Papyrus 90
Great Khorasan Road 171
Greco-Roman world 42–49, 75–83, 104–12, 129–38, 152–9, 177–85, 203–10, 226–35
Greek culture and Greek philosophy 3–4, 9–13, 52, 138
'Greek fire' 156
green vitriol 68, 76
Greenland ice cores 155
Gregory of Nyasa 133
Gudea Cylinders 63, 172–3
guilds 15, 205–7, 210
Gwinnett, A. John 34
gypsum 59

Halleux, Robert 20, 49
Hathor 32
Hedvall, Arvid 48
Heimpel, Wolfgang 95–6
Henderson, Julian 71, 73, 84, 98
Hephaestus 129–30, 226–9, 235
Heraclitus 7
Heraclitus' fire 42–3
Herakleides 205
Hermes and Hermeticism 42–3, 119, 132, 192, 231
Herodotus 85–6, 120, 153, 161, 167
heurematography 130, 138
hieroglyphic system 52–4, 214
hierogrammats 23
Hinke, William 221
Hippocrates (and Hippocratic writing) 76–7, 108, 133–4, 153, 156, 178, 181, 207–8
Hippolytus of Rome 110
history and historians of chemistry 2–4
Hittites 13
Hoefer, Ferdinand 3–4
Homer and Homeric hyms 129–30, 167–8, 185, 235
homologous alloforms 124
Hong, Sungmin 155
Horus 132
Houses of Life 24

Hughes, J. Donald 152–3
Hyperides 180, 204

iconography 163, 211, 215, 234
    associated with the womb 220–3, 234–5
ideograms 53–4
the *Iliad* 162, 226–7, 235
Imhotep 119–20
industrial parks 95
infrastructure 169
ink-making 76
iron and ironworking 13, 25, 44, 87, 130
Irtysen 190
Isimud, vizier 123
Isthmeos, John 234

John of Antioch 159
Johnson, Cale 21
Johnson, Mark 222
Jones, Horace Leonard 158
Julian of Ascalon 157
Justinian 159
Juvenal 157

Kērotakis (alchemical equipment) 110–1
Khaemouaset 141
kilns 99–100
kilns and furances terminologies 98
Kirrha, city of 156
kitchen chemistry 122
kitchenware, repurposed 103
kyanos see 'Egyptian blue'
knowledge
    mastery of 114
    preservation in workshops 210
    traced back to the gods 122
    transmission of 210
König, W. 103
Kopp, Hermann 3
Kultmittelbeschwörungen 219
Kusu and Kubu 219–21

"laboratories' 93–4, 104, 187–9, 214; *see also* Edfu laboratory
lakes contaminated 155
Lakoff, George 222
Lambert, W.G. 128
Landsberger, Benno 63, 68, 73

INDEX

lapidaries 20
lapis lazuli 13–15, 34, 59–60, 92, 150, 170
Larsa Goldsmith's Hoard 97
Late Uruk 34, 64
Laurium (or Laurion) 107, 178
lead 78–9
    white lead (*psimythion/cerussa*) 77–8, 82, 136
leather 67–8
leavening 77
Lesis 204
Levantine paradigm 102
Levey, Martin 75, 103, 216
lexical lists 39
lexical studies 225
Libavius, Andreas 104
Liebig, Justus 3
Lincoln, Bruce 124
Lippmann, Edmund 3, 110
Lipšur Litanies 173
literary transmission of chemical knowledge 3–4, 17–21
litharge 77–9, 107, 136, 178
Liu, Jinyu 205
London Medical Papyrus 114
Louvre Museum 215
Lucretius 5–7, 44, 131, 154
Lugalbanda Epics 172

magic and magicians 115–16, 167, 190–1
Malalas 234
Mallowan, Max E.L. 98
Manetho of Sebennytos 52, 120–1
manumission inscriptions 204
the *Marcianus* 154
Marduk 128
Marduk-Ea formula 195–6, 199
Martelli, Matteo 21, 80, 202
Martial 157
Mary the Jewess 16, 108–10, 112, 159, 232
mass production 88
measuring devices 173
medicine and medicinal plants 13, 123–4, 134–5, 167
Mediterranean civilizations 4, 9, 166
memorization of sacred texts 192, 195
Memphis 162
Menkheperreseneb II 164

mercury 80–1
Mereruka tomb 217
Mesopotamia 32–42, 63–75, 94–103, 122–9, 137, 143–52, 168–77, 185, 195–203, 216–26, 234
metals, formation and composition (metallogenesis) 47–9, 129, 134
metal-working 115, 148, 152, 226
metallurgy 5–10, 13, 58, 62, 79, 99, 130, 143–4, 148, 152, 155, 166, 213, 235
metaphor, use of 225, 234–5
Metropolitan Museum, New York 227
Michalowski, Piotr 219–20
Micon 179
military camps 218–19
military equipment 115, 140, 166
military force, supplies acquired by means of 143
mineral resources and mineralogy 10, 164, 178
mining 143–5, 153–6, 160
Minoans 163
Mithridates, King 13
mixis 47–8
Moorey, Peter S. 97–102
Moses 2
mud-bricks 140
Museum of Fine Arts, Boston 230
Mycenaeans 163
myrrh 192
mytho–poetic reflection 30
mytho-scientific rites 31–2

naked craftsmen 227–8
naruqqu partnerships 170, 175
natural philosophy 7, 10, 84
natural resources, depletion of 152
Nebros 156
Neilos 159, 232
Neri, Antonio 18
Nicholson, Paul T. 73
Nile Delta 85–6
Ninhursag Temple 217
Ninkasi Hymn 223–4
Nissen, Hans J. 36, 64
Nubians 163

Oates, David 98
obsidian 170
ointments 29–30, 187, 190
Olympiodorus 16, 42–3, 48, 155
onomastica 55
Oppenheim, Alan 71, 73, 220
Orsenouphis 206
Osiris 25
Osiris-Khentymentiu statues 194
Ostanes 131, 207
Ostia 158
Ovid 130

Paestum 105, 180
Palestine 164
Pan of the Desert 115
Panopolis 30
Papinian 206
Papyrus Jumilhac 25
Papyrus Leidensis and *Papyrus Holmiensis* 18
papyrus, manufacture of 89–90
Paracelsus 20
Paraphrase of Shem 133
Partington, James R. 3–4, 11, 42
Pattinson, Hugh Lee 78–9
perfume 46, 61–3, 74–5, 83, 105–6, 116, 180, 187–91, 202, 208, 214–16, 230
Peryplus of the Red Sea 181
Petosiris 205, 214, 216
Petronius 44, 82
pharmaceutical effectiveness 123
pharmakon, meanings of the term 135
Philo of Byzantium 76
Phoenicians 8–9, 164–5, 168
Photius 154
physical characteristics, embarrasing 150
Physiognomic Corpus 200–1
pigments 81–2
Place, Victor 67
"plant" stone 125
Plataia, city of 156
Plato 42, 45–7, 76, 130, 152, 157
Pliny the Elder 6–9, 14, 19–21, 61, 76–83, 89–90, 107, 130, 153–6, 180–1, 215
plunder 145, 164
Plutarch 51–2, 119, 121, 178, 208
poisoning 142, 135–6, 156, 160

pollution 142, 148–9, 156–7, 160
  effects on health 149
  mitigating the effects of 160
Polydamna, Queen 167–8
Polygnotus 179
polymaths 114
Pompeii 104–6, 158, 180, 184–5, 208, 231
Posidonius 7, 49, 131, 231
Postgate, Nicholas 38–9
Potts, Daniel T. 67, 145
priestly culture 121, 130, 132
Principe, Lawrence M. 80
procedure texts 65
Proclus 47
Prodicus of Keos 130
production quotas 174
professions, humorous 150
Properties (of substances) 39
proto-cuneiform 36, 100–2
Proto-Elamite 144
Pseudo-Aristotle 7, 79
Pseudo-Democritus 13, 48, 50, 76, 81–4, 110, 158, 181–2, 207, 209
Puntites 163
purple dyeing 83, 158–9
pyramid of Giza 58

quicklime 59

Rabâ-ša-Marduk 177
Ramses II 90
ratio, of labor to finished products 95
raw materials
  sourced from macrocosm 147
  relative value 174
Reade, Julian E. 202–3
recipes, chemical (and recipe books) 17–19, 29–31, 132, 181, 187–91, 196–7, 208–10, 216
recycling
  of raw materials 147
  of Ekur Temple by Naram-Sîn 173
refining of ore 148
Reiner, Erica 73
Rekhmirê, vizier 163
"*relief du lirinon*" 215
Robson, Eleanor 202
Rochberg-Halton, Francesca 128, 200

rocks needed for building temples 87
Roman Empire 13–14, 181, 184
Ruelle, Charles Emile 3, 15–16
Rumor, Maddalena 41, 202
rust 79

Sallaberger, Walther 225
Salonen, Armas 98–9
Samosta 156
Schamp, Jacques 20
Schøyen Collection 196
science
    as a distinct form of knowledge 2
    emergence of chemistry as 1
    prestige and reputation of 2
sealed containers, representing internal states 223
secrecy 15–19, 189–91, 234
Sekhet and Serket 115–16
semantic alternatives 50
semantic determinatives 36
Senac, Jean Baptiste 2–3
Seneca 7, 19–20, 82, 131, 231
Seth 25, 31
Shanaka, King 189–90
šiknu lists 41
Silk Road 4–5
silver 144–5, 149–50
    as a general standard of value 174
skills, cultural and technical 126
slag 147
slavery 21, 205
smelting 178
snakebite 115–16, 167–8, 230; *see also* venom
soda 12
Soldt, Wilfred H. 68–9
sourdough 225
spoils of war 146
standard of living 139
standardization 94
Statius 157
Steinert, Ulrike 220–3
Steinkeller, Piotr 145–6
Stephanus of Alexandria 159
stewards 141
Stoicism 7
Stol, Marten 220, 224
"stone, plant, tree" lists 39

stones
    "from before the flood" 127
    imitation 74
    within the human body 125
storage economies 94
Strabo 153, 155, 161
substances linked with the gods 23–4, 32
sulfur 156, 166–7
    'sulfur water' (or 'divine water' 80, 110
Sumerian language 125
Suphis, King 120
Surgon II 38
Synesius 16, 42, 48, 110

tablets foundation 64
tanning 68, 157–8
taxation 205
textiles 66–8; *see also* fabrics
Texts of the Pyramids 24–5
Thales 42–3
Theophrastus 8, 14, 20–1, 41, 46, 77, 80–3, 90, 179, 208, 234
Theopompus 80
Theosebeia 159, 207
Thompson, Campbell 70, 202
Thoth 116–19
Thutmosis I 91
Tiberius, Emperor 82
the *Timaeus* 45–6, 132–3
tin and tin metallurgy 47–8, 57–8, 63–4, 79, 144, 160
tools, use of 216
Tosi, Maurizio 97, 102–3
Tosorthros, King 119
Totelin, Laurence 230
toxicity 135–6, 149, 154, 160
tournette 102
trade and trade routes 9, 144–5, 160–5, 169–71, 175, 185
tradition, literary 18
training 205
translucent solid, man-made 71
transmutation of metals 3, 11–12, 234
transport costs 178
Treatise of Sympathies 26
"tributes" 163
tuyère 102
Tyre 158
Tyrian purple 83, 184

Udjahorresnet 167
Ulrich, Roger 227–8
United States Environmental Protection
    Agency 149
urine 157–8
Ursin, Johann Heinrich 2
Uruk Cycle 171–2
Uttu 122–3, 222

valuable materials 12
value, standard of 174
Van Minnen, Peter 206–7
Veldhuis, Nick 39–40
venom 116, 135, 193
Viano, Cristina 47
vinegar 77–9
Vitruvius 77, 79, 82, 107, 131
viziral administrations 85
Vulcan *see* Hephaestus

Waetzoldt, Hartmut 67
Wasserman, Nathan 151–2
waste material 142, 147–9, 160
water supplies 153
waterways 169
weapons, manufacture of 58

white earth 67
white lead *see* lead
Wilson, Anne 110
wine-making 212–13
women, role and status of 158–9, 212,
    215
wood, supplies of 140, 152, 160, 165
Woolley, C. Leonard 147
"Workshop-of-the-Goldsmiths" 27
workshops 14–15, 92–3, 104, 157–8,
    177–8, 184–5, 203–9, 235
  books kept in 208–9
  diagnostic features 97
  family workshops 203–4
writing systems 216

Xenophon 177–8

Zaccagnini, Carlo 177
Zagros mountains 144–5
Zamkow, Martin 224
zigurat of Assur 38
zinc and zinc metallurgy 45, 80, 147
Zoroastrianism 2, 16, 27, 42–3, 75, 80,
    108–11, 121, 132–3, 154–5, 159,
    207, 232–3